AUTOMOBILE ENGINES

MOTOR MANUALS VOLUME ONE

AUTOMOBILE ENGINES

IN THEORY, DESIGN, CONSTRUCTION, OPERATION AND TESTING

Arthur W. Judge

*Associate of the Royal College of Science, London; Diplomate
of Imperial College of Science and Technology (Petrol Engine
Research); Whitworth Scholar; Chartered Engineer; Member
of the Institution of Mechanical Engineers; Member, Society
of Automotive Engineers (U.S.A.)*

EIGHTH AND REVISED EDITION

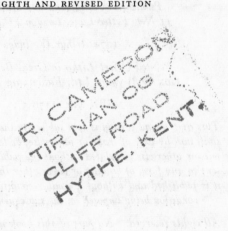

LONDON

CHAPMAN AND HALL

First published 1925
Second edition 1931
Third edition 1937
Reprinted 1940
Fourth edition 1942
Reprinted 1944, 1946
Fifth edition 1950
Reprinted 1953
Sixth edition 1956
Reprinted 1960, 1961
Seventh edition 1963
Reprinted 1965, 1967, 1969
Eighth edition 1972

Published by Chapman and Hall Ltd.
11 New Fetter Lane, London EC4P 4EE

© 1972 *Arthur W. Judge*

Printed Offset Litho in Great Britain by
Cox & Wyman Ltd, Fakenham, Norfolk

SBN 412 11170 5

PREFACE TO THE EIGHTH EDITION

WHEN first published in 1925 this book was the first of a series of Motor Manuals planned in particular for the motor engineer, mechanic, apprentice and student, so that the treatment adopted was of an elementary and practical nature, but with a minimum of theory and mathematics.

Since then, due to the public response for these Manuals, they have been revised and reprinted many times—in the present case no less than seventeen times—so that their scopes have been extended beyond the original plan.

The present edition relates to some of the important developments which have occurred since the date of the last one and are given in Chapter XI.

The new material includes accounts of engine performances, the standard methods of rating horse power, exhaust pollution and remedies, modern combustion chambers, cross-flow heads, toothed-belt drives, accounts of engines with vee-four, six, eight and twelve cylinders, automatic engine cooling fans, valve clearance adjusters, etc. In view of the marked improvements made in the Wankel type engine and the fact of its adoption by certain engine manufactures for their automobiles, an account of the more recent developments of this engine is included. It should be pointed out that in a book of this nature, produced at an economic price and for the type of reader mentioned earlier, it has been necessary to deal somewhat briefly with the selected subjects. As these Motor Manuals are distributed in the U.S.A. and in countries to which American cars are imported, an English-American Glossary of the more widely used names is included on page 8.

In conclusion, the present opportunity is taken to acknowledge the co-operation of certain manufacturers in the provision of information and illustrations, in particular Jaguar Cars Ltd.,

Vauxhall Motors Ltd., Standard Triumph International Ltd., Austin-Morris (British Leyland), AC-Delco and to the *Automobile Engineer*.

A. W. JUDGE

Farnham, Surrey
1972

CONTENTS

ENGLISH - AMERICAN GLOSSARY

English	*American*
Aluminium	Aluminum
Anti-clockwise	Counterclockwise
Bumper	Bumper
Bush (metal)	Bushing
Carburettor *or* Carburetter	Carburetor
Car Bonnet	Hood
Colour	Color
Contact Breaker	Breaker
Dismantling	Disassembling
Dynamo	Generator
Earth (metal)	Ground
Fibre	Fiber
Gauge	Gauge *or* Gage
Gear Lever	Gearshift Lever
Gudgeon Pin	Piston Pin
Inlet Manifold	Intake Manifold
Licence Plate	License Plate
Lb. per sq. in.	PSI *or* psi
Motor Car	Automobile
Moulded	Molded
M.P.H.	MPH
Paraffin	Kerosene
Petrol	Gasoline
R.P.M.	RPM
Screwed	Threaded
Spanner	Wrench
Sparking Plug	Spark Plug
Starting Motor	Starter *or* Starting Motor
Silencer	Muffler
Windscreen	Windshield
Wing	Fender
Vice	Vise

CHAPTER I

THE COMBUSTION PROCESS IN THEORY AND PRACTICE

Introduction. —Modern mechanical transport on land, sea and in the air has been made possible only by the development of the steam and internal combustion types of engine. The use of steam engines is now confined to the heavier class of transport and includes steamships, locomotives and road rollers, although motor-cars propelled by steam engines attained a considerable degree of success in the past, and there were, up to about 1930 one or two satisfactory makes of steam car still in service in America. The Doble steam motor-car, externally indistinguishable from a petrol engine type, silent in operation, extremely flexible and economical in fuel, represented the modern development of this type of engine after the first World War.

Although the present volume is confined to that extensive class of engine designated the internal combustion engine, yet a few remarks on the steam type may not be out of place here, if only to keep before the reader the past successes and the potential possibilities, of this engine even to-day. Invented by James Watt towards the end of the seventeenth century for stationary engine purposes, the steam engine was apparently first applied to road locomotion (or automobile) purposes by a French engineer, Cugnot, in 1769, when a four passenger vehicle attained a speed of $2\frac{1}{2}$ m.p.h. for a period of about 15 minutes—the time corresponding to the capacity of the small boiler. Cugnot, a year or two later, built, with the help of the French Government, a steam tractor for artillery haulage purposes; it was designed to haul a load of $4\frac{1}{2}$ tons, but was not used owing to the French Revolution.

Adverse popular opinion, the prejudice of stage coach interests and legislation (the tolls on steam cars were ultimately raised to £2 each against the three shillings for stage-coaches) practically killed the steam automobile until in 1896 conditions changed.

Prior to this no less than 54 Bills were introduced into Parliament on the subject of road vehicle taxation, and the mechanically propelled vehicle was only allowed to proceed along the road under the guidance of a man, who preceeded it, carrying a red flag; its maximum speed, therefore, could not exceed about 4 m.p.h. The passing of the Locomotives on Highways Act in 1896 removed these restrictions, and real progress in automobile development has continued from this date.

In 1902 a Serpollet steam car attained a speed of 75 m.p.h. at Nice, thus beating the record of 65·79 m.p.h. previously made by Jenatzy on an electrically-driven car in 1899. Serpollet's record was subsequently beaten several times by petrol cars, but in 1906 the latter's records were swept off the board by those set up by a Stanley steam car, which attained a speed of 127·5 m.p.h. on the Ormonde Beach sands. In 1909 a huge Benz petrol-engine raised this figure to 140·9 m.p.h., but from this period the steam car has lapsed into obscurity as a speed record breaker, although it is but fair to add that the subsequent success of the steam car in hill-climbing competitions was so pronounced that ultimately it was debarred by the Automobile Clubs from competing. Here we must leave it and turn to its confrère, the petrol-type engine, but in passing it should be added that there is little doubt that if all of the design, skill, research work and energy which has been devoted to the latter had been similarly applied to the former, a different state of affairs might now exist.

Types of Combustion Engine.—The steam engine may be termed an "external combustion" engine, for the fuel (coal or heavy oil in this case) is burnt in a separate boiler, and the steam generated at a pressure is led into the cylinder of the steam engine.

In the case of the "internal combustion" engine, which comprises all types of explosive engine—petrol, heavy oil, gas and Diesel types—the fuel is burnt, or exploded within the engine itself.

Internal Combustion Engines.—An examination of the history and development of this type reveals the interesting fact that it is possible, although not always practical, to employ a very wide range of solid, liquid and gaseous fuels, in conjunction with air, or oxygen as the combustion medium. Thus the earliest authentic record credits Huyghens, in 1680, with the invention of an engine in which gunpowder was exploded in order to drive the air out of a cylindrical vessel, closed at one end and fitted with a piston in its barrel. The result of forcing the air out, and of the cooling down of the combustion products, caused a partial vacuum within the cylinder, so that the piston was forced into it by the external atmospheric pressure. This inward travel of the piston was utilized to raise weights and to work mechanisms. Coal dust is another fuel which can be used for internal combustion engines and much experimental work has been in progress with this type of engine in recent years.

Hydrogen gas was employed by Cecil in 1820 as the exploding agency, whilst Lenoir—to whom we owe a great debt for the marked progress he made—in 1860 contructed several hundreds of engines using coal gas. These engines belonged to the "non-compression" class, and were very wasteful in fuel—using about seven times as much fuel as a modern gas engine of the same power. It is interesting to note that electrical ignition was employed on these engines.

In 1876 Otto invented his famous "Otto-cycle" engine, in which the charge of gas and air to be exploded was compressed before it was ignited. This pre-compression of the charge is the fundamental secret of success of the modern automobile and other engines. By compressing the charge a very considerable increase in the explosive effort (and therefore the power output of the engine) and also a marked reduction in the quantity of fuel used, result. Gaseous fuels applicable to internal combustion engines include hydrogen (H), carbon monoxide (CO), acetylene (C_2H_2), coal-gas (a mixture of various combustible gases such as hydrogen, marsh-gas, carbon monoxide and hydrocarbons), water-gas (carbon monoxide and hydrogen mixture), methane, butane, blast-furnace, producer and other gaseous mixtures. Liquid fuels include the heavier products obtained after the fractional distillation of coal and of petroleum; these fuels are employed in heavy-oil compression-ignition type engines. The lighter fuels include petrols, paraffins, benzole, alcohol and similar fuels of a spirit nature. Coal dust as previously stated has also been used, successfully.

Generally speaking, any fuel which contains either hydrogen (H) or carbon (C) is capable of combustion in oxygen; this fact, well known to chemists, accounts for the wide range of fuels now available for internal combustion engines.

The Process of Combustion in Engines.—It is as well, at the outset, to have a clear idea of the process of combustion since this constitutes the basic principle of internal combustion engine operation.

Combustion is the name given to the process of chemical combination of a combustible, such as any of the fuels previously mentioned, and oxygen. In this process, heat is evolved, and is utilized in expanding the gaseous products of combustion; the expansive effort of these gases is utilized in engines to give the "working" or "power" stroke.

Confining our remarks to the case of liquid fuels, these as we have seen are organic substances composed of the chemical elements, carbon and hydrogen (often with the addition of oxygen). These elements are usually defined by the capital letters C, H and O respectively, and the composition of the fuel is expressed by a chemical formula including these.

Thus, the hydrocarbon fuel _hexane_, which is the principal constituent of a pure petrol has the formula C_6H_{14}. This formula expresses the fact that one molecule of hexane contains 6 atoms of carbon and 14 of hydrogen. Since the atom of carbon weighs 12 times that of hydrogen,* it follows that the carbon and the hydrogen weigh 6×12 (or 72) and 14×1 (or 14) respectively. Expressed as percentages, these weights are 83·6 and 16·4 respectively, so that one pound, say, of petrol, of this type, contains 0·836 lb. carbon and 0·164 lb. hydrogen.

The carbon, in combustion with oxygen, forms carbonic acid gas, denoted by the formula CO_2; the hydrogen forms water in the form of steam, or H_2O. The combustion products are therefore carbonic acid gas and water (or steam) in the case of petrol and air combustion. The full statement, in chemical language, of the process is as follows:

---

* The oxygen atom is 16 times, and the nitrogen atom 14 times, as heavy as that of hydrogen.

$$C_6H_{11} + \left(6 + \frac{14}{4}\right) O_2 = 6CO_2 + \left(\frac{14}{2}\right)H_2O$$

$$\text{or } C_6H_{14} + \frac{19}{2} O_2 = 6CO_2 + 7H_2O$$

Interpreting this relation in ordinary language, 1 cubic foot of petrol vapour will require $9\frac{1}{2}$ cubic feet of oxygen to burn it completely, and the resulting combustion products (or exhaust gases) consist of 6 cubic feet of carbonic acid gas and 7 cubic feet of water vapour.

It can be shown from this result that 1 lb. of a typical petrol will require about 15 lb. of air to burn it completely, a fact which is of importance to carburettor designers and users of automobile engines.

Benzole requires less air (about 13·3 lb.). Alcohol also requires less air (about 9·0 lb.), per lb. of fuel, for complete combustion.

Detonation.—If pure oxygen is used in a petrol engine instead of air (which is a mixture consisting of 76·8 per cent. nitrogen and 23·2 per cent. oxygen, by weight), the explosion will be almost instantaneous and very violent; exceedingly high pressures will result and unless the engine is specially designed to withstand these, disastrous results may occur.

Again, if acetylene gas be used in a petrol engine, the explosion is also most violent; many engine cylinders have been wrecked by inexperienced persons experimenting with acetylene.

In both cases this extremely rapid explosive effect is termed *detonation*.

Fortunately, the atmospheric air contains the neutral or inert gas, nitrogen, in the proportion of nearly 80 per cent. This gas will not burn with oxygen, and its presence damps down the detonation tendency. It is possible, in the case of petrol engines, to obtain detonation by using too high a compression with a given fuel and air mixture. Thus if the compression pressure is much over 90 lb. per sq. in. when paraffin vapour is used, detonation occurs. For petrol mixtures of the high octane type, benzole and pure alcohol, the limiting compression pressures are approximately 170, 180 and 200 (min.) lb. per sq. in., respectively. An engine with a badly designed combustion chamber or one which normally runs "hot," will begin to detonate at lower compressions than these.

Although many theoretical explanations have been advanced to account for the phenomenon of detonation, it is only more recently that experimental investigations have given a more satisfactory indication of the causes. Thus it has been established that detonation occurs when the rate of inflammation or burning of the mixture exceeds a certain critical value; the latter depends upon several factors including the nature of the fuel used, compression ratio, combustion chamber design, ignition timing, etc.

It is now believed that the earlier stages of the burning of the charge after the spark occurs are normal in character, but that towards the end of the combustion process, namely, at about the last 25 per cent. of the flame travel in the combustion chamber, a sudden burning effect of the remaining part of the charge, or "end gas" occurs, which gives rise to a very rapid rate of pressure rise which produces impact effects on the combustion chamber walls, usually termed detonation or knock.

The subject of combustion chambers and detonation is referred to later on page 100.

That this explanation is substantially correct is indicated by the fact that it is possible to design combustion chambers based upon the results of its application which enable higher compression ratios to be employed with a given grade of fuel without detonation, than are possible in previous engines.

Some Facts About Detonation.—In recent years a considerable amount of investigation work has been carried out on combustion processes and detonation, a summary of which will be found in the footnote reference*. Apart from theoretical and laboratory results, the following practical aspects have been established, namely:

(1) In certain overhead valve engines, there is a greater tendency towards detonation at lower engine speeds. This is believed to be due to increased temperature of the mixture and greater

* *Modern Petrol Engines*. A. W. JUDGE. 3rd Edition. (Chapman and Hall Ltd., London).

formation of peroxides—chemicals which are associated with detonation conditions.

(2) Detonation occurs at maximum power output, so that the anti-knock fuel is only needed at larger throttle openings; a lower grade of fuel could be used for cruising conditions.

(3) To obtain the greatest benefit from a given anti-knock fuel an engine should be run with a higher compression ratio than that giving no detonation, but the ignition should be retarded at full throttle.

(4) Detonation is more likely to occur with weaker than with rich mixtures, within certain limits. Detonation is a minimum for very rich mixtures, namely, 9 to 11:1 (air-fuel ratio) and a maximum for rich mixtures of 12 to 14:1.

(5) Combustion chamber design and sparking plug location have a very important influence on detonation so that while some engines, using a given fuel, mixture strength and compression ratio will detonate, others with different designs of combustion head cannot be made to detonate. Overhead valve engines on account of their higher volumetric efficiencies and higher mean pressures tend to detonate more readily than side-valve engines under similar conditions of mixture, load, speed, etc.

Pre-Ignition.—Another cause of engine knocking, which is quite distinct from detonation effects, since it *always occurs before the ignition spark*, whereas the latter always occur afterwards, is that known as pre-ignition. It is often due to the inefficiency of the engine cooling arrangements, so that all of the surplus heat cannot be disposed of, with the result that the temperature within the combustion chamber increases progressively until some projecting part, such as the electrodes of the sparking plug or a protruding piece of carbon deposit, becomes so hot that it ignites the charge before the spark occurs at the sparking plug. The piston is actually ascending when this happens and it therefore receives a retarding force. Generally, once pre-ignition commences, its effects become more and more enhanced with the result that the engine loses power and eventually stops.

The chief causes of pre-ignition may be summarized, as follows:

(1) Carbon deposits.

(2) Unsuitable type of sparking plug, i.e., one that runs too hot or has too long a reach.

(3) Overheated exhaust valve head, or edge.

(4) Ignition too far retarded.

(5) Mixture much too weak or rich, giving too slow a burning rate.

Post-Ignition Effect.—There is another mixture ignition effect that may occur in a modern car engine, namely, running-on of the engine after the actual ignition has been switched off. This effect was a fairly common one with the early air-cooled rotary type aircraft engines, which often continued to operate after the ignition was switched off. The effect—known as post (or after) ignition—appears to be due to the presence, in the combustion chamber of a metal or carbon projection, or edge, which becomes heated during the initial combustion stages, to a temperature above the ignition point of the mixture and tends to supplement the electric ignition. During the exhaust, suction and compression strokes this "hot spot" cools down below the ignition temperature of the charge, but as soon as the electric spark ignites the charge, it heats up again. When the engine is at its working temperature and the ignition is switched off, this "hot spot" continues to act as an auxiliary igniter of the charge, until the temperature of the combustion chamber is reduced, sufficiently, to prevent this effect.

Engine Rumble.—When the compression ratio of certain types of petrol engine is raised, progressively, above about 10:1, a low-pitched thudding noise, accompanied by engine roughness in operation occurs. This effect is caused by high rates of cylinder pressure rise, after ignition. This excessive pressure rise occurs at or near to the top dead centre of the piston and causes crankshaft vibration which gives the "rumble" effect. It can be reduced by fine adjustment of the ignition timing, or by using a highly aromatic fuel with T.E.L. and phosphorus additives.

Fuels.—Before passing on to a consideration of the manner in which the process of combustion of a fuel with air is applied and utilized to its best advantage in modern engines, it will be necessary to say a few words concerning the principal properties of suitable fuels.

The best fuels are those which contain the highest proportions of hydrogen in their composition, since this latter element has the greatest heating value. By *Heating* or *Calorific Value* is meant the heat generated during combustion of the fuel with air or oxygen. It is usually defined as the total number of heat units (in this country, British Thermal Units*) evolved during the combustion of a given weight of the element or fuel.

In the following table is given the principal properties of various fuels. Hydrogen has heating value of 62,000 B.T.U.'s.

TABLE 1
Properties of Fuels

Name	Chemical Symbol.	Heating Value. B.T.U.'s per lb. (Lower.)[1]	Density. Wt. per gallon at 15° C.	Maximum Compression Pressure which can be used.	Fuel Consumption[2] in Engine of Efficient Design, with 5:1 Compression Ratio.
			lbs.	lbs. per sq. in.	lbs. per I.H.P. hour.
Light Low Octane Petrol (Aromatic, Free)	C_6H_{14}	19,100	7·18	105·5	0·415
Medium Octane Petrol	Mixture	19,000	7·27	118·0	0·421
Heavy Petrol ..	Mixture	18,790	7·67	140·5	0·425
Paraffin (Illuminating Oil) ..	Mixture	19,000	8·13	86·0	0·523
Benzene	C_6H_6	17,300	8·84	180	0·458
Alcohol (Ethyl) ..	C_2H_6O	11,470	7·98	> 204	0·663
Methylated Spirits ..	Mixture	10,200	8·21	163·5	0·740
Ether..	—	—	7·35	47·5	—
Carbon	C	14,540	—	—	—
Hydrogen	H	52,500	Gaseous	—	—
Carbon Monoxide ..	CO	4,329	,,	—	—
Ethylene	C_2H_4	21,350	,,	—	—
Methane (Marsh Gas)	CH_4	23,510	,,	—	—

(ETHANOL (C_2H_5OH))

[1] This is the total heat of combustion, less the latent heat of the steam formed by the combination of hydrogen in the fuel to water vapour.

[2] Values obtained from Ricardo's variable compression engine tests.

* The British Thermal Unit is the quantity of heat required to raise the temperature of 1 pound of water at 60° Fahrenheit through 1 degree Fahrenheit and is equivalent to 778 foot pounds in work units.

From the motorist's point of view, the best fuel at a given
price per gallon is the one having the greatest density with a good
heating value combined with a high octane rating.

Petrol.—This is a light highly volatile spirit obtained from
petroleum by distillation.   Its weight varies from 7 to $7\frac{3}{4}$ lb. per
gallon.   It contains about 84 and 16 per cent., repectively, of
carbon and hydrogen by weight.   About 15 parts, by weight, of
air to 1 of petrol are required for complete combustion, but modern
engines will work on mixtures of from about 9 to 1 (rich in
petrol) up to about 20 to 1 (weak in petrol).   Inflammable vapour
is given off at ordinary atmospheric temperatures.   Petrol attacks
most varnishes, and dissolves rubber, oils and fats.

Highest Useful Compression Ratio.—Most commercial
non-premium petrols are mixtures of other fuels belonging principally to certain groups or series, known as Paraffins (these include
pentane, hexane and heptane), Naphthenes, Olefines and
Aromatics.

Each of these series confers properties upon the petrol.   Of
these, the more important is that of the Aromatic Series (including benzine, toluene and xylene) which enables the working
compression ratio of the engine in which it is used to be raised.

For each series of fuels there is a limiting compression ratio,
known as the *Highest Useful Compression Ratio* (H.U.C.R.) above
which the engine will not work satisfactorily owing to detonation
effects.

For aromatic petrols the H.U.C.R. ranges from about 6 to 7,
whilst for petrols containing paraffin and naphthene series fuels
the value seldom exceeds 5·5.

The petrols containing aromatic fuels usually have a density
of ·74 to ·76, and will enable compression pressures up to 150 lb.
per sq. in. to be used, whilst with pure aromatic fuels compression
pressures of at least 180 lb. per sq. in. are possible.

Benzole.*—The commercial benzole is a mixture of benzene
(C_6H_6) with other spirits such as toluene and xylene   The 90 per
cent. grade used in automobile engines boils at 80° C. to 100° C.,

---

* A previously used mixture of equal parts of benzole and petrol, known
as "50/50 Mixture," possessed advantages over pure petrol in the matter
of fuel economy and reduced "knocking" tendency.

and weighs about 8·7 to 8·9 lb. per gallon at 15° C. At higher temperatures it weighs less. It has a heating value of 16,700 to 17,500 B.T.U.'s per pound (lower value). Its principal advantage when used in petrol engines is that in the case of engines prone to **knock** or pinking under heavy load conditions, as in the case of a car climbing a hill on top gear, benzole will lessen this detonation and subsequent overheating effect. This is due to the fact that benzole has a higher H.U.C.R. than ordinary petrols, so that to obtain the full benefit, the compression ratio can be raised above its value for the petrol engine.

When benzole was substituted for petrol, in earlier car engines it was found that the performance was appreciably better and the fuel consumption from 15 to 20 per cent. lower.

Modern benzole fuel is a mixture of benzole with high grade petrol blends, sometimes having additives to oppose carbon formation and corrosion in the cylinder. Owing to the fact that benzole is heavier by about 15 per cent. than petrol, it becomes necessary to weight the carburettor float, otherwise the jet level will be raised. A smaller jet can be used with benzole.

Benzole requires about 13½ lbs. of air per lb. for complete combustion, which is rather less than for petrol; it has a mixture range of about 11 to 1 to 19 to 1, and with the richer mixtures is apt to deposit much carbon and to gum up the valves. Benzole attacks most varnishes strongly, so that varnished cork floats (which have been used in certain American cars) should not be used.

Paraffin.—This fuel has sometimes been used in low compression petrol engines, but for the best results an efficient vaporizing and mixture heating device is required. The heat of the exhaust gases is generally used for this purpose. Paraffin is heavier than petrol (8·10 to 8·2 lb. per gallon). When used in petrol engines, it is generally necessary to have a petrol supply for starting purposes.* Its heating value is about 19,000 to 10,500 B.T.U.'s per lb., and inflammable vapour is given off at about 80° Fahr. (260° C.).

Alcohol.—This spirit is obtained from vegetable and organic

* Fuller information on this subject is given in Volume No. 2 of the Motor Manual Series.

sources, such as sugar- starch- and cellulose-containing products. Sugar-beet, mangolds, potatoes, artichokes, grain, molasses, and mahua flowers are amongst the more important alcohol sources.

Alcohol has the chemical formula C_2H_6O, and in the pure state has a density of 7·95 lb. per gallon, and heating value of 11,500 B.T.U.'s per lb. (higher). Typical alcohols that can be used as fuels include ethyl-alcohol, methyl-alcohol and methylated spirits. Methylated spirits usually consist of ethyl-alcohol (80 per cent.), wood spirit (10 per cent.), petroleum naphtha (0·5 per cent.), and water (9·5 per cent.).

When benzole is used as a denaturant it increases the heating value; thus a 50 per cent. alcohol-benzole mixture would have a heating value of about 15,300 B.T.U.'s per lb.

An important property of alcohol, when considered as a fuel, is that it is capable of being used at much higher compression ratios than petrols without detonation effects, its highest useful compression ratio, in the pure state, being rather more than 15 to 1, with a corresponding octane rating above 100.

Methyl-alcohol, theoretically requires, $6\frac{1}{2}$ parts of air to 1 part alcohol, by weight for correct combustion, while ethyl-alcohol needs 9 parts of air to 1 part alcohol. As the ordinary petrol carburettor is adjusted to supply about 13 to 15 parts of air to fuel, it will be seen that the alcohol carburettor will require a larger main jet. Alcohol has about three times the latent heat of evaporation of petrol so that *more heat* is required to vaporize alcohol in the carburetted air-fuel mixture, but owing to the greater cooling that results, the mixture is cooler and a greater weight is taken into the cylinders than for an air-petrol mixture; a greater maximum power output results, on this account. Further, there is a marked gain in output due to the much higher compression ratio that can be used, but owing to its low heating value and relatively high cost per gallon, *alcohol fuels* are relatively *much more expensive than petrol fuels*.

Alcohols are not used, in their pure state as fuels in petrol engines, but are mixed with other fuels, or fuel mixtures, so as to impart some of the important properties to the resulting mixture, namely, higher compression operation without engine knock, greater fuel economy and lower exhaust temperatures. The high latent heat of the alcohol content plays an important part in the effective cooling of the cylinder, under the high compression and engine speed conditions; in this respect the alcohol-blended fuels have an advantage over those which derive their high octane rating by addition of tetra-ethyl lead compounds.

When alcohol is used in petrol engine carburettors it is necessary to heat the induction manifold (or inlet air supply), to ensure proper vaporizing of the fuel.

Anti-Knock Fuels.—Fuels, such as petrols and paraffins, are known to detonate, when the compression pressures used exceed a certain limiting value. By adding a small quantity of another special chemical or fuel it is possible to raise the compression pressure appreciably without experiencing detonation. Thus if about 1 to 2 per cent. of aniline be added to petrol, very much higher compressions can be used. Ethyl iodide, xylidine, and tetra-ethyl lead are also effective in this respect; these liquids are termed "stabilizers," and their solutions with petrol are known as *anti-knock* fuels.

Octane Number of Fuels.—Considerable research has been undertaken with the object of improving petrol engine fuels, to enable higher compressions to be used without detonation occurring.

New blends of fuels made by special refining methods and synthesis have been produced, whilst the performances of engines using ordinary fuels have been improved by adding stabilizers, or "anti-knock" constituents to them.

In order to compare these fuels from the point of view of their anti-knock qualities it is now the custom to use a standard fuel, having two constituents, viz., *Iso-octane* and *Heptane*; both are hydrocarbon fuels.

The former is a fuel having a higher anti-knock value than all ordinary fuels used in modern petrol engines. The latter, on the other hand, is a very poor fuel of bad "knocking" properties.

For this reason iso-octane is given the arbitrary rating of 100, and heptane, O. Any other fuel will have an *Octane Value or Number*, intermediate between those of these two fuels.

The octane value of any fuel represents the percentage of iso-octane in the mixture of heptane and iso-octane which has

similar knocking properties—when tested in a standard engine—
to that of the fuel.

The usual motor octane values of commercial motor petrols
range from 75 to 80 for the cheaper petrols to 85 to 100 for the
higher anti-knock fuels used with "high-compression" range of
production engines.

Research and Motor Octane Ratings of Fuels.—The
Octane number, or O.N. of a fuel as determined by the method
described, using a standard laboratory engine operating under
closely controlled conditions of speed, air intake temperature,
speed and ignition timing is referred to as the *Research Octane
Number*. These O.N. values do not necessarily apply to automobile engines so that a closer approximation is obtained by
altering the text conditions on the same standard engine, e.g., by
increasing its speed, air temperature range and ignition timing.
The O.N. rating obtained by this method is known as the *Motor
Octane Number*. For the higher anti-knock fuels the O.N.
rating is lower for the Motor than the Research method due to the
higher temperature and speed for the text.

There is no standard relationship between the Research and
Motor octane ratings, that can be applied to the wide range of
fuels, engine designs, compression ratios and speeds but it may
be mentioned that fuels which show relatively large differences
between the Research and Motor test methods are stated to be
Temperature Sensitive and this sensitivity is defined, as follows:

Sensitivity = Research Octane Number—Motor Octane Number

Compression Effect upon Octane Rating.—It is now well
known that as the octane rating of a fuel is raised, it becomes
possible to increase the compression ratio of an engine, without
knock effects, but with increased power output and reduced fuel
consumption. Whilst it is not possible to give a definite formula
or graph showing this relationship that could be applied generally
to various engines and fuels, an approximate relationship between
these two quantities is that shown in Fig. 1.

Since the limiting value of the octane rating by the previously
described method is 100 the results shown for the compression
ratios of approximately 10·25:1, corresponding to pure iso-octane

fuel, represent the limit of the scale, so that some other method must be used for fuels having higher O.N. ratings than 100 as described later.

Octane Rating and Performances.—If the power output of an engine under standard operating conditions, e.g., those em-

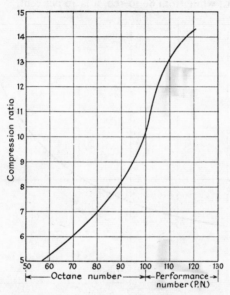

Fig. 1.—Relationship between Octane Number, Performance Number and Compression Ratios.

ployed in the Motor Research O.N. rating tests, is measured for various compression ratios using suitable fuels of known O.N. ratings it is possible to establish a relationship between these quantities; further, it is also possible to extend the results for fuels above the standard 100 O.N. rating as shown in Figs. 1 and 3.

The results of certain tests made by J. D. Davis* using petrols made by different processes, to which increasing small amounts of tetra-ethyl lead (T.E.L.) had been added, are shown in Fig. 2,

* *Proc. Inst. Mech. Engrs.*, 1953.

in which the power outputs are plotted against the corresponding
O.N. ratings. It will be seen, from the graph, that whilst there
is a definite power gain for fuels up to about 60 O.N. rating,
namely, about 20 per cent., there is a much greater gain at the
higher O.N. ratings. Thus, from 60 to 80 O.N. the gain is about
19 per cent., while from 80 to 100, it is 40 per cent.

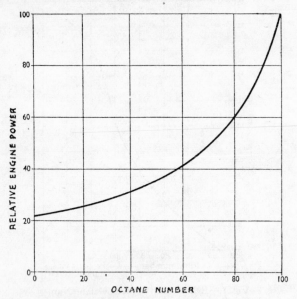

Fig. 2.—Showing effect of Octane Rating of Fuel on
Engine Power Output.

The Higher Octane Fuels.—In seeking to improve the power
outputs of aircraft engines during the 1940-45 period, various
fuels and additives were employed to raise the octane ratings above
the 100 O.N. values. Blending agents, including iso-pentane,
neo-hexane, iso-octane, etc., with small proportions of T.E.L.
were employed to raise the anti-knock properties above the 100
O.N. rating. Instead, however, of using a chemical fuel scale,
the new method adopted was to use the power increase or *performance number* (P.N.) so that by starting with the 100 octane

number as the standard or reference, a fuel that produced a power increase or P.N. of 30 per cent. would be given a rating of 130, and so on.

Fig. 3 shows how the P.N. values of fuels are related to the octane numbers. It may be here mentioned that aircraft engine fuels of 150 to 170 P.N. rating have been made and used in piston engines.

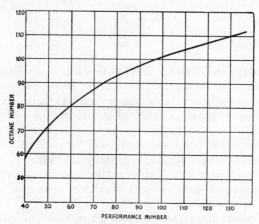

Fig. 3.—Relationship between Octane Number and Performance Number.

To readers who are interested in these high O.N. and P.N. ruels, the following empirical relationship is employed for fuels above the 100 O.N. rating:

$$\text{Octane Number} = 100 + \frac{\text{P.N.} - 100}{3}$$

Commercial Petrols or Gasolenes.—Modern petrol engine fuels are blends of several hydrocarbon fuels, obtained by refining petroleum. They consist of certain liquids having different physical properties and octane ratings and include straight-run, reformed, thermally-cracked and catalytically-cracked petrols blended in selected proportions.

The earlier fuel used, during and after the 1939 War was known as Pool Petrol and had an octane rating between about 70 and 73. Later, the octane ratings were raised and other grades of fuel, of still higher ratings became available, commercially. The general octane ratings*, covering various petrol pump grades, in 1962, were, as follows:

Grade	Octane Number	
	Research Rating	Motor Rating
Regular	83– 84	79– 80
Premium Grades	95– 99	84– 90
Super Grades	100–102	88– 91

Advantages of High Octane Fuels.—It is shown later in this chapter that the thermal efficiency of an engine increases progressively as the compression ratio is raised, so that by using the higher compression ratios rendered possible by the modern high octane fuels this efficiency can be raised, with the result that the fuel consumption per horse-power will be reduced. Further, the maximum power output for a given cylinder capacity increases with the compression ratio, but the maximum combustion pressure also increases and so the pistons and cylinder units must be made stronger on this account.

Fig. 4 shows the performance curves of a certain automobile engine fitted with a variable compression device capable of altering the compression ratio from about 5:1 to 14:1. The values, shown for the maximum and mean cylinder pressures, and for the fuel consumptions are relative to those for an engine with a compression ratio of 5:1. It will be seen from these graphs that over the compression ratio range shown, the maximum pressure increases by 320 per cent.; the mean pressure—upon which the power output depends—increases by about 38 per cent., while the fuel consumption per brake horse-power decreases by no less than 28 per cent.

In connection with these results it should be mentioned that as a result of modern engine research with combustion chambers of various shapes and with different arrangements of the valves and sparking plugs, that each particular shape of chamber and

* Associated Octel Company.

disposition of the inlet and exhaust valve and placing of the sparking plug, corresponds to a given octane value of fuel, such that when using this fuel detonation will just be avoided.

The values of octane numbers and equivalent highest useful compression ratios previously given must, therefore, be regarded

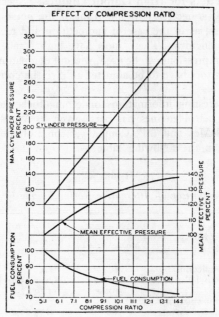

Fig. 4.—Performance Curves for Modern Petrol Engines.

as applying to one particular shape of combustion chamber and valve and plug disposition and are therefore applicable, generally, to other shapes.

It has been shown that with the plain L-headed chamber shown at *F* in Fig. 35, on page 101, the highest compression ratio that can be used with petrol of 80-octane value is about 5·8:1, whereas with other more efficient designs, compression ratios up to 7·8:1 to 8·0:1 can be used without knocking or any other ill-effect for the same fuel.

As mentioned previously the earlier pool petrol had an octane rating of about 72, but with the efficient combustion chambers used in post-war engines, it was possible to employ compression ratios up to about 6·5:1 to 7·2:1, and to obtain brake mean pressures of 118 lb. to 125 lb. per sq. in.

Special Fuels for High Output Engines.—Special fuels are used when it is necessary to obtain maximum horse-power from given engines, as in the case of competition and racing-car and motor-cycle engines. These fuels have high octane ratings, so that high compressions can be used, without detonation occurring. The carburettors have to be adjusted to give the correct mixture strengths.

Some interesting information concerning fuels developed for racing-type petrol engines has been given by W. B. Rowntree of the Shell-Mex and B.P. Company,* some of the data being reproduced in the following table:

TABLE 2

Comparison of Alternative Racing Engine Fuels

Type of Fuel	Octane Number by Motor Method	Maximum Compression Ratio, without knock	Air-Fuel ratio for Maximum Power
Ethanol (ethyl alcohol) ..	Over 100	Over 15:1	6·5:1
Methanol (methyl alcohol)..	Over 100	Over 15:1	4·5:1
Benzole	100	11 to 12:1	11:1
50/50, 70 octane spirit and benzole	80	9 to 10:1	11·5:1
70/25, 70 octane spirit and benzole	75	7 to 7½:1	12:1
Aviation spirit 100–130 grade	100	12 to 13:1	12·5:1
Ether 50 per cent. in motor spirit	—	3·9:1	10:1

In the past a considerable number of fuel mixtures have been used to improve petrol engine performance and a few of these fuels have become commercialized.

The fuels used for blending with petrols include alcohols,

* "Trends and Developments in Racing Fuels." W. B. ROWNTREE. *The Autocar.* September 25th, 1953.

acetone, benzole, ether and nitrobenzene. Some fuels contain a very small percentage of an oil, such as castor oil or a chlorinated oily compound, somewhat similar in its effect to the Halowax oil used in tetraethyl lead compound to lubricate the upper piston rings and rotors of superchargers.

In regard to hydrocarbon fuels, *i.e.*, petrol blends, these are often used with alcohols, to improve their calorific values. If, however, the alcohol contains water the petrol will not mix freely, benzole is useful in such cases as a blending agent.

Another fuel additive is *nitromethane*. Although rather expensive, when this compound is added to methanol and certain fuel mixtures, it gives considerable gains in power. It contains oxygen, some of which is used in the combustion of the fuel, so that less air is needed than for most fuels. Therefore, the overall air-fuel ratio is reduced and a greater weight of fuel can be taken into the engine, the actual amount being, of course, controlled by the total available oxygen in the fuel itself and the air drawn into the cylinder. The net result, however, is a much increased heating value of the air-fuel charge, with increased combustion pressure and power output.

Obtaining Power from Fuels.—We have seen that every fuel, when exploded or burnt with oxygen or air, is capable of generating a very large amount of heat in proportion to its bulk. Now it is the primary object of internal combustion engines to utilize this heat to the best advantage, and for this purpose it is necessary to explode the fuel-air mixture in a closed vessel, or cylinder, fitted with a close-fitting plug, or piston, capable of sliding backwards and forwards as shown diagrammatically in Fig. 5. Let us suppose that the piston shown is a heavy one, and at first is near to the bottom of the cylinder. If, now, a mixture of petrol vapour and air in the proportions of 1 part to 15 parts be introduced through the tap T into the chamber C, the tap turned off, and the mixture exploded by causing a high voltage spark to pass between the ends of two wires, or electrodes S, a relatively large amount of heat will be developed. This heat raises the temperature of the gases formed during combustion, namely carbonic acid gas and water vapour, and causes them to expand. This expansion will tend to raise the weight of the piston, if it is not too heavy, forcing it upwards, until its lower edge uncovers the outlet orifice E, when the gases escape into the atmosphere. The piston will then fall, and in doing so will compress the remaining exhaust gases in the cylinder.

If the piston has a diameter of, say, 3 inches, and the mixture is at atmospheric pressure before explosion, the instantaneous pressure generated when the spark passes will be about 80 lb. per sq. in., which is equivalent to a total force (=pressure × area of piston) of 570 lb. As the piston moves upwards the pressure diminishes until, when the orifice E is uncovered, it falls to atmospheric value (14·7 lb. per sq. in.).

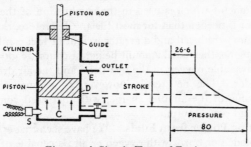

Fig. 5.—A Simple Type of Engine.

Let us now consider what the pressure, within the cylinder, will be when the piston has moved upwards to any given position. There is a physical law, known as Boyle's law, which states that the pressure of a gas at a given temperature varies inversely as its volume; if the volume is doubled, the pressure will be halved, if trebled, it will be reduced to one-third, and so on.

If we can keep our cylinder warm, it is easy to estimate what the pressure will be at any piston position. If the volume of the cylinder under the piston is, say, 1 cubic foot in the position shown in Fig. 5 and 3 cu. ft. when E is about to be uncovered, then since we have seen the explosion pressure is 80 lb. per sq. in., this value will drop at E to one-third, that is 26·6 lb. sq. in. In an intermediate position, such as at D, where the volume is 2 cu. ft., the pressure will be $80 \times \frac{1}{2} = 40$ lbs. per sq. in. The manner in which the pressure varies throughout the piston's stroke is shown graphically in the R.H. diagram, the lengths of the horizontal lines representing the pressures.

The area of such a pressure-stroke diagram represents the work accomplished by the expanding mixture: it is from such diagrams, known as *Indicator Diagrams*, that the power developed can be ascertained. It is only necessary to measure the average pressure value of the diagram, and to multiply by the length of stroke in order to obtain the mechanical work done during the piston's stroke. An example will serve to make this clear.

Example.—If the piston in Fig. 5 has a diameter of 3 inches, and an effective stroke of 3 inches, estimate the average work done per explosion, assuming the mixture before explosion to occupy 3 inches depth of the cylinder. If the piston makes 1,000 such strokes every minute estimate the horse-power.

$$\text{The piston area} = \pi \frac{(3)^2}{4} = 7 \cdot 06 \text{ sq. in.}$$

The initial pressure is 80 lb. per sq. in. and the final value 40 lb. sq. in., as we have stated before.

The average pressure measured from a diagram similar to that shown in Fig 5 is about 58 lb. sq. in.

Hence the work done per stroke

$$= \text{Total force on Piston} \times \text{Stroke.}$$
$$= 58 \times 7 \cdot 06 \times \tfrac{3}{12} \text{ ft. lb.}$$
$$= 102 \cdot 5 \text{ ft. lb.}$$

The effort exerted during the explosion and expansion of the piston would be sufficient to raise a weight of 102·5 lb. through a distance of a foot.

Now power=rate of doing work, i.e., work done per unit time, and 1 horse-power=work done at the rate of 33,000 ft. lb. per minute.

In the present example the work done per minute × 1,000 = 102·5 ft. lb.

$$\text{Hence the equivalent horse-power} = \frac{1000 \times 102 \cdot 5}{33,000} = 3 \cdot 1$$

In passing it should be mentioned that in practice the hot gases would cool down as the piston moved upwards, due to the piston and cylinder metal extracting some of their heat, so that the expansive force would be less than we have estimated.

Application to Engines.—We have now shown that the heat energy of combustion of a fuel with air is capable of being employed for doing useful work, although our example was a somewhat crude one.

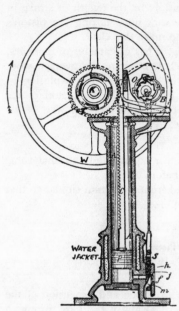

Fig. 6.—The Otto and Langen Engine of 1886.

Otto and Langen, in their famous gas engine of 1866, utilized this energy of combustion in the following manner. They employed a long vertical cylinder, as depicted in Fig. 6, fitted with a piston P, the rod of which consisted of a toothed rack C, engaging with a gear wheel I attached through a ratchet device to a flywheel W on the shaft K. A mixture of gas and air was admitted to the cylinder through the lower R.H. passage by means of a slide valve S worked by an eccentric O and rod. This was simply a device to admit the gas and air mixture at the proper time. The mixture in the cylinder was ignited by means of a flame f in the cover of the slide valve. The explosion forced the piston upwards, and during its pas-

sage the rack C engaged with the toothed wheel T, but the latter, in virtue of its free-wheel device, did not turn the fly-wheel. The power stroke in this case was the downward, or return one, due to the weight of the piston and rack, together with the atmospheric pressure on top of the piston (there being a vacuum underneath, due to the cooled exhaust gases). The ratchets in the free-wheel engaged during the down stroke, so that the piston rack turned the fly-wheel in the direction shown. The explosion was very violent and there was a good deal of noise and rattling of the gear, but nevertheless many of these engines were installed in this country and in France and worked satisfactorily.

Similar to Lenoir's gas engine of 1860, this type belonged to the non-compression class, in which the explosive charge was ignited at atmospheric pressure. All modern engines work on

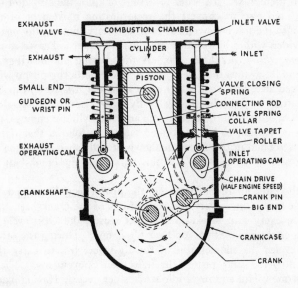

Fig. 7.—Illustrating the Components of a simple Petrol Engine and their names.

the *compression* principle, the charge being compressed to several atmospheres pressure before ignition. The result is a very considerable increase in power, reduction in size of engine, more rapid working, and much better fuel economy.

The Otto Principle.—The majority of modern engines work on a principle of operation first enunciated by M. Beau de Rochas (in 1862) and applied by Otto, in which compression of the charge was the important feature. In order to understand this principle it will be necessary to outline briefly the mechanism of a simple type of engine in order to familiarize the reader with the common automobile engineering terms employed, and to show how the method is carried out. Fig. 7 represents in outline the mechanism of the simple one-cylinder type internal combustion engine. It consists of a cylinder, having a closed end, in which a hollow piston can slide inwardly and outwardly, by a fixed amount, termed the *Stroke*. In order to define, or limit the stroke of the piston, and also to convert the reciprocating movement of the latter into the more convenient form of rotary motion, the piston is provided with a hinge-joint and pin; the latter is known as the *Gudgeon, Piston or Wrist Pin*. A rod, having pin bearings at each end, known as the *Connecting Rod*, connects the piston with a cranked arm, or *Crank*, attached to a shaft (*Crankshaft*) capable of rotating in fixed bearings. A fly-wheel is usually attached to the crankshaft, in the case of one and two cylinder engines, for a specific reason, to be mentioned later. It follows that if the crank is rotated in the direction shown, it will cause the piston to reciprocate in the cylinder and this will cause the crankshaft to turn.

In the Otto method of working, the piston receives one power impulse stroke once every two revolutions of the crankshaft, that is, once every four strokes. For this reason the Otto cycle is known as the *Four Stroke* or *Four Cycle* one. During the other three cycles the momentum of the fly-wheel has to carry the connecting-rod and piston through their movements. There is thus one useful and three idle strokes per cycle. Having made this clear, we can now consider a complete cycle in more detail, and follow each successive stroke, commencing with the one termed the *Suction Stroke*, and illustrated in Fig. 8 (*a*). In the head, or *Combustion Chamber* of the cylinder, two orifices or ports are arranged; each orifice can be completely sealed from the inside by Valves. One valve, known as the *Inlet Valve*, regulates the admission of the explosive mixture, or charge, whilst the other, or *Exhaust* or *Outlet Valve*, controls the exit of the burnt or exhaust gases. These valves are operated by simple mechanical devices, driven at one-half engine speed.

In Fig. 8 (*a*) the piston is moving downwards, and the inlet valve is open. The suction created by the piston draws the explosive charge into the cylinder. When the piston reaches its lowest position (or *Bottom Dead Centre*), the inlet valve is arranged to shut down on its seating, thus sealing the cylinder contents. On the upward stroke, Fig. 8 (*b*), the charge is compressed into the combustion chamber end of the cylinder.

Just at the moment when the piston reaches the upper end of its stroke, an electric spark is caused to pass inside the combustion chamber at the *Sparking Plug*, indicated by the coiled wires and spark-gap shown on the upper right hand side of cylinder. (Fig. 8). The resulting explosion forces the piston downwards, the burnt gases expanding at the same time. This is the *Firing Stroke*, Fig. 8 (*c*). Just before the end of the firing stroke, the exhaust valve, which until this moment has remained closed, opens, and allows the exhaust gases to flow out into the atmosphere or a silencing device. The final upward stroke of the piston forces the exhaust gases out of the cylinder. This is the *Exhaust Stroke*, Fig. 8 (*d*). At the end of this stroke the exhaust valve closes, leaving the combustion head full of burnt gases at about atmospheric pressure. These gases have no real detrimental effect, but dilute the incoming charge.

The Practical Aspect of the Four-Stroke Cycle.—Hitherto we have been dealing with the theoretical aspect of the Otto cycle; it is now proposed to consider what actually happens in the cylinder of a modern engine during the cycle of operations. In this respect it must be emphasized that unlike the steam engine with its slower speeds of 150 to 300 r.p.m., and its large cylinders and pistons, the petrol type engine has comparatively diminutive cylinders (from 2 to 5 inches diameter in the case of automobile engines), so that its individual impulses are relatively small. Owing, however, to the rapidity of the combustion and expansion of the gases, the piston moves very quickly when working normally and to facilitate these movements the crankshaft revolutions are high, namely, from 1,000 to about 5,000 r.p.m. The petrol type automobile engine gives only a fraction of its full power at low speeds, so that for slow road speeds and for hill climbing gear-boxes or variable gears (which enable the torque at the road wheels to be kept high) become essential.

Commencing with *the suction stroke*, the entering mixture of

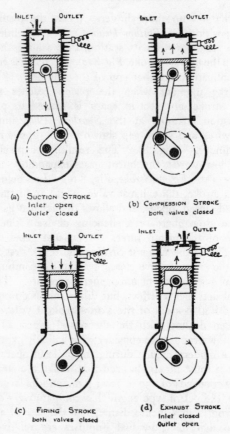

(a) SUCTION STROKE
Inlet open
Outlet closed

(b) COMPRESSION STROKE
both valves closed

(c) FIRING STROKE
both valves closed

(d) EXHAUST STROKE
Inlet closed
Outlet open

Fig. 8.—Illustrating the principle of the Four-Cycle
Engine.

air and fuel is comparatively cool and at a pressure slightly below
atmospheric. This charge enters the cylinder with a very high
velocity; in normal cases the speed of the fresh charge through the
inlet pipe and valve port varies from 120 to 240 feet per second,
that is, from about 82 to 164 miles per hour.

The high velocity charge swirls around the combustion chamber in a state of agitation which persists not only during the suction, but also the compression stroke. Owing to this *Turbulence*, as it is termed, the combustion of the fuel particles is extremely rapid, for their movements spread the flame much quicker than would otherwise be the case. The degree of turbulence depends partly upon the design or shape of the combustion chamber, and also upon the position of the valves, and the engine speed. In certain cases the shape of the combustion chamber is specially designed so as to promote turbulence. But for this turbulence effect, petrol engines would only work satisfactorily at low speeds.

Compression Ratio.*—At the end of the compression stroke, the pressure of the charge at full throttle opening varies from 120 to 190 lb. per sq. in. in most automobile engines. This pressure varies with the *Compression Ratio*; this is the ratio of the combustion chamber, plus the piston swept-volume to the combustion chamber volume. In Fig. 9 it is the ratio of clearance volume (G)+stroke volume, to the clearance volume. (G).

This ratio in the smaller mass-produced European cars is usually about 6·5 to 7·2 for ordinary commercial petrols and from about 7·8 to 9·0 for premium fuels; the average British engine of 1·0 to 2·0 litres capacity has a compression ratio of 8·0 to 8·7:1. With super-grade fuels, compression ratios of 9·0 to 10·1:1 can be used. In the case of certain American car engines, in 1962, ratios of 10:1 to 10·5:1 were used. For special racing car engine fuels, compression ratio of 10·0 to 12·0:1—and ever higher, are employed. Immediately after the spark passes at the plug and the mixture ignites, the pressure rises rapidly to about 3 to 4 times the compression pressure value, after which it falls rapidly as the piston descends, until at the moment the exhaust valve opens its value is usually from 60 to 85 lb. per sq. in., dropping to slightly above atmospheric pressure as the valve opens fully.

During the exhaust stroke, the burnt gases are at a pressure above atmospheric value, and there is thus a certain load on the piston, acting against it, or opposing its motion. With large exhaust valves, opening for the longest possible period, and an efficient silencer this loss of power due to *Back Pressure* of the exhaust gases can be minimized.

* See Footnote on Page 47.

Exhaust Pressure Waves.—The effect of the sudden release of high pressure gases into the exhaust system is to cause a series of pressure waves which usually vary from positive to negative (suction) values. Use can be made of these fluctuations to get rid of the exhaust gases in the cylinders, by a scavenging effect, whereby there is a negative pressure in the vicinity of the exhaust valve when the inlet valve is about to open.

The pressures and temperatures within the cylinder vary according to the type and design of any engine and depend upon a number of other factors, including the compression ratio, fuel, mixture strength, ignition timing, etc. The following are approximate values for a typical engine:

TABLE 3

Pressures and Temperatures in Petrol Engine

Compression Ratio	Compression Temperature °C.	Compression Pressure[1] lb. sq. in.	Mean Pressure in Cylinder[2] lb. sq. in.	Maximum Cylinder Pressure lb. sq. in.
4	280	93	140	300
5	327	125	160	400
6	357	159	172	650
7	397	195	184	800
8	427	224	190	950
9	457	256	197	1090
10	487	288	205	1195

[1] Absolute Values.   [2] Indicated Mean Effective Pressure.

Engine Temperatures.—The temperature of the charge just before the ignition depends chiefly upon the compression ratio; its value ranges from about 330° C. for a ratio of 5·0:1 up to 410° C. for one of 7·5:1.

The temperature during ignition is very high for a short period, and ranges from 1500° C. up to over 2000°.* Owing to these very high temperatures it becomes necessary to provide means for cooling the cylinder and piston, otherwise these would

* Iron melts at 1530° C., and platinum at 1750° C. The temperatures of explosion only occur for a small fraction of a second, so are ineffective in melting the metal.

soon become too hot causing piston seizure. The two methods adopted in the case of automobile engines are the water-jacketed cylinder, and the air-cooled radiating fins ones; there was also an engine which employed the lubricating oil to cool the cylinder walls and barrel.

The inlet valve is kept relatively cool by the petrol vapour which flows past it; its temperature seldom exceeds 250° C., whereas that of the exhaust valve, which is exposed to the hot exhaust gases, varies from 600° C. to 700° C. The top of the piston also gets fairly hot, since the heat cannot readily get away; it varies from about 300° C. to 350° C. in average engines. It is owing to this high "crown" temperature that the lubricating oil splashed up on the under side usually becomes carbonized.

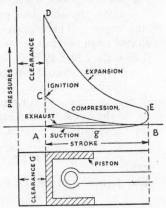

Fig. 9.—Graphical method of illustrating Cylinder Pressures.

Graphical Representation of Cylinder Pressures.—To those accustomed to reading graphs, the cylinder pressure—or indicator—diagram will be found useful in interpreting the Otto, or four-stroke cycle of operations, upon which most automobile engines work. Referring to Fig. 9, the piston and cylinder are shown in outline below, and the pressure diagram above. The base AB of the pressure diagram corresponds to the piston stroke. During the suction stroke AgB the pressure in the cylinder G is slightly below atmospheric, and during the next inward or compression stroke it rises along the line BC to the compression pressure represented by the height AC corresponding to the inner dead centre piston position. The pressure at any intermediate position is represented by the height of the pressure ordinate at that place.

At C the spark occurs, and the pressure rises suddenly to the value AD. As the piston is forced outwards, the pressure falls along the expansion line *DE*, until at *E* the exhaust valve opens, and the pressure drops quickly to a value slightly above atmospheric during the exhaust stroke *BA* (upper line).

Horse-Power.—If the pressure scale is expressed in lb. per sq. in., and the stroke *AB* in feet, the area of the diagram *BCDE* will represent the work done per cycle, that is per two revolutions of the engine. The average height of this diagram gives the indicated mean effective pressure, or, as it is termed, the I.M.E.P.

If P_m=the I.M.E.P. in lb. sq. in.

l=length of stroke in feet.

A=piston area in sq. in.=0.7851 (diameter)^2

and N=engine speed in r.p.m.

then the horse power developed in the cylinder, or *Indicated Horse Power* (I.H.P.) is given by the relation

$$\text{I.H.P.} \times \frac{P_m.l.A.N.}{66,000}$$

This formula can also be written as follows

$$\frac{9.91664\, P_m l d^2 N n}{10,000,000}$$

Where d=cylinder diam. l= stroke in ins. and n=number of cylinders.

For a two-cycle engine the constant becomes 19.8333, but the value of P_m is lower than for the four-cycle engine.

The horse-power available at the crankshaft, known as the *Brake Horse-Power*, or B.H.P., is less than this owing to the power absorbed in piston friction, bearings friction, operation of the valve gear, fuel feed pump, ignition drive unit, water pump cooling fan, oil pump, etc. As is explained later these losses usually amount to 10 to 20 per cent. of the power developed in the cylinders, i.e. the I.H.P.

Gross and Net Horse-Power.—When the power output of an engine is measured, by means of a dynamometer, separate water-cooling and exhaust systems are employed. The horse-power figures at various engine speeds are then known as the *Gross B.H.P.*[*] values. When the engine is installed in its chassis,

---

[*] Except where stated otherwise, B.H.P. figures given in this volume are gross values.

certain accessories, including the cooling water circulating pump, radiator fan, dynamo, fuel feed pump and the ignition distributor shaft and contact-breaker mechanism have to be driven from the engine. The power then available for driving the car is therefore less than the Gross B.H.P. and is termed the *Net B.H.P.*, so that:

Net B.H.P. = Gross B.H.P. − H.P. for driving the accessories

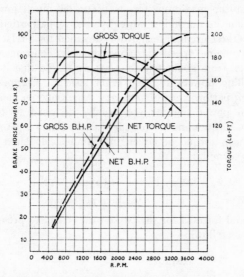

Fig. 10.—Gross and Net Horse-Power and Torque for Vauxhall Bedford Engine.

At the full engine output the Net B.H.P. is usually 15 to 20 per less than the Gross B.H.P., falling at fast idling speeds to about 7 to 10 per cent.

Engine Torque.—A knowledge of the torque or turning effort on the crankshaft of an engine is of special importance from the car's performance aspect. Whilst the crankshaft revolves, this torque is variable reaching a maximum value once every two revolutions for the four-cycle engine. As the number of cylinders of an engine are increased, so do the torque variations diminish until for engines of 8, 12 or 16 cylinders they are practically of little or no importance.

The average, or mean torque of an engine provides an indication of a car's pulling effort or performance, so that it is usual for the manufacturer to provide figures or a graph for engine torques at stated speeds. Fig. 10 shows the average torques for a Vauxhall engine at both the gross and net outputs. It will be seen from these that the maximum torque occurs at much lower engine speeds than the maximum power speeds. The particular engine for which these graphs are shown is designed for a relatively wide range of engine speeds at approximately constant maximum torque, i.e. from about 600 to 2,600 r.p.m., so that it possesses good pulling powers at these lower speeds. Competition and other high performance car engines, which operate near to their maximum power outputs are usually designed to have their maximum torques and power curve peaks closer together.

The mean torque of an engine can be calculated from the following formula:

$$\text{Mean Torque} = 5,252 \times \frac{\text{B.H.P.}}{N}$$

Where N = engine revolutions per minute

The general formula for the brake mean effective pressure of an engine is as follows:—

$$\text{B.M.E.P.} = \frac{\text{Constant} \times \text{B.H.P.}}{N}$$

Where N = revolutions per minute

so that the graph showing the relationship of the B.M.E.P. and engine speed is of identical general form to the engine torque-speed curve, but its scale, is different.

The Two-Cycle Engine.—There is another cycle of operations, upon which the petrol engine can work, in which only two strokes, or one revolution of the crankshaft, are necessary for a complete cycle.

This method, first utilized by Sir Dugald Clerk in 1880, consists in partly compressing the mixture in a separate cylinder, or in the crank-case of the engine itself, and admitting this mixture into the cylinder when the piston is at the outer end of its stroke.

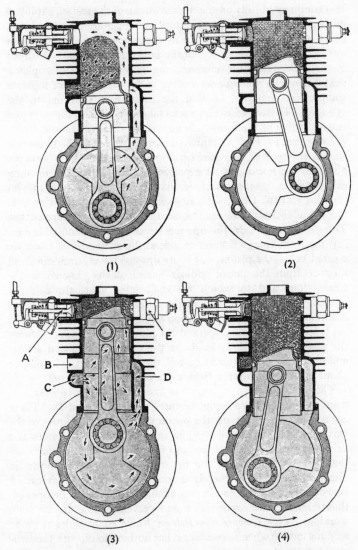

Fig. 11.—Illustrating the operation of the Two-Cycle Engine.

It is compressed fully on the inward stroke of the piston, exploded in the usual way and expanded until the upper edge of the piston, when nearing the outer end of its expansion stroke, uncovers an orifice, or port, through which the burnt gases escape. In most motor-cycle two-cycle engines, the lower edge of the piston next uncovers a second port leading to the crank-case, or separate compression cylinder, so that the mixture is admitted to the cylinder. In such cases the piston takes the place of the inlet and exhaust valve of the four-cycle engine for these operations.

Referring to Fig. 11 which illustrates the Enfield two-cycle engine, the cycle of operations is as follows: Diagram (1) shows the piston at the bottom of its expansion stroke, with the remaining exhaust gases flowing out of the exhaust port B (Diagram (3)) on the left, and the fresh, slightly compressed mixture flowing into the cylinder from the crank-case through what is termed the *Transfer Port D*. On the upward stroke of the piston (Diagram (2)) the fresh charge is compressed; at the same time the suction created below the piston, due to its upward travel, draws in fresh mixture from the petrol and air mixing device, known as the *Carburettor*, and to which we shall refer later, through the opening C (Diagram (3)). In the latter diagram the piston has reached the top of its compression stroke, and is just about to be exploded by the electric spark at the plug E. Diagram (4) shows the piston moving downwards on the ensuing expansion stroke, and uncovering the exhaust port B, the transfer port D being just about to uncover when a fresh cycle of operations occurs.

This common type of engine is known as the *Three Port* one, from the fact of there being the three ports B, C, and D. There are no valves in this type, the piston performing the duties of the valves in the four-stroke type, so that the absence of valves and valve gearing enables such engines to be made more economically than four-stroke ones. The inwardly opening valve shown in Diagram (3) at A is merely a motor-cycle control device. A Bowden lever on the handle-bar enables this valve to be opened, thus releasing the compression and preventing the engine from working. This *Compression Release Valve* enables the rider to stop his engine when necessary; it has nothing to do with the two-stroke cycle of operations.

It is not proposed to consider the mechanical details, or the different types of two-cycle engine, here, but to deal only with the theoretical principles. The manner in which the pressures in a two-cycle engine vary is illustrated in Fig. 12. The diagram of pressures is somewhat similar to that shown in Fig. 9, except for the fact that there are no suction and exhaust lines (these are the two "strokes" missing in the present case), and that the "toe" of the diagram is different.

The pressures developed in the case of a two-cycle engine are appreciably lower than in the four-cycle one of the same dimensions, and at the same speed. This is due to the fact that it is difficult to fully charge the cylinder during the short inlet port opening interval, and also that the engine tends to run much hotter, since it has twice as many explosions in the same time as the four-cycle engine. Moreover, since both exhaust and inlet ports are open together, a part of the fresh charge blows right through into the exhaust, and is lost; this results in a smaller effective charge for compression and firing. In order to minimize this charge loss, it is usual to shape the piston as shown

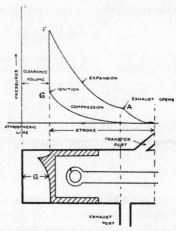

Fig. 12.—Graphical Method for Two-Cycle Engine Pressures.

in Fig. 7, namely, with a baffle on the crown, so as to deflect the incoming charge, and to obstruct its passage to the exhaust. The results of scientific tests made on three-port engines, fitted with deflector pistons, indicate a loss of charge through the exhaust port, however, varying from 15 to 25 per cent., in spite of the deflector. The net result of the disadvantages mentioned are that the ordinary two-cycle engine gives only about 50 to 60, instead of

100 per cent. more power for a given cylinder capacity, than the
four-cycle engine. Further information on two-cycle engines
is given in Chapter V.

Brake Horse-Power of Engines.—The automobile engineer
is more concerned with the actual horse-power delivered at the
crankshaft (and available for propulsion purposes) than with the
power developed in the cylinder itself. It has, therefore, become
customary to deal with the shaft, or brake horse-power (B.H.P.)
in making comparisons of engines.

As shown in Chapter X, the B.H.P. can be measured accurately
with a power brake or dynamometer, and in this respect the following relation is of interest when studying the possible limits of
performance of four-cycle petrol engines:

$$\text{B.H.P.} = \frac{\text{B.M.E.P.} \times \text{Displacement} \times \text{R.P.M.}}{792,000}$$

Where B.M.E.P. = Brake mean effective pressure in lbs. per sq. inch.
 Displacement = Total cylinder volume swept by pistons. (cu. inch.)
             = Piston area (sq. in.) × piston stroke (in.) × number
                 of cylinders.
 R.P.M. = Revolutions per minute of engine.

As it is usual, in this country to express cylinder capacities in
litres, the following relationship will be found useful:

$$\text{B.H.P.} = \frac{p, N.V}{12,950}$$

where p=B.M.E.P. in lb. per sq. in.; V=cylinder swept volume,
or capacity in litres, and N=revolutions per minute.

It will be seen that for an engine of fixed cylinder dimensions the
horse power that the engine develops will depend upon (1) The
B.M.E.P. or average effective pressure that can be obtained and
(2) The engine r.p.m. The B.M.E.P. is usually about 85 to 90
per cent. of the indicated pressure, or I.M.E.P., developed within
the engine cylinders during a complete cycle of operations.

To obtain *the maximum B.M.E.P.*, the *combustion chamber*
should be designed correctly and the engine valves and ports
should be ample, to ensure the greatest weight of mixture being
taken into the cylinder during the piston's suction stroke. The
compression ratio should be increased to the highest value, for the

fuel used, that will avoid detonation under full load conditions. Losses of power in engine friction should be reduced to a minimum, e.g., the lowest bearing friction should be aimed at.

There are other factors contributing to high B.M.E.P. values. These include correct valve and ignition timing; proper sparking plug location; efficient carburation under maximum loads; good mixture distribution; correct cylinder temperatures, etc.

Automobile Engine Developments.—Since the Second World War there has been a constant increase in the power output for a given cylinder capacity, accompanied by reduced fuel consumption per horse-power. In regard to production car engines, these improvements have been due to the adoption of higher compression ratios, increased engine speeds, greater use of overhead camshaft-operated valves and improved designs of combustion chamber.

Compression Ratios.—In regard to compressions, the pre-War engines used ratios of 5·0:1 to 6·0:1, followed in the earlier post-War years by ratios of 6·5:1 to 7·0:1, using the improved "pool" grades of petrol then available. With the later introduction of higher octane fuels, compression ratios have been steadily increased, until in 1962, these ratios have extended to the range of 7·5:1 to 10·5:1, with an average value, in this country of about 8·5:1 and in the U.S.A. of 9·0:1. It may be mentioned that there are at least twenty different makes and models of American cars having compression ratios of 10·0:1 to 11·0:1.

Engine Performances.—The performance of any petrol engine can be expressed by its output per given cylinder capacity. Thus, in this country it is usual to compare performances by the B.H.P. per litre (1,000 c.c.'s) and in the U.S.A. by the B.H.P. per cubic inch of cylinder capacity, or piston displacement.

Over the past twenty years or so, for the average medium production car engine, the output has increased from about 27 to 34 B.H.P. per litre, to 38 to 45 B.H.P. per litre. For average American car engines, from 1938 to 1962 the output has increased from 0·412 B.H.P. per cubic inch, to about 0·70 to 0·80 B.H.P. per cubic inch.

For higher powered expensive touring cars, e.g., the Jaguar,

* For later information see Chapter XI.

Ferrari, Facel-Vega and Maserati, the more recent outputs ranged from about 60 to 100* B.H.P. per litre, with corresponding outputs of 1·04 to 1·65* B.H.P. per cubic inch. It should be pointed out that these engines were in the 250 to 380 B.H.P. class with 6, 8 (vee) and 12 (vee) type engines. Certain American production cars are fitted with optional higher powered engines and these develop from about 0·88 to 0·95 B.H.P. per cubic inch.

High Performance Engines.—It is possible to increase the output of a petrol engine considerably above that of the normal car engine, previously mentioned by such expedients as: (1) Using larger valves, with streamlined ports and special valve timing methods. (2) Employing still higher compression ratios, e.g., from 10·5:1 to 12·5:1—or even higher. (3) Much higher engine speeds. (4) Multiple carburettors to give a higher volumetric (or charge) efficiency. (5) Separate exhaust pipes, or systems for the individual cylinders, attuned to the exhaust pressure system. (6) Special "racing engine" fuel mixtures.

In this way it is possible to obtain maximum outputs, for normally-aspirated (as distinct from supercharged) engines of 100 to 135 B.H.P. per litre, i.e. 1·64 to 2·213 B.H.P. per cubic inch. In this connection smaller bore engines of the vee-six, vee-eight and vee-twelve types have given the greatest outputs for a given capacity; such engines have run at speeds up to and often above 10,000 r.p.m.

For still higher outputs supercharging systems, giving up to at least 50 per cent. more power, in suitably designed engines must be used.

Engine and Piston Speeds.—As mentioned earlier, engine speeds have been increased as one means of obtaining greater power. Thus, if the mean cylinder pressure (B.M.E.P.) can be kept constant, the B.H.P. will increase in proportion to the engine speed increase, as shown by the formula on page 46. Over a 20-year period the average speeds of British engines have increased from about 3,500 to 4,000 r.p.m. up to 4,200 to 5,500 r.p.m. (for the 750 to 1,000 c.c. engines).

In the case of the higher powered expensive cars mentioned in

* Ferrari Superamerica touring car.

the previous section the maximum power engine speeds range from 5,500 to 6,750 r.p.m.

In regard to American production car engines over the period 1938–62, the average engine speeds of the whole car range have increased from 3,580 r.p.m. to 4,400 r.p.m., with maximum speeds (1962) of 4,800 r.p.m. to 5,800 r.p.m. (Ford 390 B.H.P. Galaxie). It may be mentioned that the special vee-six, vee-eight and vee-twelve high powered engines developed solely for track and road racing purposes operate at maximum speeds of about 6,000 to 9,500 r.p.m.

Piston Speeds.—The mean speed of travel of the piston in its cylinder is limited by friction and lubrication considerations. This speed can be estimated as follows:

$$\text{Mean piston speed} = \frac{S \times N}{6} \text{ (feet per minute)}$$

Where S = piston stroke in inches and
N = engine revolutions per minute

When the stroke is given in millimetres the formula becomes:

$$\text{Mean piston speed} = \frac{S \times N}{152 \cdot 4}$$

For the ordinary British production car engine of 1 to $2\frac{1}{2}$ litres capacity the mean piston speeds range from about 2,400 to 2,500 ft. per min., but for high performance engines fitted to high road speed cars, the mean piston speeds range from about 3,000 to 3,800 ft. per min.

The mean piston speeds of the average of all American cars has increased from about 2,500 ft. per min. in 1936 to 2,730 ft. per min. (1957–58), down to 2,560 to 2,750 ft. per min. (1959–62).

Engine Stroke-Bore Ratios.—Previously to 1947, the taxable horse-power of a car engine was based upon the area of the cylinder only, so that it was possible to obtain the greatest power output, within certain limits, by using the greatest piston stroke for a given cylinder bore. Usually, the earlier stroke-bore ratios ranged from about 1·4 to 1·6. After 1947, the taxable horse-power was first based upon the cubical capacity of the engine and, finally, a fixed annual charge per car was made which was independent of engine size or horse-power.

With this latter change, engine manufacturers began to reduce the piston stroke in relation to the cylinder bore until, in 1962, the stroke–bore ratios of British cars ranged from about 0·6 (Ford Anglia) to 1·25. The greater number of engines had ratios between 0·88 and 1·10.

In the case of American engines, over the period 1936–62, the stroke–bore ratios decrease from about 1·3 to 1·00 in 1955, and then down to about 0·94 in 1962.

The tendency to increase the engine bore in relation to the stroke results in *lower piston speeds* for the same maximum engine r.p.m. The larger bore enables larger valves to be used, with *increased charge* in the cylinders and, therefore, an increase in output from the same amount of fuel. Lower piston speeds mean *longer life* for the cylinder bores and piston rings. Engines can be made smaller in height and crankshafts—with their shorter crankpin throws—made stiffer.

As an example of the benefits of the larger bore-shorter stroke engine, the results of some actual tests made by Vauxhall Motors Ltd. on earlier engines of 2·736 in. bore and 3·74 in. stroke (stroke-bore ratio, 1·37) and later "square" engine of 3·125 in. bore and 3·00 in. stroke (ratio 0·96) are given herewith. In this connection, both engines had the same compression ratio. The "square" engine gave 40 B.H.P. at 4,000 r.p.m. and the earlier engine, 35 B.H.P. at 3,200 r.p.m. Both engines had maximum torques of 51 lb. ft. at 2,000 and 1,800 r.p.m., respectively.

Apart from the increased power of the "square" engine, road tests showed that it gave a greater mileage per gallon, while at a road speed of 60 m.p.h. the average piston speeds of the earlier and "square" engine were 2,334 and 1,872 ft. per min., respectively. The much lower piston speed of the "square" engine results in a lower piston travel per road mile and therefore, as stated earlier, longer cylinder bore, piston and piston ring lives.

If the formula for mean piston speeds, strokes and engine r.p.m. is studied it will be seen that for the same maximum power r.p.m., the shorter stroke gives the lower mean piston speed, while for the same mean piston speed the shorter stroke engine must operate at higher r.p.m. than the longer stroke engine.

Performances of Modern Engines.—Typical examples of

engine performances for car engines ranging from the smaller 848 c.c. (51·7 cu. in.) to the larger 7,046 c.c. (429·8 cu. in.) models are given in Table 4. While there are smaller and larger capacity engines in the complete range of available engines, those selected enable the trend of engine power output, maximum torque and B.M.E.P. to be studied. Some typical examples of current American vee-eight car engines have been included in the table.

TABLE 4

Performances of typical British and American Engines

Cylinder Capacity c.c.'s	cu. in.	Compression Ratio	Maximum B.H.P. (gross)	at r.p.m.	Maximum torque lb. ft.	at r.p.m.	Maximum B.M.E.P. lb. sq. in.	at r.p.m.
848 [1]	51·7	8·3	37	5,500	44	2,900	128	2,900
948 [1]	57·9	8·3	40	5,000	·50	2,500	130	2,500
1,172 [1]	70·4	7·0	38	4,500	53	2,500	111	2,500
1,300 [1]	79·3	8·8	72	5,200	76	3,400	145	3,400
1,500 [1]	91·5	8·3	50	4,100	77	2,000	127	2,000
1,600 [1]	97·6	8·3	66	4,800	84	2,800	131	2,500
2,000 [1]	122·0	7·5	83	4,500	105	2,500	136	2,500
2,600 [2]	158·6	8·1	113	4,800	138	1,600	129	1,600
2,785 (A)	171·9	8·7	101	4,200	138	2,000	144	2,000
2,900 [3]	176·9	8·3	120	4,750	163	2,500	138	2,500
3,168 *	193·4	9·0	140	4,800	163	2,000	127	2,000
3,523 (A)	214·9	8·8	155	4,600	220	2,400	154	2,400
3,781 [2]	230·6	9·0	265	5,500	260	4,000	172	4,000
4,561 [3]	278·1	8·0	220	5,500	283	3,200	150	3,200
4,638 (A)	283·0	8·5	185	4,600	275	2,400	146	2,400
5,918 (A)	361·1	10.0	265	4,400	380	2,400	173	2,400
6,572 (A)	400·1	10·2	325	4,400	445	2,800	167	2,800
7,046 (A)	429·8	10·0	315	4,100	465	2,200	163	2,200

(A) American (vee-eights). [1] Four cylinder. [2] Six cylinder. [3] Vee-eight.

The engines with capacities up to about 2,000 c.c. (122 cu. in.) are all of the four-cylinder vertical type.

Power-to-Weight Ratio.—The performance of an automobile depends upon its weight, in laden order and the power available from its engine. Thus, a light car with a relatively powerful engine will have a much better acceleration and top road speed than a heavier car with a lower powered engine.

The acceleration of a moving body is expressed, as follows:

$$\text{Acceleration } f = \frac{\text{Force } F}{\text{Mass } M}$$

Where f is in feet per second, per second, F is in pounds and $M = W \times g$, where W = weight of body in pounds and g = the acceleration due to gravity, i.e. 32·19 feet per second, per second.

In the case of the automobile the force F is derived from the engine, *via* the transmission and road wheels, while the weight W is that of the laden vehicle, so that the acceleration will be pro-

portional to $\dfrac{F}{W}$, or $\dfrac{\text{B.H.P.}}{W.}$

Therefore, the greater this ratio of $\dfrac{\text{Power}}{\text{Weight}}$ the better will be

the performance, assuming that the car is of sound design, with suitably chosen transmission ratios. The top road speed can also be shown to depend upon the same ratio.

An examination of the power-to-weight ratios of current model automobiles shows that for the smaller cars, up to about 1,200 c.c. engine capacity the ratio is between about 30 and 45 B.H.P. per ton (laden) weight.

For production cars with engines of 1,400 to 1,600 c.c., the ratio is from about 45 to 55. This ratio also applies to certain larger four-cylinder engine production cars. For the rather higher performance cars with engines of 1,500 to 2,500 c.c., power-to-weight ratios from about 60 to 75, corresponding to top road speeds of 90 to 100 m.p.h. are employed.

In the larger and higher performance car range, in which car accelerations, from 0 to 50 m.p.h., can take only 6 to 10 seconds and maximum speeds up to 150 m.p.h. are attainable, the ratios usually extend from about 80 to as much as 200 B.H.P. per ton. Thus, the Jaguar E-type car with a maximum speed of about 150 m.p.h. has a power-to-weight ratio of 195·4. and the Aston-Martin D.B.4 car, 227 for a somewhat similar maximum speed. For current standard American cars the ratios range from about

120 to 160. The power-to-weight ratios of racing cars are, of course, appreciably higher than these figures.

Greatest Possible Output of Any Engine.—It is of interest to designers and others to know what is the greatest possible power output that can be obtained from any unsupercharged engine. It is not a difficult matter to ascertain this from considerations of the quantity of air passing through the engine at any given

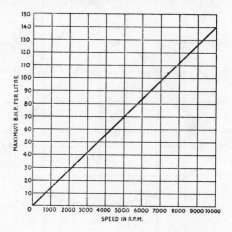

Fig. 13.—Maximum B.H.P. per litre Graph.

engine speed, per minute, since the amount of oxygen used determines the power output. If this is done it can be shown that the maximum output, in B.H.P. per litre, will vary with engine speed in the manner indicated in Fig. 9, namely, from about 14 at 1,000 r.p.m. to 140 at 10,000 r.p.m., and in direct proportion at higher speeds. It should, however, be made clear that these values represent the limiting outputs theoretically attainable using certain assumptions concerning volumetric, mechanical and thermal efficiencies, etc.

From these results it follows that the other possible methods to increase the output per litre, above the values given in Fig. 13 are, as follows: (1) To operate at still higher engine speeds. (2) To increase the charge (volumetric) efficiencies so as to induce more air into the cylinders during the suction stroke. (3) To employ special fuels which contain their own oxygen contents, e.g., Nitromethane in Methanol, and (4) By increasing the inlet air pressure, i.e. by supercharging.

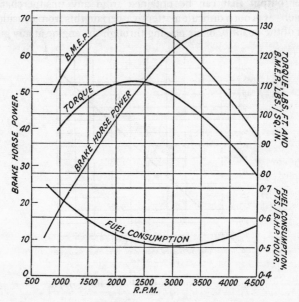

Fig. 14.—Typical Engine Performance Graphs.
(Four-cylinder engine.)

Typical Engine Performance Curves.—To illustrate the improvement that has occurred in the performances of production automobile engines, the curves shown in Fig. 14 for a typical four-cylinder mass-production engine of 1·6 litre cylinder capacity.

Referring to the brake horse-power curve, the scale for which is given on the left, it will be seen that the maximum B.H.P. is about 68, at 4,000 r.p.m. The maximum B.M.E.P. occurs at about 2,300 r.p.m. and is 131 lb. per sq. in., whilst the maximum torque, which also occurs at the same engine speed is about 111 lb. ft. per litre.

Charge or Volumetric Efficiency.—During the suction stroke of a four-cycle engine the piston, in moving from its top to bottom dead centre, should theoretically draw into the cylinder a volume of mixture at atmospheric pressure equal to its area multiplied by its stroke, i.e., its working volume. In practice the quantity of mixture drawn into the cylinder is always appreciably less than this theoretical amount, so that it is usual to term the ratio of the actual to the theoretical amounts the charge or *Volumetric Efficiency*. Thus:

$$\text{Volumetric Efficiency} = \frac{\text{Actual Charge Weight}}{\text{Theoretical Charge Weight}} \times 100 \text{ (per cent.)}$$

The greater this efficiency the more power will be developed from a given size of engine. The usual value of the volumetric efficiency in the case of a well-designed engine is from 80 to 85 per cent. at full throttle opening. The principal reasons for the inability of the engine to obtain a full weight of charge during the suction stroke are as follows: (1) Throttling effects due to presence of carburettor restrictions to mixture flow, presence of bends in inlet manifold, changes of section of manifold and inlet port, etc. (2) Heating of the incoming charge by the hot inlet port, cylinder combustion chamber, walls, piston, hot exhaust valve, remaining hot exhaust gases in clearance space, from previous exhaust stroke, etc.

Methods of increasing the volumetric efficiencies of existing engines include: (1) The use of one dual-barrel carburettor per two engine cylinders. (2) Separate carburettors to each cylinder. (3) Substitution of fuel injection into each cylinder, instead of using carburettors. (4) Using an air compressor (supercharger). (Methods (2) and (3) give appreciable increases in charge efficiency while method (4) can give considerable power increase.

In the case of supercharged engines, whilst it is of equal importance to minimize the results due to (1) and (2), the fact that the mixture is forced into the cylinder under pressure, instead of relying upon engine suction, ensures that a greater quantity of mixture will enter the cylinder during the suction stroke.

Horse-Power and Air Consumption.—It has been shown by

H. Ricardo that there is a definite relation between the *weight of air* entering the cylinder per minute and the *indicated H.P.* developed and that this relationship is, within fairly small limits, the same for all hydrocarbon fuels. This statement, however, does not mean that all engines of similar dimensions will give the same power output, since other factors—such as the thermal and volumetric efficiencies, mixture strength and mechanical efficiency—must also be taken into account.

It is, however, useful to note that for engines having the same mechanical efficiency and operating upon rich mixtures, using either petrols, benzole or alcohol, the indicated horse-power can be estimated, approximately, from the following relation:

$$\text{I.H.P.} = k \times E \times A.$$

Where A=lbs. of air consumed per I.H.P. per hour and k=a constant which, for various hydrocarbon liquid fuels varies between the limits of 1·94 and 2·00, and E=thermal efficiency of the engine, expressed as a decimal. The deduction that can be made from this relationship is that for the *maximum power the engine should consume the greatest amount of air per unit time*; as shown later, the method of supercharging an engine to obtain greater outputs is based upon this principle.

It is also of interest to remember that the results of tests by Ricardo show that, irrespective of the fuel employed in the engine, the amount of energy liberated for each pound weight of air consumed is approximately 1,300 B.T.U.'s, for rich mixtures.

Estimating the Mean Effective Pressure.—For engine design purposes it is useful to be able to estimate, approximately, the mean effective pressure that will be developed. This can be done as follows:

Assuming that the engine is working on petrol having a calorific value of 18,290 B.T.U.'s per lb. and with the correct air-petrol proportions of 15:1 by weight, it can be shown that the heat energy of the mixture works out at 45 ft. lb. per cu. in.

If the volumetric and also the thermal efficiencies were at their ideal values, 100 per cent., the theoretical indicated M.E.P. would be given by the following relation:

$$\text{Theoretical I.M.E.P.} = 12 \times 45$$
$$= 540 \text{ lbs. per sq. in.}$$

In practice the indicated thermal efficiency of a well designed petrol engine would be of the order of 30 to 35 per cent., say, 32·5 per cent. for a compression ratio of about 6:1. The volumetric efficiency would be about 80 per cent. for such an engine.

The actual I.M.E.P. is obtained by multiplying the theoretical I.M.E.P. by the actual thermal efficiency and also by the volumetric efficiency. Taking the values previously mentioned we have:

$$\text{Actual I.M.E.P.} = 540 \times 0.325 \times .80.$$
$$= 140 \text{ lbs. per sq. in.}$$

The corresponding brake M.E.P. is then obtained from the last value by multiplying it by the mechanical efficiency of the engine, which is about 85 to 90 per cent., say, 87·5. So that we have:

$$\text{B.M.E.P.} = 140 \times .875$$
$$= 122.5 \text{ lbs. per sq. in.}$$

From the latter value, the probable horse-power output of an engine of a given bore, stroke, speed and number of cylinders can be estimated by the use of the formula given on page 46.

The Efficiency of Petrol Engines.—It is interesting to consider how the high speed petrol engine compares with other types of prime mover in the matter of power output and fuel consumption, and to see what proportion of the fuel is usefully employed. The best method of comparison, perhaps, is that in which the power output for a given supply of combustion heat is considered. The method, which gives comparative results for any type of engine, whether working with steam, heavy oils, gas or petrol, involves a quantity known as the *Thermal Efficiency*. The thermal efficiency of any type of engine unit is simply the ratio of the useful work or power obtained to the heat supplied from the fuel. The useful work is defined by the Indicated Horse-Power and the heat supplied, by the product of the quantity of fuel used per given time and its calorific value.

If an engine uses w lb. of fuel, of calorific value C, per hour and the Indicated Horse-Power is denoted by I.H.P., then:

$$\text{Thermal efficiency} = \frac{\text{Useful work done per minute}}{\text{Mechanical equivalent of heat supplied per minute}}$$

Now, the useful work per min = I.H.P. $\times$ 33,000 $\times$ 60

(Since 1.H.P. = 33,000 ft. lbs. per min.)

and: Mech. Equiv. of heat supplied per min. = $C \times w \times \mathcal{J}$.

Where $\mathcal{J}$ = Joule's equivalent, viz., 1 B.T.U. = 778 ft. lbs.

So that Thermal Efficiency $= \dfrac{\text{I.H.P.} \times 33{,}000 \times 60}{C \times w \times 778} \times 100$ per cent.

$$= 2{,}545 \cdot \frac{\text{I.H.P.}}{C \times w}$$

Example. An engine with a mechanical efficiency of 85 per cent. consumes 0·55 lb. of petrol, of calorific value, 18,500 B.T.U.'s per lb., per B.H.P. per hour. Find the indicated and brake thermal efficiencies.

Since Mechanical Efficiency $= \dfrac{\text{B.H.P.}}{\text{I.H.P.}}$

then I.H.P. $= \dfrac{\text{B.H.P.} \times 100}{85} = 1 \cdot 177$ B.H.P.

Then (1) Brake thermal efficiency $= \dfrac{2{,}545 \times 1 \times 100}{18{,}000 \times \cdot 55} = 25 \cdot 7$ per cent.

(2) Indicated thermal efficiency $= 1 \cdot 77 \times 25 \cdot 7 = 30 \cdot 22$ per cent.

The thermal efficiencies of petrol and Diesel engines are much higher than those of reciprocating steam engines, due to the fact that the latter types require steam boilers in addition to the steam engines. In each case there is a loss of combustion, or heat efficiency, so that the overall thermal efficiency, which is the product of the two separate efficiencies, is relatively low in comparison with the petrol and Diesel engines, in which combustion of and work done by the fuel takes place within the cylinder itself, so that heat losses are much lower.

The brake thermal efficiency of a locomotive or stationary reciprocating engine is of the order of 6 to 10 per cent., using ordinary boiler pressures, but with such refinements as feed water heaters, condensers, superheated steam, etc., the efficiency can be increased to 15 to 23 per cent.; power station and marine reciprocating steam engines come into this category. The steam turbine is the most efficient of steam engines and with modern aids to higher efficiency, such as feed water heaters, high degrees of superheat, high steam pressures, etc., brake efficiencies of 27 to 35 per cent. are attained.

The efficiency of a modern petrol engine, using suitable fuels, depends very largely upon the compression ratio; the higher this ratio is, the higher will be the efficiency. The following table gives some typical values of brake thermal efficiencies for various kinds of petrol engines and for typical Diesel engines.

TABLE 5

Brake Thermal Efficiencies of Petrol and Diesel Engines

Type of engine	Compression ratio	Brake thermal efficiency per cent
Small petrol two-cycle (three-port)	5·0:1	15 to 20
Larger air-scavenged two-cycle engine ..	6·0:1	18 to 24
Automobile four cycle, at 2,000 r.p.m. {	4·0:1	23
	5·0:1	27
	6·0:1	30
	7·0:1	32
High speed racing type	10·0:1	35
Aircraft supercharged, four cycle .. {	7·0:1 to 8·5:1	31 to 36
High speed automobile Diesel engines {	10·0:1	36
	14·0:1	39
	18·0:1	42
High speed sleeve valve Diesel engine (1,300 r.p.m.) (Ricardo)	15·0:1	44

Efficiency of Automobile Engines.—The thermal efficiency of automobile engines varies within certain limits, due to several factors; these and their influences are enumerated, briefly, below:

The *Thermal Efficiency*

(1) Is generally less at part loads, and a maximum at full loads, so that the fuel consumption at part loads is heavier. Thus at quarter to half full load, the fuel consumption is from 50 to 30 per cent. greater than when the engine runs on open throttle.

(2) Is higher as the compression ratio is increased. Detonation and other combustion considerations fix the upper limit of this ratio.

(3) Is higher for weaker mixtures of air and petrol. The usual ratio for maximum efficiency is about 17 to 18 parts of air to 1 part, by weight, of petrol; the ratio for complete combustion is about 14·5 to 15·0.

(4) Is dependent upon the shape of the combustion chamber. The best efficiencies are usually given by those engines having the smallest combustion chamber surface for a

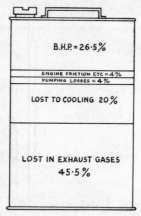

Fig. 15.—Showing how the Fuel Heat Energy is utilized in a Petrol Engine.

given volume. The ideal shape is spherical; the over-head valve, "pocket-less" engine, with inclined valves and dished piston, is the nearest approach to this ideal. The T-headed engine is about the least efficient.

(5) Increases with engine speed up to a certain limiting value, after which it decreases. The maximum efficiencies are usually obtained at speeds of from 70 per cent. to full load speed values.

(6) Depends upon the type of fuel used. For the same compressions and speeds, fuels, such as alcohol, benzole and high octane petrols, give the better efficiencies.

How the Fuel's Heat is Utilized.—Due to the Otto cycle of operation of the petrol engine and the limiting temperature range that can be employed, in practice, only a certain percentage of the fuel's total heat energy can be obtained during the combustion process in the engine. Thus, for an engine operating with a compression ratio of 9:1 only about 58 per cent. of the fuel's energy is available, theoretically. Of this heat quantity, which is initially in the form of highly heated combustion gases, some is lost through the combustion chamber and cylinder walls by conduction to the cooling water (or fins); some is lost by friction of the mechanical parts of the engine; some by oil heating, carburation, radiation and other minor causes, while the greatest loss is that due to the heat carried away by the hot exhaust gases. When the heat losses from these various causes is totalled up, it is found that in the case of a modern car engine, having a compression ratio of 9:1 and operating at 3,000 r.p.m., only about 26·5 per cent. of the original fuel's energy is available at the crankshaft for power supply purposes. The following is an analysis of the various heat losses in the example chosen:

TABLE 6

Showing how the Fuel's Heat Energy is Utilized in a Petrol Engine

Heat available as useful B.H.P. at the crankshaft	26·5	per cent.
Heat absorbed, due to engine friction and cooling fan	4·0	,,
Heat absorbed in 'pumping' the mixture and gases	4·0	,,
Heat carried away by the cooling water or (fins)	20·0	,,
Heat lost by the exhaust gases and minor items	45·5	,,
	100	,,

It has been shown that as the engine speed increases, the *frictional losses*, due to the piston, crankshaft bearings, etc. increase more rapidly than the speed. Also, as the compression ratio is increased the *exhaust losses* decrease and the B.H.P. increases. In connection with the values given in Table 6, as due allowance has been made for the power absorbed in driving the cooling air fan, cooling water pump, the oil and fuel feed pumps and ignition distributor shaft, etc., the figure given for the useful B.H.P. is the net value.

Diesel Engines.—Assuming the brake thermal efficiencies of a high speed Diesel engine to be between 38·2 and 42·5 per cent., the corresponding percentages of heat energy of the fuel converted into useful work as B.H.P. would therefore be the same, namely, 38·2 and 42·5, respectively. Thus, by using the compression-ignition principle instead of the Otto cycle one the amount of heat wasted can be reduced from 10 to 15 per cent. Another method of reducing the waste energy is to utilize the hot high speed exhaust gases to drive a gas turbine of the Rateau type, coupled to a centrifugal pump for supercharging the mixture supplied to the engine. This method has been applied successfully to aircraft piston (and certain Diesel) engines and has resulted in a marked increase in power output, so that aircraft engines thus equipped could maintain their ground H.P. rating up to much higher altitudes than gear-driven supercharged type engines.

Supercharging.—The power obtainable from a given engine is determined by the weight of mixture which fills the cylinders at the beginning of the compression stroke. If we could double the quantity of air, we could obtain twice the quantity of mixture, and much higher cylinder mean pressures and powers would be realized. If we previously compress the mixture to one-half its volume, before compression in the cylinder, about twice the charge weight can be obtained.

The power obtained from this doubled charge will not, however, be twice as much, due to combustion considerations into which we cannot enter here, and to mechanical power losses in connection with the pre-compression of the charge.

Supercharging is the name given to the process of increasing the charge weight in the cylinder so as to obtain more power from a given cylinder capacity.

The subject of supercharging is dealt with more fully in Chapter IV.

CHAPTER II

THE PETROL ENGINE AND ITS COMPONENTS

Development of the Petrol Engine.—Although there is some doubt as to the original inventor of the petrol engine, it appears fairly certain that the first road car driven by benzene vapour with air was constructed by a German named Siegfried Marcus, of Mecklenburg, in 1864 to 1868, according to a photograph taken at about that time. In 1873, Marcus's improved vehicle was exhibited at the Vienna Exhibition; this vehicle appears to have been constructed from a hand-cart, the two rear wheels of which were replaced by the fly-wheels of the benzene engine used for driving it.

There was little progress to record between 1862 and 1875, but several inventors appear to have been working on the subject of road-car design. In 1889, a vehicle was actually built by Butler, an Englishman, and was able to travel along the roads at 4 to 12 m.p.h. This three-wheeled vehicle was driven by means of a small two-cylinder engine. of $2\frac{1}{4}$ ins. bore, driving direct on to the hub of the rear wheel, through an epicyclic gear with a reduction ratio of about 6 to 1. The first ignition system was a magneto, with a low tension make-and-break sparking device in the cylinders. Ultimately, this was abandoned in favour of a primary battery and induction coil, with a wipe contact. A float feed carburettor was used; this appears to be the first instance in which this type was employed.

On the Continent, Gottlieb Daimler, in 1886, constructed a benzene engine-driven vehicle, which was fairly successful. This was a motor-cycle and employed carburetted air in an engine working on the Otto, or four-stroke, principle. Another pioneer who was working independently and on somewhat parallel lines, was Benz, who in 1885 produced a satisfactory motor-tricycle using electric ignition. Daimler, and later Levassor, in France, used the hot tube ignition system.

These early engines were virtually modified horizontal gas-engines, running at relatively low speeds, namely, from 400 to 700 r.p.m.; they were also very heavy and somewhat erratic in their running. It was only when Count de Dion and M. Bouton, two French engineers, realized that the future development of the automobile engine was along the lines of reduced weight, and increased power output, that any real advance was made. They constructed the famous De Dion Bouton engine running at 1,500 r.p.m.—a considerable speed increase over the engines of their rivals. Later events proved conclusively the soundness of their foresight and judgment. To Messrs. Panhard and Levassor also belongs much credit for constructing one of the first successful motor-cars*, in 1901, from which date progress was, relatively speaking, very rapid.

It will be seen from this somewhat brief survey that the automobile engine has been developed from the early gas engines of Lenoir and Otto, by lightening the construction, improving the valve and cylinder design, using the compression principle, employing the vapours of light spirits such as benzine and petrol, by utilizing electric in place of hot-tune ignition and by increasing the engine speed.

The Complete Engine.—If the principle of the four-stroke type of engine, explained in the preceding chapter has been properly understood, the beginner will experience little difficulty in following the application of this principle to modern petrol engine practice. Although the sectional views of engines at first sight may appear to be complicated and a little bewildering, he should bear in mind the fact that these engines work on a similar system to that shown in Fig. 8, and so should not find it difficult to understand engine designs. For this reason it is proposed to take our first step from the outline drawing of Fig. 8 to the rather more detailed one shown in Fig. 16. This latter illustration shows all of the essentials of a modern air-cooled petrol engine, but devoid of the details which might at this stage confuse the reader.

Keeping the Cylinder Cool.—It has been stated in the pre-

---

* Many of the earlier motor-cars and motor-cycles, with their original engines (1896-1914) can be seen at the Science Museum, South Kensington, London.

vious chapter that the heat of combustion is relatively, very great, and that there is an appreciable proportion of this heat conducted through the cylinder walls whilst the engine is working, namely about 20 per cent. of the combustion heat. In the case of an

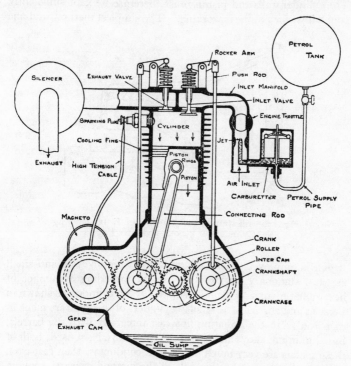

Fig. 16.—Showing the Components of a simple Air-Cooled Petrol Engine.

engine of $3\frac{1}{2}$ inches cylinder diameter and 4 inches stroke (about the dimensions of a motor-cycle engine) running at 2,000 r.p.m., no less than 2,500 B.T.U.'s of heat are given to the metal of cylinder and piston every minute. This quantity of heat is sufficient to raise $1\frac{1}{2}$ gallons of water from freezing to boiling-point every minute. If the cylinder of the engine was merely a plain metal cylinder, it would be raised to a red heat after a very short period of running if it did not seize up before this.

Further, the incoming charge is liable to explode whilst it is still entering the cylinder, and thus to fire back into the carburettor. The cylinder walls and piston must therefore be kept sufficiently cool whilst the engine is working. The simplest method, and one

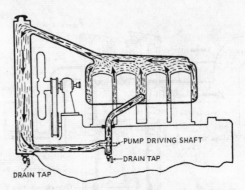

Fig. 17.—Illustrating the principle of Water-Cooling the Cylinders of a Four-Cylinder Engine.

which is employed upon certain aircraft, motor-cycle and small car and stationary engines, consists in providing a number of heat-radiating fins on the hottest parts of the cylinder, as shown in Figs. 11 and 26. Most air-cooled cylinders of this type are made of cast-iron, with the radiating fins cast integrally with the barrels, but aluminium alloys and also copper fins have been used; both of these metals are very much better heat conductors than cast-iron.

Fig 17 illustrates the principle of the method of water-cooling, and shows how the heat of the cooling water is dissipated or carried away by means of a *Radiator* (shown on extreme left). The latter consists of a large number of thin metal tubes, of great surface area, through which the hot water passes, and these tubes are cooled by radiation due to the large tube surface, and also by a current of air blowing past the outside surfaces of the tubes.

Further information on engine cooling is given in Chapter IX.

Supplying the Mixture to the Cylinder.—Petrol is contained in the tank or reservoir shown in Fig. 19 and is fed to the petrol-air proportioning and mixing device known as the *Carburettor*, along the petrol supply pipe; a tap is usually provided near the tank to shut off the supply when required.   Referring to Figs. 16 and 18, the petrol enters the base of the carburettor *Float Chamber*, past a tapered needle valve.   The float chamber contains a hollow metal float, the purpose of which is to lift the pivoted levers shown, so as to depress the needle valve and shut off the supply of petrol as soon as the petrol in the float chamber reaches a given height.   In this manner a constant level of petrol can be maintained in the float chamber, so that flooding is prevented; moreover, it is essential always to maintain a constant level of the petrol in the small jet for the correct working of the carburettor.   The jet referred to consists of a vertical pipe of small diameter fitted with a plug at its upper end, having a very fine hole; in the case of a motor-cycle engine this hole is only about $\frac{1}{30}$ to $\frac{1}{50}$ inch diameter. The open end of the jet is arranged in the centre of a constricted tube, throat, or *Venturi*, the purpose of which is to increase the speed of the air which enters at the lower end, so that it will flow very quickly past the jet, and in so doing will cause a reduction of pressure below that of the atmosphere at the jet outlet.   Since the pressure on the surface of the petrol in the float chamber is atmospheric, whilst at the jet it is less than this, it follows that the petrol will flow out of the fine jet as a spray and will be carried along with the air which is sucked in during the suction stroke of the piston.

The dimensions of the air supply pipe and of the jet are carefully proportioned so as to give about the correct ratio of air to petrol, namely about 15 parts of air to 1 of petrol, by weight, but means are provided for varying the proportions between the limits of about 8 (rich) and 20 (weak) to suit certain running conditions.

The *Throttle* shown between the carburettor and inlet valve is simply a valve provided for the purpose of governing the engine speed and power; it acts by regulating the quantity of mixture admitted to the engine.   For slow running and light loads it is nearly closed: for high speeds and full power it is opened wide. The throttle shown in Fig. 16 consists of a cylindrical member, capable of rotating in the outer cylindrical casing, and it is provided with a cylindrical hole of the same diameter as that of the inlet pipe. It will be seen that if this inner piece is rotated in its casing it will gradually restrict the area through which the mixture can pass to the engine.

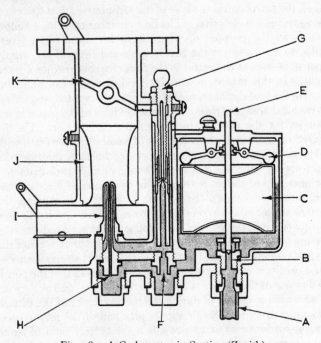

Fig. 18.—A Carburettor in Section (Zenith).
A—Petrol union-nipple. B—Float-needle seating.
C—Float. D—Counter-weight. E—Float needle.
F—Compensating jet. G—Slow-running device.
H—Main jet. I—Main jet cover. J—Choke tube.
K—Butterfly throttle.

Fig. 18 illustrates a typical carburettor in detail. In this case there are two petrol jets, namely, the *Main Jet* on the left, which corresponds to the one previously described, and the *Pilot Jet*, *G* on the right. The purpose of the latter is chiefly to supply the very small quantity of mixture required to start from the cold or keep the engine running slowly, when there is no load on it, as when an automobile engine is running idle. There is also a compensator jet to keep the mixture strength approximately constant when the engine is working at various speeds, since it is known that the simple type of carburettor illustrated in Fig. 16 can only give the correct mixture proportions at one definite speed. Above this speed the mixture obtained is richer in petrol, whilst below it is weaker.

All modern carburettors are provided with additional means, or devices, to render the mixture proportions approximately constant over the working speed range; the compensator device referred to in Fig. 18 comes under this heading.

It will be noticed that another type of throttle, namely, the *Butterfly Valve*, is employed in this case; it is shown closed, and is opened by rotating with the pin shown in the centre. This type is now almost universally used. The air inlet is on the lower left hand side, whilst the choke, or venturi tube for increasing the velocity of the air, surrounds the jet. Plugs are provided below the jets to facilitate their removal for cleaning or changing purposes. It is usual to fit a petrol strainer, or filter, between the tank and the float chamber needle valve in order to trap any solid particles of matter which might otherwise block up the fine petrol jet. It will also be observed that the upper end of the needle valve is accessible; by raising the needle by hand, the level of the petrol can be raised. This procedure, which is known as "flooding" or "tickling" the carburettor, enables a rich mixture to be provided for starting the engine more readily.* It is not now employed in modern carburettors.

Modern Carburettor Improvements.—The later types of carburettor are mostly of the "down draught" pattern, in which the air for combustion is drawn vertically downwards through the carburettor, the latter being placed near the top of the engine. This arrangement enables the heavier mixture to assist, by its weight, in the induction process, whilst the number of passages and bends to the combustion chambers is minimized; it also renders the carburettor more accessible.

---

* A full account of Carburettors and Fuel Injection Systems is given in Vol. II of this series.

Carburettor main air inlets are fitted with *air cleaning devices* and, since in the past there was a certain amount of noise made by the passage of the air through the carburettor unit, many modern carburettors are also fitted with *air silencers* of the gauze or perforated plate pattern. In some cases *flame arresters*, of copper gauze have been fitted over the air inlets to stop back-fire flames.

The *starting of car engines* from the cold has been improved by the fitting of automatic devices for enriching the mixture at the time of starting. One typical starting device has a *thermostat* which, when cold, operates an air valve known as a *choke* for reducing the main air supply; as soon as the engine warms up, however, the air-valve is opened by the device.

The mixing chamber above the jet is usually heated by exhaust gases; there is often a throttle valve for shutting off the gases when the engine's throttle is fully opened.

Acceleration pumps are also fitted for the purpose of ensuring an adequate supply of petrol for acceleration, when the accelerator is suddenly pressed down.

In modern carburettors, means are provided for making the mixture weaker in petrol for all normal running purposes, but to give a richer mixture, for full power when the throttle is fully opened; this method results in a saving in fuel consumption.

Finally, a *mixture control* is usually fitted for the purpose of altering the mixture proportions to suit climatic conditions, i.e., extreme heat and cold atmospheres.

Direct Fuel Injection.—As an alternative to the carburettor method, fuel can be sprayed into the cylinder or inlet manifold, so that the manifold then becomes an air charge intake, only. In this way a greater charge weight and better distribution of the charge to the various cylinders are obtainable. This method is described more fully in Chapter III.

Supplying Petrol to the Carburettor.—The petrol supply is carried in a tank at the rear of the car, although earlier models were fitted with a dashboard or bonnet tank to allow the petrol to feed the carburettor by gravity.

The favoured method a few years ago was to employ a system known as the *vacuum feed*, whereby the reduced pressure in the

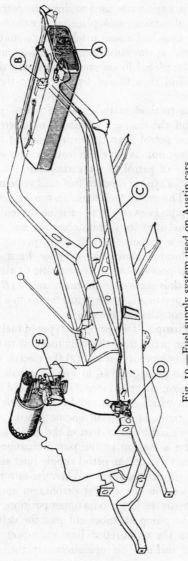

Fig. 19.—Fuel supply system used on Austin cars.

inlet pipe of the engine was used to draw the petrol from the rear tank into a small auxiliary tank placed under the bonnet above the engine. This tank had a float mechanism to shut off the suction from engine and at the same time to open a valve allowing the outside air to be placed in communication with the petrol in the tank. The petrol then flowed by gravity to the float chamber of the carburettor.

The modern method is to employ a petrol pump, operated mechanically off the engine's camshaft, or electrically from the battery, to draw petrol from the main tank and force it into the carburettor; most pumps have automatic release valves to obviate excessive supply of petrol to the carburettor.

Fig. 19 shows a typical modern fuel feed system as used on the Austin cars. The main fuel tank, shown at (A) supplies fuel along the fuel pipe lines (C) to the engine-driven fuel pump (D), which draws fuel into its suction chamber and pumps the fuel upwards, under a pressure of 3 or 5 lbs. per sq. in. to the carburettor (E), located above the cylinder head of the engine The fuel tank is provided with a pneumatic or eletric petrol level gauge unit, which is situated in the main tank at (B), and transmits its readings by pneumatic or electrical means to a gauge mounted on the instrument panel of the car.

Fuel Feed Pump.—Two principal types of fuel pump are used for supplying the petrol from the main fuel tank to the carburettor, namely, (1) the engine-driven and (2) the electric solenoid pump. The former type pump, shown in Fig. 20 consists basically of a flexible diaphragm, clamped around its edges and flexed by means of a rocker arm which is actuated by the engine camshaft cam, shown on the right, at one-half engine speed. The cam lever and pull rod, move the central part of the diaphragm downwards, thereby creating a suction in the pump chamber above. This causes a suction valve on the petrol supply inlet side to open, and allow petrol to fill the chamber. When the cam moves round to its "no-lift" position the central diaphragm spring, below the diaphragm, returns the latter to its upper position, thereby forcing the petrol in the pump chamber out past the delivery valve and union outlet to the carburettor float chamber; the inlet valve remains closed during this operation. If the outlet pressure exceeds a certain value—usually only a few pounds per sq. in.—then a release valve on the outlet side opens and allows the petrol to flow back into the suction side again, so that the float chamber of the carburettor is thereby relieved of excess fuel pressure. The electric-type pump operates off the car battery and employs a solenoid device to flex the diaphragm of the pump member. The

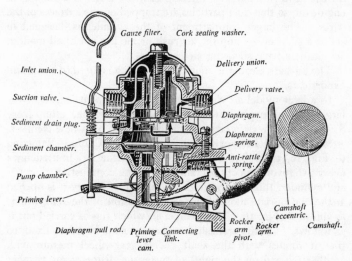

Gauze filter. *Cork sealing washer.*

Inlet union.

Delivery union.

Suction valve.

Delivery valve.

Sediment drain plug.

Diaphragm.

Sediment chamber.

Diaphragm spring.

Anti-rattle spring.

Pump chamber.

Priming lever.

Camshaft eccentric.

Rocker arm.

Rocker arm pivot.

Camshaft.

Diaphragm pull rod. *Priming lever cam.* *Connecting link.*

Fig. 20.—Illustrating the operation of the A.C.-Delco Mechanical Petrol Pump.

pump works as soon as the ignition key is switched on, so that no initial priming of the pump is required, as is sometimes the case with the engine-driven pump. In this connection, the hand priming lever for the latter type of pump is shown on the left in Fig. 20. The S.U. fuel pump used on all B.M.C. car engines is a typical example of the electric type of pump.

Air Cleaners.—The cleaning of the air taken through the carburettor has a marked effect upon the useful life of the pistons and cylinders, so that it is now usual to fit cleaners to the air-intakes of the carburettors. These cleaners operate either upon the direct filtering principle, using felt or hair elements, to stop the dust particles, or upon the centrifugal one, whereby the air is rotated at high speed before it enters the carburettor, thus throwing any solid matter outwards into a dust collecting chamber.

A popular air filter is the large capacity cylindrical type mounted on top of the engine. It has a short cylindrical compartment at one end, filled with fine wire twisted and partly compressed to fill the space. This wire filtering element is given a coating of engine oil, so that dust particles are trapped on the surfaces of the wires. The larger compartment of the cylinder is designed to act as a silencer, to stop the air rush or hissing noise; all modern air cleaners have silencing elements.

The oil-bath cleaner is another type which is fitted to home and export cars.

In a later model air cleaner the filtering element is a relatively large pleated paper concentric type filter which, after every 8,000-10,000 miles is taken out and replaced by a new element.

Mechanical Interpretation of the Otto Cycle.—Referring to Fig. 16, the mixture from the carburettor is admitted once every two revolution of the crankshaft to the combustion chamber and cylinder, through the *Inlet Valve Port*. This port is opened at the appropriate moment by the depression of the *Inlet Valve*, as shown in Fig. 16. The manner in which this is carried out is as follows. On the crankshaft a gear-wheel is keyed or fixed, so that it rotates with the shaft. This gear-wheel meshes with another (shown on the right) of twice the diameter, or number of teeth: the latter wheel therefore makes one revolution to every two of the crankshaft. Keyed to this wheel is a *Cam* (the *Inlet Cam*) which is so designed that it is cylindrical over about three-quarters of its surface, and has a curved projection on the other part. Thus during one complete rotation of the larger gear-wheel, the cam lifts the *Push-Rod* shown, for about one-quarter of its revolution, and allows it to remain stationary during the other three-quarters (i.e., when the inlet valve is closed during the compression, firing and exhaust strokes). During the inlet stroke the cam raises the push-rod, tilts the *Rocker Arm* about its central pivot or pin, and opens the inlet valve. When the roller of the push-rod returns to the plain portion of the inlet cam, the inlet valve spring closes the valve, and keeps it closed. The valve, it will be observed, opens inwards, and closes on to a conical seating; the pressures of compression and explosion therefore force the valve on to its seating and tend to render the joint more gas-tight. The *Exhaust Valve* is operated by its own cam, push-rod and rocker in exactly the same way, at the correct moment and for the appropriate period. It will be observed that the exhaust and inlet cams are not in the same positions relatively to the crank-shafts; this is because these cams have to

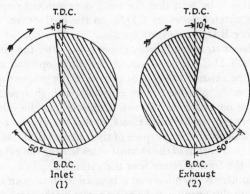

Fig. 21.—Typical Inlet and Exhaust
Valve Timing Diagrams.

open the valves at different relative times, in order that the Otto cycle of operations may be carried out.

Valve Opening Periods.—The sequence of valve opening periods can also be shown graphically in the cases of both the two and four-stroke engines by denoting the corresponding positions of the crank on the crankshaft at which these operations occur. Such diagrams are known as *Valve Timing Diagrams*.

Referring to Fig. 21 (1,) this shows, on a crank-angle circle, the moments at which the inlet valve opens and, later, closes, once every two crankshaft revolutions, for a typical car engine. It will be observed that the valve opens at 8° before the crank and piston reach the highest positions, or top dead centre (T.D.C.). As the piston descends during its suction stroke, the inlet valve remains open throughout this stroke and continues open until the piston has moved up its compression stroke, until the crank angle is 50°, when the valve closes. The object of this early opening and late closing of the inlet valve is to obtain the maximum quantity of fresh mixture at higher engine speeds by taking account of the momentum or inertia of the mixture, which effect continues for a time after the piston has reached its bottom dead centre (B.D.C.).

Fig. 21 (2) shows the exhaust valve timing diagram for the same engine. It will be seen that the valve commences to open well before the piston reaches its B.D.C. on the final stroke, namely, at 50° before B.D.C.; remains open during the whole of the following exhaust stroke and finally closes at 10° after the piston has reached its T.D.C. The object of this early opening and late closing of the exhaust valve, is to get rid of the greatest quantity of exhaust gases, possible—more especially at high speeds.

It will be noticed from the two diagrams that the inlet valve opens before the exhaust valve closes, so that both valves are open, for an angle of 18°. The purpose of this *Overlap* is to take advantage of the momentum of the exhaust gases leaving the cylinder, to help draw the fresh mixture into the cylinder.

It should here be mention that it is usual to show both the inlet and exhaust valve opening and closing angles on the same diagram.

More Recent Valve Timings.—An examination of the valve timings of modern cars (1962) shows that, with a few exceptions the valve timings for cars ranging from about 0·75 to 3·5 litres vary but little between the limits shown in Table 7 opposite.

The variations in the valve timings shown are due to various factors, e.g., valve cam design, inlet and exhaust manifold design, maximum speed, compression, etc. This accounts for the occasional unusual timings that are used. Thus for the recent Volkswagen 1492 c.c. opposed cylinder engine, the valve opening and closing angles, using the order of Table 7, were: 1, 32, 38, 0, while for the Buick 4600, 325 B.H.P. engine the corresponding angles were: 33, 77, 75, 44, giving an overlap of 77°, so that the valve opening periods were appreciably longer than for the usual engines.

High Performance Engine Valve Timings.—The valve timing diagrams in Fig. 21 are for a typical and popular production car engine. For higher performance engines different timing diagrams are used, the main object being to give rather longer inlet and exhaust valve opening periods, to ensure the maximum amount of fresh mixture at the higher engine speeds, together with the greatest possible clearance of the exhaust gases— to keep the cylinder temperatures down to the permissible limit.

In these engines it is usual to open the valves earlier and to close them later, so that the total opening periods are greater than those shown in Table 7.

TABLE 7

Valve Timings of Car Engines

British Cars				
Inlet Valve		Exhaust Valve		Overlap of Values
Opens Before T.D.C.	Closes After B.D.C.	Opens Before B.D.C.	Closes After T.D.C.	
(Degrees) 5–20 15 (Aver.)	(Degrees) 40–65 50 (Aver.)	(Degrees) 40–60 50 (Aver.)	(Degrees) 10–25 16 (Aver.)	(Degrees) 15–38 30 (Aver.)
American Cars				
12–26 20 (Aver.)	50–70 56 (Aver.)	48–68 56 (Aver.)	10–28 20 (Aver.)	23–50 37 (Aver.)

An example of the valve timing used in one of the successful Climax racing car engines has been selected. The Climax FPF four-cylinder engine, of the overhead twin camshaft type had a bore of 81·2 m.m. (3·2 in.) and stroke of 71·1 m.m. (2·8 in.) giving a cylinder capacity of 1475 c.c. (90 cu. in.). The compression ratio was 10 : 1. Using 100-octane fuel this engine developed a maximum output of 146 B.H.P. at 7,300 r.p.m., maximum torque of 108·5 lb. ft. and B.M.E.P. of 188·5 lb. sq. in. both at 6,500 r.p.m. The dry weight of the engine was 255 lb., i.e. 1·75 lb. per H.P.

The valve timing used on the 1·5 litre engine, and on larger models up to 2·5 litres was as follows:

Inlet valve opened at 40° before T.D.C. and closed at 70° after B.D.C.

Exhaust valve opened at 65° before B.D.C. and closed at 45° after T.D.C.

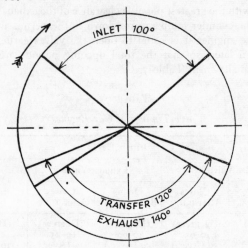

Fig. 22.—Port opening Diagrams
for Two-cycle Engine.

The inlet valve opening was 290° and, for the exhaust valve, 290° also.

The inlet-exhaust overlap was 85°.

Two-Cycle Engines.—The simpler type of two-cycle engine, with crankcase compression, has no valves, the mixture introduction, or charging and the exhaust operations being governed by ports in the cylinder walls which are opened and closed by the upper edge of the piston. The latter controls three operations, namely, (1) Opening and closing of the carburettor port to the crankcase. (2) Opening and closing the transfer ports from crankcase to cylinder and (3) Opening and closing exhaust ports.

These three operations can be shown on the engine crank angle diagram as indicated in Fig. 22, for a Levis engine. The inlet port to the crank case opens near the end of the (underneath) suction stroke of the piston at 50° before T.D.C. and closes 50° after T.D.C., giving an inlet opening period of 100°. The transfer ports open at 60° before B.D.C. and close at 60° after B.D.C., giving a 120° transfer period. The exhaust ports, which open before the transfer ports, at 70° before B.D.C., close after the transfer ports at 70° after B.D.C., giving 140° of exhaust opening.

When the two-cycle engine is of the compressed air scavenging kind, this air is admitted during the transfer period (Fig. 22) and gets rid of the exhaust gases, leaving the cylinder filled with cool air at a higher pressure than atmospheric. This air is compressed and at some given point during the compression stroke fuel is injected under high pressure through a spraying nozzle. The mixture is ignited by an electric spark in the usual manner. This gives a much more efficient combustion process, with higher power output, than for the three-port engine of Fig. 22.

The Piston.—The function of the *Piston* is to receive and transmit the force of the burning and expanding gases to the crankshaft, *via* the *Connecting Rod*; it must also act as a gas-tight plunger for the suction, compression and exhaust operations. The piston receives the thrust of the connecting rod; this causes it to press heavily upon the cylinder wall during the firing stroke. The piston is given a good bearing area for this reason and means are provided for lubricating the cylinder walls. Split rings, made of cast-iron, but which are flexible, known as the *Piston Rings*, are provided to ensure gas-tightness. The piston also houses the *Gudgeon*, *Piston* or *Wrist Pin*, upon which the upper end of the connecting rod rocks.

The connecting rod, as we have mentioned, is the connecting link between the piston and the crankshaft, whereby the reciprocating motion of the former is converted into the rotary motion of the latter; the lower or *Big-End* bearing of the connecting rod is on the *Crank Pin*, or *Journal*, of the crankshaft.

The camshaft, which rotates at one-half engine speed, is also known as the half-speed shaft.

Referring to Fig. 16 again, on the extreme left is shown a

gear drive for a high voltage electricity generating device, or *Magneto*, which produces the high voltage electric spark for the ignition of the charge. In modern engines another system, known as *Coil Ignition* is used.

Finally, the whole of the mechanism and gearing below the piston is enclosed in an oil-tight casing, known as the *Crankcase*, or *Crank Chamber*. This latter is usually in the form of a cast-iron, sheet-metal or aluminium alloy casting and not only houses the bearings for the crankshaft, camshaft, and other moving parts, but contains the lubricating oil, or "oil spray", so that the moving parts can be adequately lubricated. The crankcase is also provided with projections or lugs to bolt the engine to the frame of the automobile and with brackets upon which the engine-driven accessories are mounted.

The Complete Engine.—Having considered the principles and operation of the four-stroke petrol engine, together with some of its chief components it is proposed, next, to give some typical examples of car engines, in order to show their general arrangements and working members. Since the overhead valve engine is now more favoured, a typical engine is selected as the first example. Fig. 23 shows a water-cooled engine in end cross-section. In this type the inlet and exhaust valves are arranged in the cylinder head, the latter being made detachable for de-carbonizing and valve grinding. The mechanism for operating the valves, clearly shown in Fig. 23, consists of a camshaft driven at one-half engine speed, the cams of which actuate the tappets above and the latter operate the rocker arms through long tappet rods having cups at their upper ends, engaging with ball-ended members screwed into the right-hand ends of the rocker arms or levers. The latter can rock on central bearings so that as their right-hand ends are pushed upwards by the push rods the left-hand ends depress the valves. When the cam ceases to lift the push rod the valve springs—which are of the double-type—maintain the push rod and tappet in contact with the cam and also return the valve to its seating.

The cylinder has a water-jacket which communicates with the water spaces in the cylinder head so that there is a continuous flow of cooling water through these spaces when the engine is

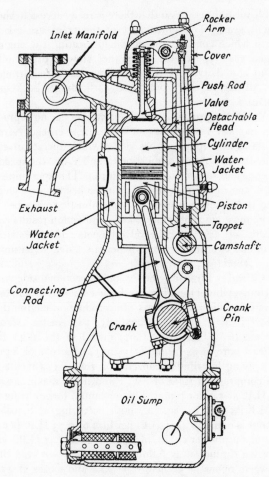

Fig. 23.—End Sectional View of Overhead Valve
Water-Cooled Car Engine.

working. The carburettor and inlet manifold are on the upper
left-hand side and the exhaust manifold seen just below. The
whole of the valve mechanism in the cylinder head is enclosed by a

detachable oil-tight cover so that there is ready access to the valve adjustment gear. The lower part of the crankcase terminates in a sump which contains the oil for lubricating the engine. A submerged gear-type of oil pump (not shown) draws oil from this sump and delivers it to the various parts requiring lubricating. All surplus oil from these parts drains back into the sump again.

Side-Valve Engine.—The side-valve engine, now practically obsolete, in most respects is similar to the overhead valve type, but—as its name implies—it has valves which operate in a pocket formed on the side of the combustion chamber, i.e., the space above the piston when on its top dead centre. The sparking plug is situated in the upper part of the cylinder head and is located in regard to the piston, valves and the combustion chamber shape, to give the most efficient ignition and combustion results.

The camshaft operating the valve tappets and valves is similar to that of the overhead valve engine, the valve actuation being direct and, therefore, more simple.

Four Cylinder Engine.—So far, only a sectional side view of a typical (four-cylinder) engine has been considered, so that it is appropriate to follow with views of a production engine that has established a high reputation namely, the Austin A.40 four-cylinder engine. The later (1962) model of the A.40 Mk. II engine has a bore of 64·6 mm. (2·54 in.) and stroke of 83·7 mm. (3·3 in.) giving a cylinder capacity of 1098 c.c. (67·3 cu. in.). It has a compression ratio of 8·5: 1 and has a maximum output of 50 B.H.P. at 5,100 r.p.m. The maximum torque is 60 lbs. ft. and B.M.E.P., 135 lb. sq. in., both occurring at 2,500 r.p.m. The output is equal to 44·4 H.P. per litre or 0·73 H.P. per cu. in. The engine output per sq. in. of piston area is 2·37 H.P.

The valve timing is, as follows:

Inlet valve opens at 5° before T.D.C. and closes at 45° after B.D.C.

Exhaust valve opens at 51° before B.D.C. and closes at 21° after T.D.C.

The Mk. II engine is also made integral with the transmission for the front wheel drive Morris 1100 car.

Referring to the *left hand* (or *nearside*) *view* in Fig. 24 (A), this

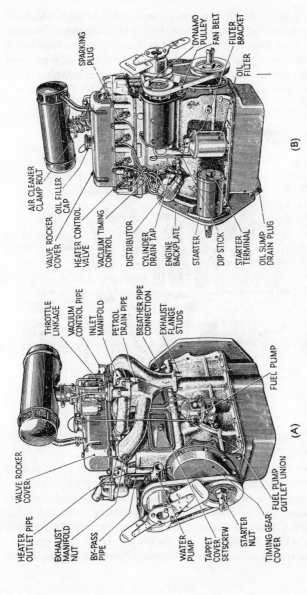

SPARKING PLUG

DYNAMO PULLEY
FAN BELT
FILTER BRACKET

OIL FILTER

(B)

AIR CLEANER CLAMP BOLT

OIL FILLER CAP

VALVE ROCKER COVER

HEATER CONTROL VALVE

VACUUM TIMING CONTROL

DISTRIBUTOR

CYLINDER DRAIN TAP

ENGINE BACKPLATE

STARTER

DIP STICK

STARTER TERMINAL

OIL SUMP DRAIN PLUG

THROTTLE LINKAGE

VACUUM CONTROL PIPE

INLET MANIFOLD

PETROL DRAIN PIPE

BREATHER PIPE CONNECTION

EXHAUST FLANGE STUDS

FUEL PUMP

(A)

VALVE ROCKER COVER

HEATER OUTLET PIPE

EXHAUST MANIFOLD NUT

BY-PASS PIPE

WATER PUMP

TAPPET COVER SETSCREW

STARTER NUT

TIMING GEAR COVER

FUEL PUMP OUTLET UNION

Fig. 24.—Showing External Features of the A.40 Engine (A) Nearside view. (B) Offside view.

shows the large air cleaner and silencer unit through which all the air is drawn (by engine suction) before entering the carburettor below. Fuel from the petrol tank is drawn into the fuel pump through a long copper pipe (not shown), whence the fuel is pumped up to the carburettor float chamber above, to mix with the air charge as it is drawn into the engine.

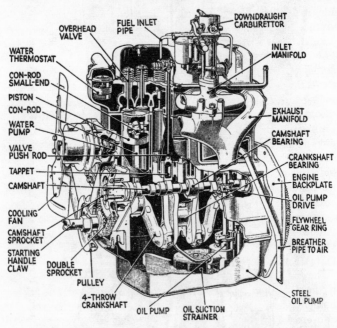

OVERHEAD VALVE
FUEL INLET PIPE
DOWNDRAUGHT CARBURETTOR
WATER THERMOSTAT
INLET MANIFOLD
CON-ROD SMALL-END
PISTON
CON-ROD
WATER PUMP
EXHAUST MANIFOLD
VALVE PUSH ROD
CAMSHAFT BEARING
TAPPET
CRANKSHAFT BEARING
CAMSHAFT
ENGINE BACKPLATE
OIL PUMP DRIVE
COOLING FAN
FLYWHEEL GEAR RING
CAMSHAFT SPROCKET
BREATHER PIPE TO AIR
STARTING HANDLE CLAW
DOUBLE SPROCKET
PULLEY
4-THROW CRANKSHAFT
OIL PUMP
OIL SUCTION STRAINER
STEEL OIL PUMP

Fig. 25.—Part Sectional View of the A.40 engine.

The external end of the crankshaft is fitted with a vee-pulley, which with the aid of a vee-belt drives two other pulleys, namely, one for rotating a two-bladed fan which draws cooling air through the radiator; it also drives, within, a circulating water pump for the cylinder water jacket and head. The other pulley drives a dynamo which supplies current to the car battery, in order to keep the battery charged at all times. Referring, again, to the crankshaft, to the left of the crankshaft pulley there is a projection with a claw end, for the starting handle engagement.

The purpose of the breather pipe, shown in Fig. 24 (A) is to release any compression or suction that may occur within the crank chamber, due to the piston motions. This pipe is connected between the space in the overhead valve chamber and the carburettor air cleaner. The vacuum control pipe of the upper right side, in Fig. 24 (A), is employed to operate the vacuum ignition advance mechanism.

Referring, next, to the *right* (or *offside*) *view* in Fig. 24 (*B*), this shows, clearly, the valve cover, sparking plugs, starting electric motor, the oil filter (that is used to clean the engine oil as it circulates), the ignition distributor and the belt-driven dynamo and cooling-fan (and water pump) unit.

Some Internal Details.—Having considered the exterior features of a typical car engine, it may be of interest to describe some of the internal details of the same engine, as shown in the part-sectional view, in Fig. 25. Commencing with the crankshaft, this has two end and one centre bearing, to support it. The two inside and the two outside cranks are in the same planes, the inner pair being at 180° to the outer pair. The cylinders "fire" at equal intervals of 180°. Thus if the left hand (or front) cylinder is numbered (1), and the other in the order (2), (3) and (4), then the firing intervals for the cylinders will be, as follows: (1), (3), (4), (2).

The crankshaft has a pair of equal chain sprocket wheels—shown on the left—which, by means of double roller chains, drives another double sprocket wheel of twice the number of sprocket teeth, fixed to the end of the camshaft; the latter has three bearings in the crankcase unit. The camshaft, therefore, rotates at one-half engine camshaft speed, to give the four-stroke cycle. Fig. 25 shows also the oil-pump vertical drive shaft and its driving gears at the camshaft; the connecting rod and piston arrangement; operating mechanism for one valve, viz., the third from the left; the water circulating pump impeller; cooling fan; carburettor; parts of the inlet and exhaust manifolds; valve cover on cylinder head; breather pipe, which in this case opens direct to the air below; starting motor engagement gear teeth on fly-wheel; oil sump casing and the thermostat unit. The latter device is located in the hot water outlet from the cylinder head to the top of radiator; it serves to disconnect the radiator from the engine when the water in the system is cold, thus allowing the cooling water in the cylinder jackets to heat up quickly.

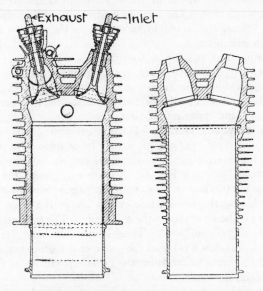

Fig. 26—Examples of Air-cooled Cylinders. Left—Aluminium Alloy with Steel Barrel and shrunk-in alloy steel Valve Guides and Valve Seatings. Right—Steel Cylinder with Cast Aluminium Alloy Head, and Bronze Valve Seatings.

Air-Cooled Cylinder Materials.—The majority of air-cooled cylinders are made of a hard grade of grey cast-iron the fins being chill cast, with the result that they are extremely hard. An average composition of cast-iron suitable for motor cylinder and piston castings is as follows: Iron from 93 per cent. to 95·5 per cent., Carbon (combined) 0·5 per cent. to 0·8 per cent. (total) 2·7 per cent. to 3·3 per cent., Silicon 1·2 per cent. to 1·8 per cent., Manganese 0·6 per cent. to 1·0 per cent., Sulphur (max.) 0·12 per cent.,

Phosphorus (max.) o·85 per cent. This cast-iron gives an average tensile strength of 16 tons per sq. in. The thickness of the walls is usually from $\frac{5}{32}$ to $\frac{7}{32}$ inch, and the fins taper down to a fine edge. Aluminium alloys have also been used for air-cooled cylinders, more particularly on aircraft engines, although the use of these alloys for the cylinder heads of motor-cycle engines is now favoured. The alloys themselves are not hard enough to withstand the hammering action of the valves and the wearing action of the piston, so that it is necessary to insert cast-iron or bronze valve seatings for the valves (these are usually cast in place), and cast-iron, alloy steel or Nitralloy hardened liners in the case of the cylinders.

A typical aluminium alloy for cylinder and piston castings is as follows: Aluminium 91 per cent., Copper 7 per cent., Tin 2 per cent. Other well-known alloys are the British Engineering Standards Alloys 2L8, 3L11 and the "Y" and R.R. Hiduminium Alloys. Elektron, a magnesium alloy, weighing only 60 per cent of aluminium, has been used for pistons, crankcases, etc.

Air-cooled automobile engine cylinders are occasionally made of steel. In this case a hard carbon steel (40 to 60 tons per sq. in. tensile stress) forging is used, the fins and barrel being machined from the solid. A separate aluminium or cast-iron cylinder head is generally employed. In one or two cases copper cooling fins have been employed, the separate fins being threaded, or pressed on to the steel or cast-iron barrel; the difficulty in this case is to obtain a good contact between the two metals; a film layer of dirt acts as a partial heat insulator. Copper is the better heat conductor, aluminium and its alloys come next, whilst steel and cast-iron are much inferior, but possess the best wearing and strength qualities.

Bonded Aluminium-Cast Iron Cylinders.—By the combination of cast-iron or steel cylinder barrels with aluminium-alloy cooling fins, the superior wearing qualities of the former metals and the excellent heat conduction of the latter metal are obtained. Hitherto, the methods of bonding aluminium alloys to cast iron or steel, namely, by shrinking, screwing or caulking spiral sheet aluminium fins into steel cylinder grooves, have proved uneconomical in manufacture and not very efficient in the thermal conduction sense. More recently, however, a method for obtaining an almost perfect bond between aluminium and ferrous metals, known as the Alfin process has been commercialized* and is widely used for high efficiency aluminium-finned cylinders and their heads. The difficulty of getting rid of the aluminium oxides at the joints has been overcome, so that there is a definite bonding layer between steel (or cast iron) cylinder or its head, and the aluminium alloy cooling fins.

Water-Cooled Cylinders.—Water-cooled automobile engine cylinders are, generally speaking, more intricate in design, since the cooling water ports and passages have to be arranged in the castings. The examples given in Figs. 23 and 25 will enable some idea to be obtained of the general type of casting employed.

Previously, plain cast-irons have been used for cylinders, as they gave sharp clean castings in the case of the complicated multiple cylinder blocks. The irons suitable for such purposes had not, however, the best wearing properties. Later, the earlier rather softer, but sharp casting irons were replaced by the much harder and therefore longer wearing plain cast irons using steels or refined pig iron, in their manufacture. More recently alloy cast irons, such as chromium (Chromidium), with nickel and nickel low-silicon irons have been used. Although more expensive than plain irons these alloy irons give from two to three times the mileage distance, before cylinder reboring becomes necessary. It may be mentioned that copper cast-irons have also given very good results for automobile cylinders.

More Recent Progress in Cylinders.—The principle changes in cylinder block and head design have been those connected with better cooling of the valve ports and sparking plug bosses. In some cases positive delivery of cooling water to these parts is provided by directed flow of the water from the pump through carefully designed ports or channels. The exhaust valve—which is the hottest part of the combustion chamber and is, usually, the limiting factor from the viewpoint of the highest compression ratio that can be employed without the occurence of detonation—has received particular attention in the matter of valve seat cooling. The usual arrangement is now to cool

*Wellworthy Ltd., Lymington, Hants.

the regions mentioned by pump-directed flow, but to employ thermo-syphon cooling for the cylinder jackets. In some instances rather longer water jackets around the cylinder barrels have been provided for improved cooling of the metal and the lubricating oil. In general, there is now better positioning and dimensioning of the water passages in the cylinder block and head, allowing a freer flow of cooling water; further, the thickness of the head metal tends to be more uniform in order to avoid the *distortion* that occasionally occurred in previous models.

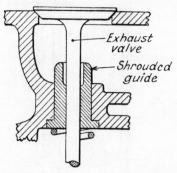

Exhaust valve

Shrouded guide

Fig. 27.—A Shrouded Valve Guide.

The use of *aluminium alloy cylinder heads* has increased; this is on account of their better heat conduction and, therefore, the use of higher compressions than for cast-iron heads. One notable disadvantage of such heads is that they are apt to corrode around the cylinder studs and become difficult to remove for decarbonizing and valve grinding purposes. The *use of graphite* on the cylinder studs before the head is placed in position will obviate this trouble.

In regard to the *valve guides* in the cylinder block. When *leaded fuels* have been used over an appreciable period, deposits of lead salts on the exhaust valve stems have often been the cause of *sticking valves*. To avoid this some valve guides are shrouded in the manner indicated in Fig. 27, so that the valve stem is protected from the lead-salt laden exhaust gases.

Chromium-plated Cylinders.—In order to prevent the

appreciable wear in the upper parts of cylinders due to a combination of corrosion and frictional influences, this portion or the whole of the bore is sometimes electroplated with *porous chromium*, an extremely hard metal that contains fine surface cavities which act as oil reservoirs.   Cylinders thus coated will operate for periods of 90,000 miles and upwards before reconditioning is needed. The Van der Horst process of porous chrome-hardened cylinder

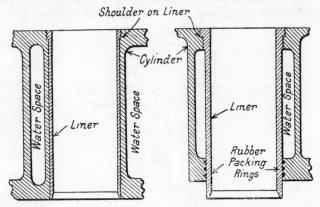

Fig. 28.—Showing (left) Dry Cylinder Liner and (right), Wet Liner.

bores that is much used on automobile petrol and Diesel engines is an example of this method.

In one Rolls Royce unit engine a short hardened alloy cast iron sleeve is used.   This is arranged at the top of the cylinder barrel, where the maximum wear occurs.

Cylinder Liners.—It is now also customary in certain engines, to employ plain cast-iron or aluminium cylinders, and to fit the cylinder barrels with liners machined from an alloy cast-iron. These irons are of the oil-hardening type, and give considerably longer useful service than plain cylinder irons.   Many of the more expensive car and commercial compression ignition engines are fitted with hardened and ground alloy cast-iron liners.

Some mass production car engines, e.g., the Standard Vanguard and the Citroen engines are fitted with detachable cylinder liners.

When the liner forms the complete cylinder barrel, with the cooling water in contact with its outer surface it is termed a *Wet Liner*, but when the liner is inserted into the existing band it is called a *Dry Liner* (Fig. 28). The former type has the advantages of better cylinder cooling and ease of insertion and removal. The dry liner is more difficult to insert and is now used mostly for reconditioning worn cylinder barrels; the latter are bored out to take thin section dry liners of alloy cast-iron. Usually, after the cylinders have been rebored twice, they are fitted with dry liners when the third rebore becomes due.

The use of the wet-type liner is increasing; the favoured metals are the nitrided steels (nitralloys), nitrided cast-irons (Nitro-castirons) heat-treated centrifugally cast-irons in loded cast-irons, chromium and other alloy cast irons, suitably heat-treated. These metals are much harder than ordinary cylinder block irons and give at least 50 per cent. greater wear resistance. Moreover, when the cylinders are eventually reconditioned, standard replacement liners of similar dimensions are employed, so that a standard size of replacement piston can be used; this is obviously an advantage over the method of reboring worn cylinders and using oversize pistons of various step-up dimensions. Wet-type cylinder liners used in car engines are, relatively, thicker than dry liners; a typical wet liner would be from $\frac{3}{16}$ in. to $\frac{1}{4}$ in. thick and a dry liner from $\frac{1}{16}$ to $\frac{1}{8}$ in.

Cylinder Head Gaskets.—In the case of detachable cylinder heads, the joint between the two machined surfaces has to be made both water- and gas-pressure-tight, by means of a special gasket, which was usually, made of two very thin sheets of copper, with asbestos packing between; this liner, when the cylinder head nuts are screwed up, beds down and conforms to the machined surface shape, thus making a good joint. The previous *copper-asbestos gasket* had the drawbacks of altering the compression ratio when screwed down and also tending to distort the cylinder barrel. Thinner gaskets, of copper and asbestos, previously compressed to their final thickness offered a marked inprovement, until more recently *the steel-asbestos-copper gasket* was introduced. This thin gasket has a steel face rim around the combustion spaces to resist the high cylinder pressures and temperatures, and a copper face

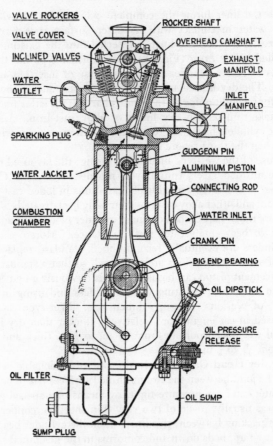

Fig. 29.—Typical Overhead Camshaft Engine,
showing principal components.

VALVE ROCKERS
VALVE COVER
INCLINED VALVES
WATER OUTLET
SPARKING PLUG
WATER JACKET
COMBUSTION CHAMBER
ROCKER SHAFT
OVERHEAD CAMSHAFT
EXHAUST MANIFOLD
INLET MANIFOLD
GUDGEON PIN
ALUMINIUM PISTON
CONNECTING ROD
WATER INLET
CRANK PIN
BIG END BEARING
OIL DIPSTICK
OIL PRESSURE RELEASE
OIL FILTER
OIL SUMP
SUMP PLUG

for the water and oil passage joints. It is also advantageous for aluminium heads, in preventing corrosion of the copper.

Another recent successful gasket is the *single steel ridged* or "corrugated" one, which is much used on mass produced engines. The ridge portions act as springs, when the cylinder head nuts are tightened, so that the head, gasket and block are virtually "solid". This type of gasket does not rely upon distortion of a soft steel, but employs a hard (150 Brinell) steel designed to give the spring action mentioned.

It is usual to coat these gaskets with a special varnish which melts when the engine heats up and seals all the smaller interstices of the block and head. The *stainless steel gasket* is a still later

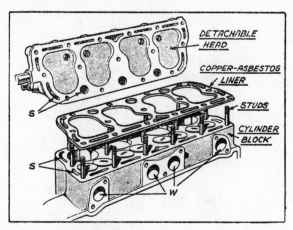

Fig. 30.—Four-Cylinder Engine showing Detachable Head and Liner.

development of the ridged type and is claimed to last for the life of the engine.

A typical detachable cylinder head is shown in the upper part of Fig. 23 in cross-sectional view, whilst Fig. 30 shows a four-cylinder engine with its detachable head removed. The copper-asbestos gasket is also shown ready to drop over the studs on the top of the cylinder block.

The cylinder heads of water-cooled engines are made hollow for the water to circulate round. For this purpose a number of slots and holes, marked S in Fig. 30 are arranged to correspond with one another in both the cylinder head and block.

The cylinder head, it should be noted, carries the sparking

plugs, top water-pipe connection to radiator and, as in the case illustrated, the cooling fan pulley boss. The four holes marked W. in Fig. 30 represent the exhaust manifold connections.

In the case of motor-cycle engines it is the practice to fit cast-iron cylinders with aluminium alloy detachable heads. This arrangement not only ensures better conduction of the heat but enables a higher compression to be employed.

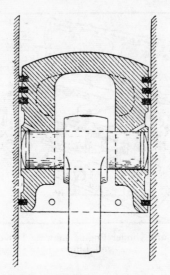

Fig. 31.—The Cross Liner-
less Aluminium Cylinder.

The advantages of a detachable cylinder head are as follows:—
(1) It enables the cylinders to be machined uniformly, and also the heads; and (2) it obviates the necessity of removing the whole (and usually heavy) cylinder block for decarbonizing and valve re-grinding.

Linerless Aluminium Cylinders.—The Cross linerless aluminium alloy cylinder was made from a special hard and strong alloy and machined all over. The piston used in this cylinder was also of aluminium alloy and prevented from actually touching the cylinder walls by three rings of a specially hardened type (Wellworthy); these rings fitted the bottoms of their piston grooves (Fig. 31). The advantages claimed for this system are that there is appreciably less cylinder wear; the cylinder and piston have the same co-efficient of expansion and therefore require very little working clearance, thus giving good gas-tightness, low oil consumption and freedom from cold piston slap ; better heat conductivity so that a rather high compression can be used and, reduced weight.

Tests have been made on motor-cycle engines fitted with this system over very long distances and the results obtained appear to justify the claims made.

Motor Car Engine Cylinder Construction.—The complete cylinder unit consists of (1) The cylinder block. (2) The cylinder head. (3) The oil sump and (4) The inlet and exhaust manifolds. These are separate components which are assembled together by bolts or studs and nuts.

The cylinder block is of a somewhat complex construction since it contains the cylinder barrels, water-cooling jackets, the bearing mountings for the camshaft and crankshaft, the compartments for the timing gears and the crank chamber; the latter is formed by bolting the pressed steel or cast aluminium oil sump to the cylinder block.

In regard to the crankshaft main bearings, the upper halves of the housings for the bearings are machined in the lower part of the cylinder block, studs being provided for the separate main bearing caps. This method of arranging for the main bearings to be mounted on the cylinder block gives a rigid construction and also, for mass production engines, saves expense.

Fig. 32 illustrates the main features of an Austin cylinder block, the cylinder head, end plate, oil sump, timing gear cover, etc., having been removed, to show the jointing faces.

The castings for cylinder blocks are usually normalized and then stored for a period of some weeks in order to allow the metal to "age" before machining. This ensures that there will be no subsequent volume changes. The lowest part of the crankcase usually terminates in a flat machined face provided with a relatively large number of studs for the purpose of attaching the pressed steel

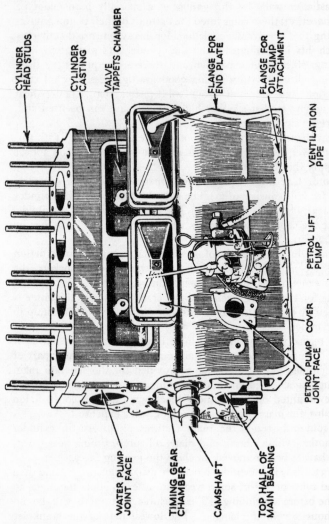

CYLINDER HEAD STUDS

CYLINDER CASTING

VALVE TAPPETS CHAMBER

FLANGE FOR END PLATE

FLANGE FOR OIL SUMP ATTACHMENT

VENTILATION PIPE

PETROL LIFT PUMP

COVER

PETROL PUMP JOINT FACE

WATER PUMP JOINT FACE

TIMING GEAR CHAMBER

CAMSHAFT

TOP HALF OF MAIN BEARING

Fig. 32.—Typical four-cylinder Engine Cylinder Block.

or cast aluminium (or magnesium) alloy oil sump. The crankcase of the cylinder block casting also includes the bearer arms for bolting the engine to the chassis frame; the front portion of the clutch housing is also made integral with the cylinder block casting. The front end section of the latter also contains the

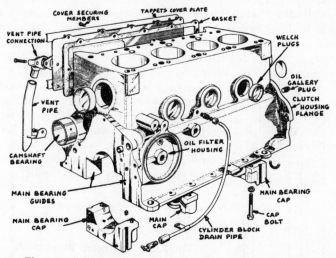

Fig. 33.—Cylinder Block and Components of Four Cylinder Vauxhall Engine.

timing gears or chains and sprockets. In one or two instances, however, these are at the rear or fly-wheel end of the crankcase, but this arrangement does not give the same accessibility as at the front end.

Another example of a cylinder block namely, that of the Vauxhall four-cylinder 92 cu. in. engine block is illustrated in Fig. 33. The overhead valve gear and sparking plugs are located in the cylinder head, which is not shown.

The block, as in the previous example carries at its lowest part the housings for the three main bearings of the crankshaft, the lower bearing caps for which are shown in Fig. 33. The tappet gear and camshaft is located on the farther side of the block, the tappets being enclosed by the cover plate and its gasket. The cylinder block is drilled longitudinal for an oil passage, from which the main bearings are fed. After drilling this passage the outer ends are sealed, so as to be pressure tight, by the end plugs shown. At the bottom of the right hand side of the water jacket a screwed hole for the cylinder drain pipe and its cock is provided.

Cylinder Welch Plugs.—When the somewhat complicated cylinder block is cast, it is necessary to provide certain holes for core extraction, machining purposes and sometimes to allow for expansion of the hot coolant. The holes for internal machining purposes are afterwards plugged up securely. The other block core holes are machined with cylindrical recesses which are recessed a little at their inner ends (Fig. 34, (C)) to receive metal discs (P) which, initially, are slightly convex but, on being driven into their machined hole recesses become flat, to bear tightly against the sides of their holes. It is the usual practice to apply cylinder head jointing compound to the holes before inserting the Welch plugs (discs).

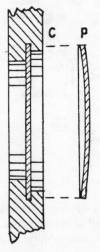

Fig. 34.—Showing typical Welch Plug.

Vee-Eight Cylinder Blocks.—It is usual in the case of the cylinder blocks for vee-eight engines to cast the two banks of four-cylinders each, as a single unit and to machine its lower horizontal face so that the oil sump can be attached to it. The main bearings housings which usually are five in number, are machined (with the bearing caps bolted in position, in the bottom part of the cylinder block). Usually, the face of the lower horizontal plane is level with the axis of the main bearing members; this provides excellent accessibility for main bearing maintenance purposes.

Aluminium Cylinders and Heads.—More recently, increasing attention has been given in the U.S.A., and to a lesser extent in European countries to the use of aluminium alloys for automobile cylinders, crankcases and heads, so that no cast-iron is used for these components. In the U.S.A. such engines have been made by semi-gravity and pressure die-casting methods, with satisfactory results. While, as shown in the earlier Cross motor-cycle engine, aluminium alloy pistons can be used in aluminium alloy cylinders, the results obtained with larger automobile engines have not been altogether satisfactory, so that the wear problem has been solved by such methods as (1) Chromium-plating the cylinder bores. (2) Using cast-iron liners pressed into the cylinder bores. (3) Using cast iron liners, with serrated outer surfaces cast into the bores, when the aluminium block castings metal is poured. (4) Cold spraying the cylinder bores with steel by the well-known metallizing process. (5) Fitting chromium-plated compression rings to the aluminium alloy pistons. The principal advantages of aluminium alloy, die-cast cylinders and heads are as follows:

(1) An appreciable saving in engine weight, up to about 30 per cent. over the cast-iron type engine, giving smaller engines for the same power output.

(2) An increased compression ratio, without detonation, due to the much higher thermal conductivity of aluminium.

(3) A greater horse-power output and lower fuel consumption per horse-power on account of the increased compression, and better cooling properties.

(4) Smaller piston-to-cylinder clearances, when chrome-plated pistons or steel-sprayed cylinder bores are used.

Against these important advantages must be offset the following possible drawbacks:

(1) Greater manufacturing costs than with cast-iron heads and blocks.

(2) Necessity to use cast-iron valve and (sometimes) sparking plug insets.

(3) Greater possibility of cylinder coolant spaces corrosion by the cooling water.

(4) Tendency of the cylinder heads to "stick" to the cylinder block, due to intermetallic corrosion between the steel studs and the aluminium alloy.

In regard to the corrosion possibility this has been practically eliminated by using high silicon aluminium alloy—which also has a lower thermal expansion than other alloys—and by the use of suitable corrosion inhibitors in the cooling water.

It may be mentioned that the "all-aluminium" engine warms up more quickly from the cold than the cast-iron one and also has a smaller cooling water capacity—usually by about 20 to 25 per cent.

Pressure-cast cylinder blocks are used in the American Motors six-cylinder-in-line and Chrysler "Slant-Six" engines. The Oldsmobile "Rockette" vee-eight engine uses semi-gravity die castings for the cylinder blocks. The more recent Rolls Royce vee-eight engine has high silicon aluminium cast cylinder blocks and heads.

Copper Cylinder Heads.—Copper has even better thermal conductivity than aluminium or its alloys. Thus, if the conductivity for copper is represented by 100, that of aluminium is about 58 and grey cast-iron, 18. Thus, if copper is used instead of aluminium for cylinder heads, still higher compression ratios can be used, with the same fuels and combustion chambers, than for aluminium. The results of laboratory and road tests on engines using cast copper heads have indicated appreciable increases in power output, accompanied by lower fuel consumptions. Even with the same compression ratios, the copper head engines showed better fuel economy than when aluminium heads were used.

Combustion Chamber Shapes.—The shape of the combustion chamber has an important bearing upon the output and thermal efficiency of the engine. This shape depends upon the disposition of the valves. If the valves are in the cylinder head, it is possible to approximate to the ideal spherical shape shown in Fig. 35(A). The inclined overhead valves and dished piston shown in Fig. 35 (B) are the nearest practical approach to (A). A more common arrangement, which is also efficient, is that shown in (C); this design can be constructed more cheaply, and is often used on high speeds and racing engines.

Most overhead valve engines give more turbulence to the fresh charge than in the other designs shown in the lower diagrams of Fig. 35, and also give less heating to the incoming charge. An arrangement, which represents a compromise between the over-head and the side-by-side valve shape, is that illustrated in (D) Fig. 35 and in which the inlet valve is over the exhaust. This arrangement enables a single camshaft to be used; the inflowing mixture also cools the exhaust valve.

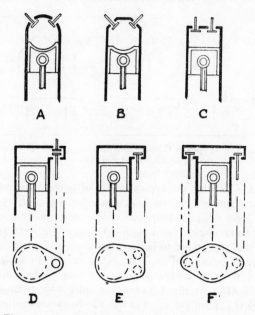

Fig. 35.—Showing various Combustion Chambers and Valve Dispositions.

The side-by-side valve engine can be constructed economically, and is accessible; it was widely employed in light and medium cars. The turbulence promoted in this type is not as satisfactory as in the overhead types, but the special combustion chamber design illustrated in Fig. 39 improves the degree of turbulence considerably, and renders this type more efficient. The T-headed engine shown at F (Fig. 35) is now obsolete; it is bulky, inefficient and expensive to produce.

Some Earlier Comparative Results.—In regard to the overhead valve type of engine with inclined valves in the head as shown in Fig. 36, the results of research work carried out by Ricardo with different positions of the sparking plug and, in one instance with the cylinder head fitted with two sparking plugs as shown at *B* (Fig. 36), show that the highest useful compression

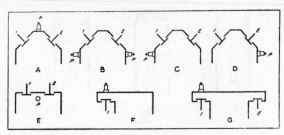

Fig. 36.—Types of Combustion Chamber, with different
Sparking Plug Positions.

ratio (H.U.C.R.) for the typical grade of petrol used was 6·0:1 for the arrangement shown at *A*, 5·9:1 for that at *B*; 5·3:1 for that at *C*, and 5·6:1 for that at *D*. The corresponding value for the side valve engine head with sparking plug over the inlet valve as at *F* was found to lie between 4·3:1 and 4·6:1.

The plain cylindrical type of head, similar to Diagram *E*, Fig. 36, with the sparking plug on the side of the combustion chamber wall, centrally between the inlet and exhaust valves gave an H.U.C.R. of 5·2:1. For the Tee-head combustion chamber, shown at *G*, gave the low value of 3·6:1.

Mention should also be made of the symmetrical pocketless combustion chambers of the single-sleeve valve, the Aspin and Cross rotary engines which are described in Chapter III; each of these combustion chambers enables relatively high values of the H.U.C.R. to be utilized without detonation effects.

Non-Detonation Type Combustion Chambers.—A considerable amount of experimental work has been carried out in recent years with the object of improving the combustion chamber so that higher compression ratios can be employed with fuels which previously required lower compressions in order to prevent

detonation.   Now that the causes of detonation are more fully
understood the shape of the combustion chamber can be designed
accordingly and improved performance thus obtained.

The general principle of the method employed is illustrated
in Fig. 37, which shows an efficient design of car engine cylinder
head.   It will be seen that the chamber is divided into three
zones.   The first zone, on the left, is the ignition area which is

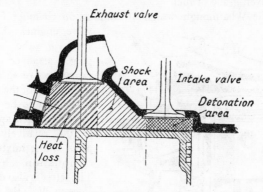

Fig. 37.—Combustion Zone.

likewise the heat loss area.   In this area the metal is exposed to
burning for the longest time and it is the region of the highest
temperature.   This area should therefore be protected against
heat loss and the exhaust valve placed within this zone.

The second zone is the shock area, because during the time this
area is burning the crankpin and piston are passing through
top-dead-centre and thus the structure is given the maximum
effect of the pressure rise.   It is desirable to reduce the pressure-
rise rate in this area and hence a portion of the volume must be
displaced from this point.   If this is placed in the third area,
which is the detonating zone, more volume-to-surface results
than is satisfactory for detonation control.   The relatively cooler
inlet valve when placed in this area helps to absorb the heat of
this super-compressed last portion of the charge to burn.

This is added to the last third, and, providing the depth of the
section through this area is not over $\frac{9}{16}$ in., detonation may be

adequately controlled for higher compression ratios. In this manner it is possible to get a lower heat absorption at the first area allowing a lowering of the flame front in the shock area and yet providing additional heat in the detonating area.

It may here be mentioned that a mechanical method was devised by Vauxhall Motors Ltd. to obtain the best shapes of combustion chamber for utilizing the highest compression ratios for a given grade of fuel.

In the Whatmough design of combustion chamber for side-valve engines shown in Fig. 38, the sparking plug is arranged over the exhaust valve so that the hottest part of the mixture is the first to be ignited, the flame then spreading to the relatively cooler parts of the mixture and combustion chamber. The latter is also designed so as to allow the mixture to flow into it on stream-line principles. In order to secure as even a distri-

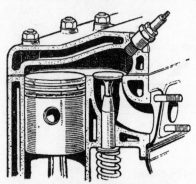

Fig. 38.—The Whatmough Combustion Chamber.

bution of temperature as possible the cooling water is arranged to give the most efficient removal of the surplus heat from the combustion chamber possible.

The previously inefficient side valve combustion chamber shown at E in Fig. 35, was improved considerably by Ricardo, who evolved the shape shown in Fig. 39. This gives increased turbulence, for when the piston nears the top of its compression stroke the charge is, as it were, squeezed into the space over the valves, thus giving it greater turbulence, and promoting more efficient combustion conditions.

More Recent Combustion Chambers.—Marked improvements have been made in the actual shapes of combustion chambers of modern engines and in the disposition of the valves and sparking plugs to give the best conditions for (1) Short flame travel path; (2) Good volumetric efficiencies; (3) Satisfactory scavenging of the exhaust gases; (4) Adequate cooling of the sparking plug points; (5) Good cooling of the exhaust valve head; (6) The use of the highest possible compression ratio for any fuel of given octane rating.

The principles discussed on page 103, in regard to combustion zones have been followed and developed with the above objects

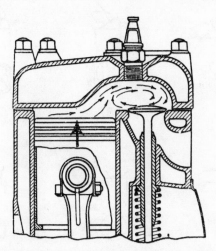

Fig. 39.—The Ricardo Turbulence Combustion Chamber Design for Side Valve Engines.

in view. It has been shown that the best results are obtained when the valves are of ample head proportions, *the inlet being rather larger than the exhaust*. The sparking plug should be located in a fairly central position so as to give the shorter flame travel paths in all directions, as far as possible. Further, a certain amount of swirl or turbulence of the compressed mixture should be provided by the combustion chamber shape, in relation to the piston at the top of its compression stroke, so that the sparking plug will be cooled sufficiently by the mixture movement.

To obtain such results it is necessary to depart, somewhat, from the combustion head and valve arrangements shown in Fig.

35 on page 101, and to use inclined valves or one vertical and one inclined valve. There is a relatively large number of alternative arrangements to those shown in Fig. 35, but if these are examined in the light of the principles described on page 103, the number of practicable designs will be found to be very limited. Here, it may be added that the L-head and T-head combustion chambers

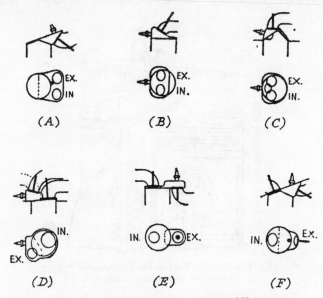

Fig. 40.—Alternative Arrangements of Valves, Sparking Plugs and Combustion Chambers.

are now obsolete, but the Ricardo and Whatmough modifications to the former type have increased its efficiency and enabled good performance results to be obtained.

Fig. 40* shows, diagrammatically, some alternative combustion head arrangements that have been developed with the object of fulfilling some or all of the conditions enumerated previously;

"Poppet Valve Cylinder Heads, etc." J. Swaine. *Proc. Inst. Mech. Engrs.*, 1948.

it will be observed that in each case the valve-actuating mechanism is more complicated than for the side-valve engine.

Referring to (A) Fig. 40, this represents a development from the side-valve engine to give a more compact combustion head and approximately central sparking plug position, but with a bias towards the exhaust valve. Owing to the valves being in an

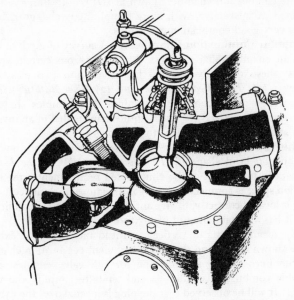

Fig. 41.—Rolls Royce F-type Cylinder Head.

offset pocket, the full amount of mixture may not be drawn into the cylinder; further, the sparking plug and exhaust valve would not be fully cooled by the mixture.

The arrangement shown at (B), namely, with two inclined overhead valves, with parallel stems, is an improvement on the vertical overhead method as it gives a wedge-type combustion zone with central plug position. The mixture during the final compression stage is squeezed out from the extreme right side and thus given some degree of turbulence which will provide valve and plug cooling; this is often referred to as a "squish" effect. The initial flame front near the plug is of wide dimensions, which is a distinct advantage.

The alternative arrangement shown at (C), to give somewhat similar results, necessitates a dished piston head, but enables vertical-type overhead valves to be employed.

The combustion chamber shown at (D) requires an overhead, inclined exhaust valve, with more complicated valve gear, but it gives very good combustion conditions.

Popular Combustion Chambers.—A survey of modern combustion chamber shapes shows that, apart from certain special types, these generally resolve themselves into two classes, as follows: (1) *The Wedge Shape* and (2) *The Bathtub Shape.* The wedge shape chamber, resembles the examples shown in Figs. 37, 38 and 40 (B). The bathtub shape is that of an inverted bath and is indicated in Figs. 39 and 40 (E).

The "F"-Head Engine.—The overhead valve disposition and combustion head shape in (E) *known, generally, as the F-head*, appears to fulfil most of the enumerated requirements; the combustion space is compact and there is directional turbulence. The plug is located correctly, to give anti-knock combustion conditions, and it has a certain amount of mixture swirl cooling effect.

The arrangements of vertical overhead inlet and side exhaust valve shown at (E), and the partly-turbulent combustion chamber on the lines of the Ricardo one for side-valve engines, gives suitable combustion conditions and tends to simplify the valve gear. It will be observed that the plug is located over the exhaust valve and that the flame front is a wide one in this region.

The more recent Rolls Royce and Bentley engines employ the F-head arrangements, corresponding to (E), in Fig. 40. In this design the inlet valves are of the push-rod operated overhead type (Fig. 41)* and the exhaust valves of the side-valve pattern. The inlet valve is of rather larger diameter than the exhaust valve, to ensure a high volumetric efficiency. The sparking plug is located over the exhaust valve.

The Rover Combustion Head.—Of the many possible alternative arrangements, that shown at (F) (Fig. 40) corresponds

*Courtesy. *The Autocar.*

to the modern Rover car engine combustion head, given in sectional view in Fig. 42. It approximates to the hemispherical shape, with the plug at the centre of the flat side, giving minimum flame travel path, but permitting large valve diameters for induc-

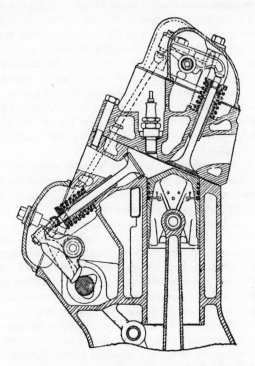

Fig. 42.—Sectional View of the Rover Car Engine
Cylinder Head.

tion and exhaust clearance. The piston at the end of the compression stroke forces out the mixture past the overhead inlet valve and across the plug points, thereby tending to cool them. This gives a high efficiency combustion chamber, such that it enables compression ratios of 7·75:1 to 8·0:1 to be employed with 80-octane fuel and about 7·2:1 to 7·5:1 for 72-octane fuel.

Brake m.e.p.'s up to 140 lb. per sq. in. are obtainable with higher compression ratios. Further, it has been shown that weaker mixtures down to about 20 parts air to 1 part petrol can be employed without misfiring and with very low fuel consumption.

The Hemispherical Combustion Head.—If the combustion chamber of a petrol engine could be made in the form of a complete sphere, with the sparking plug at the centre it would represent the ideal shape, since a sphere gives the smallest surface area for its volume. This means that there would be the minimum loss of the heat due to the combustion of the fuel, during the combustion period, than for any other combustion chamber form. Such an engine should show the highest thermal efficiency. A possible arrangement of this type is shown at (A), in Fig. 35, but it will be obvious that the sparking plug cannot be located at the centre of the spherical space.

It may be mentioned that a sphere has only about 60 per cent. of the surface area of an ordinary side-valve combustion head of equal volume and about 80 per cent. of a cylindrical head, as at (C), in Fig. 35 of the same diameter and volume.

If the conditions of minimum heat loss were the sole consideration, then the spherical combustion chamber would be ideal, but unfortunately, the various other important requirements that necessitate a different shape of combustion chamber. To mention but one of these requirements, namely, that of miminum flame travel before the detonation zone is reached, it would not be possible with a sparking plug located on the surface of a spherical chamber to obtain this result; neither would it be practical to arrange for the sparking points at the centre of the sphere.

Instead, therefore, of using a spherical combustion chamber, *a hemispherical cylinder head*, with a flat or domed piston, has found much favour with designers of high output engines. In this case overhead valves are used and there are no side pockets—as with the L-headed or T-headed type of combustion chamber.

The hemispherical head has twice the surface area of the circular cylindrical head, so that appreciably larger valves can be used, with a consequent increase in the volumetric efficiency.

The higher volumetric efficiency of the hemispherical head engine, combined with the relatively short flame travel path from the centrally located sparking plug and other favourable combustion conditions, enable this type of engine to develop very high outputs as expressed in H.P. per litre. An example of the exceptional performances obtained include that of a Norton motorcycle engine that won a Tourist Trophy race. This 500 c.c. air-cooled engine developed 49 B.H.P. at 6,750 r.p.m. and gave a B.M.E.P. of 203 lb. per sq. in. at 6,000 r.p.m.   The equivalent power output was 98 H.P. per litre for this non-supercharged engine.

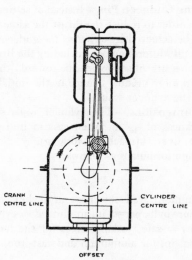

Fig. 43.—The Offset or Desaxé Cylinder Arrangement.

Typical examples of the various makes of hemispherical-head car engines include the Jaguar XK, Armstrong Siddeley Sapphire, Bristol, Aston-Martin and the American De Soto V-8 (1953) and Buick XP-300.

The Desaxé Engine.—In order to reduce the thrust of the piston on the cylinder walls during the firing stroke, the cylinders in some engines are set forward in the direction of rotation, so that their centre lines pass in front of the crankshaft centre as shown in Fig. 43. This *Offset*, or *Desaxé* arrangement gives also a rather slower piston motion at top-centre and tends towards higher combustion pressures. It does not affect the engine balance appreciably.

The usual amount of offset is from one-fifth to one-sixth of the cylinder diameter. The piston stroke is about 0·3 per cent longer with this amount of offset, and with a connecting-rod to crank ratio of 4 : 1.

The practice of offsetting the engine has been followed in the case of several automobile engines.

Offsetting the Gudgeon Pin.—Instead of setting the cylinders axis forward in order to reduce the thrust the same result can to a limited extent be achieved by *offsetting the* gudgeon pin. Thus, in certain General Motors engines (including the British Vauxhall) the gudgeon pins are offset by $\frac{1}{16}$ in. towards the thrust side. This provides a more gradual change in the initial thrust of the piston against the cylinder wall.

Cylinder Dimensions.—The cylinder walls should have a minimum thickness of $\frac{3}{16}$ in. when cast in iron or aluminium alloy and $\frac{1}{16}$ in. in steel (machined forging).

A useful design formula is as follows:

$$t = \frac{p\,d}{2f}$$

where t = minimum wall thickness in inches, p = maximum explosion pressure in lb. per sq. in. absolute, d = cylinder diameter in inches, and f = safe working stress for the metal used, i.e., 2,000 lb. per sq. in. for aluminium and cast-iron, and 3,500 for steel.

Another empirical formula is

$$t = \frac{d}{32} + \tfrac{1}{8} \text{ (inch)}$$

Cylinder Tests.—Cylinder water jackets are usually tested for soundness and cracks by subjecting them to hydraulic, steam or air low pressure tests. With the steam method the heat tends to open any cracks, making them more readily discernible. When the cylinder block design allows, the cylinder barrels can be tested hydraulically, for soundness using pressures well above the maximum peak pressures developed soon after ignition, e.g., 700 to 1000 lb. sq. in.

The Piston.—The purpose of this important component is to receive the force due to the pressure of the combustion products, during the firing and expansion periods and to transmit this force to the connecting rod.   It must operate in the cylinder with the minimum of friction and must be gas-tight, so that there are no pressure leakages past its surface during the expansion and compression strokes.   As the crown of the piston is exposed to

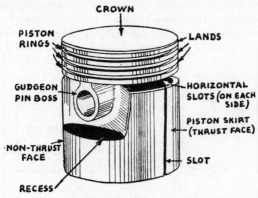

Fig 44.—A modern Split-skirt Aluminium Alloy Piston, showing the names of the principal parts.

the high temperatures due to combustion—usually about 2,000°—2,500° C.—and to a cycle of temperature changes during the working of the engine it must be both strong in construction and also as light as possible, in order to reduce the inertia forces due to its velocity changes and weight.

In regard to the maximum load on a modern engine piston this is estimated to be about 1¼ to 1½ tons in the case of a piston of 2·6 in. diameter, as used in some 1·5 litre engines.   Modern pistons are invariably constructed from strong aluminium alloys which are about 50 to 60 per cent of the weight of earlier cast-iron pistons of similar diameter.   It should be mentioned that cast-iron and alloy steel pistons have been used in modern engines on account of their better wearing qualities and of the smaller working clearances that can be used.   This is on account of the greater heat expansion of aluminium and its alloys.

On the other hand aluminium alloy, which is very much lighter is a far better heat conductor; it therefore runs much cooler, and more free from carbon deposit. Owing to its higher co-efficient of expansion than cast-iron, it has usually to be made a "looser" fit when the engine is cold; this extra clearance used to result in a knocking noise, known as *Piston Slap*, which disappeared, however, as soon as the engine warmed up. To overcome this objection, it is usual to split the *Skirt* or lower end of the piston, and to slightly spring the segments. Other designs of non-slap aluminium alloy pistons are described later in this section.

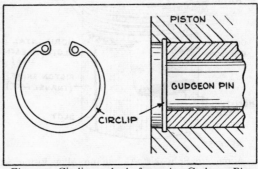

Fig. 45.—Circlip method of securing Gudgeon Pin.

The Gudgeon Pin.—The gudgeon or piston pin is usually made hollow to save weight, and to give as much bearing surface as possible. The material used is a high-grade carbon or low-nickel content steel, case-hardened and ground to size.

Various methods have been used, in the past, to secure the gudgeon pin in the piston bosses. These included the tapered pin through the piston boss and gudgeon pin and the piston ring encircling the piston in a slot machined opposite to the centre of the gudgeon pin. Of the various alternatives the Circlip method, shown in Fig. 45 is now the only one that is widely used.

The usual method employed in modern aluminium alloy pistons is to make the gudgeon pin a push fit in the small-end bearing and piston bosses. It is secured against end movement by means of spring clips, such as those shown in Fig. 45.

In certain modern designs the gudgeon pin is actually clamped tightly in the small end of the connecting rod and allowed to rock in the piston bosses. This method is used in the B.M.C., Ford and Vauxhall engines. The complete piston and connecting rod assembly of the B.M.C. engine is illustrated in Fig. 46; the gudgeon pin and locking or pinch bolt are shown in this illustration.

When the piston is made of high strength aluminium alloy, the floating gudgeon pin will operate satisfactorily in this alloy without the necessity for bronze bushes.

In some modern engines the gudgeon pin is made a *thermal fit* in the piston bosses. Thus, when the piston is heated to the temperature of boiling water, the holes in the bosses expand sufficiently to enable the cold gudgeon pin to be pushed readily into its working position.

The Small End Bearing.— The smaller end of the connecting rod is bushed with a phosphor-bronze bush and the gudgeon pin is made a rocking fit in this bush. The clearance between the pin and the inside of the bush is usually 0·0006 in. to 0·001 in.

In some instances the gudgeon pin is a bearing fit in both the small end bush and the piston bosses. Then, the clearance is ·001 in. to ·003 in. at 75° F. When the pin is clamped to the connecting rod small end, the pin extensions are a bearing fit in the piston bosses, with clearances of ·001 in. to ·0025 in.

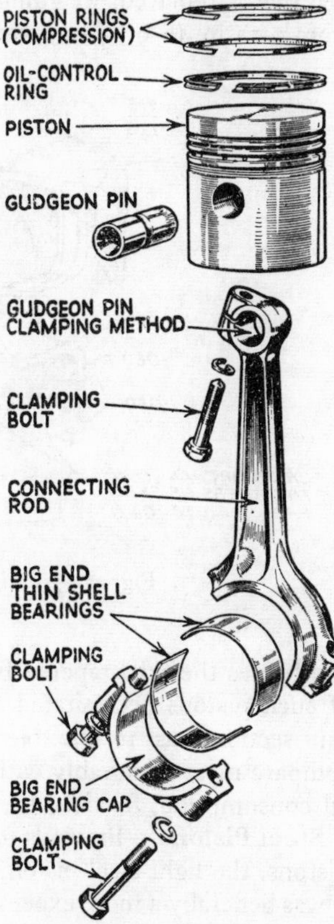

Fig. 46.—Components of Morris Oxford Piston and Connecting Rod.

PISTON RINGS (COMPRESSION)

OIL-CONTROL RING

PISTON

GUDGEON PIN

GUDGEON PIN CLAMPING METHOD

CLAMPING BOLT

CONNECTING ROD

BIG END THIN SHELL BEARINGS

CLAMPING BOLT

BIG END BEARING CAP

CLAMPING BOLT

A Cast-iron Piston.—Fig. 47 illustrates an alloy *cast-iron piston* employed on certain American engines of the late war period when aluminium was in short supply. Although cast-iron is about three times as heavy as aluminium, the $3\frac{3}{4}$ in. diameter piston shown weighed only about 10 per cent. more than the aluminium one that it replaced, weighing only 1·23 lb. The crown tapered from 0·12 in. to 0·13 in., while immediately below the gudgeon

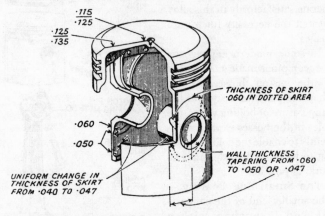

Fig. 47.—A Light Cast-Iron Piston.

pin bosses the skirt tapered from 0·06 in. to 0·04 in. The making of such pistons necessitated efficient casting methods, involving thin sections cast to fine tolerances. The pistons were stated to compare most favourably with aluminium alloy ones in regard to oil consumption, friction, freedom from scuffing and fit.

Steel Pistons.—Previous to the general adoption of aluminium pistons, the light steel piston was used in racing car engines, but it was generally a more expensive piston to manufacture.

The Lincoln Zephyr engine, made by the Ford Company, used the copper steel piston design shown in Fig. 48. This piston had a skirt thickness of only ·035 inch, and a crown thickness of ·090 inch. It was, therefore, an excellent heat disperser as compared with the heavier cast-iron pistons. It was so light that the piston shown (2¾ in. diameter) weighs only 11 oz., which is about 1 oz. more than the same diameter of aluminium piston.

The clearance between the piston and cylinder wall can, of course, be made much smaller than for aluminium alloys, so that the oil consumption is reduced and no piston-slap occurs at starting.

The pistons were cast in "gates" of eight pistons, the latter being afterwards separated by sawing.

Aluminium Alloy Pistons.—In the case of aluminium alloy pistons, the clearances are greater, as we have mentioned before.

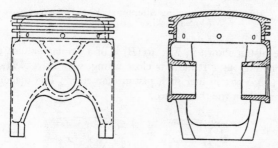

Fig. 48.—A Copper-Steel Piston.

The average clearance for plain aluminium alloy pistons working in cast-iron or steel cylinders is $\frac{3}{1000}$ to $\frac{6}{1000}$ in. per inch diameter of cylinder. An appreciably greater clearance is given at the top of the piston, however. Thus in the case of a 3 in. diameter cylinder the skirt clearance is about $\frac{25}{1000}$ in. at the bottom; that of the first land $\frac{12}{1000}$ in.; of the second land $\frac{14}{1000}$; of the third land $\frac{16}{1000}$ in.; and of the top or fourth land $\frac{20}{1000}$ in. The compensated types of aluminium alloy pistons described later in this section have relatively smaller clearances compared with those of the plain cylindrical ones.

Typical examples of the clearances for different parts of two different kinds of aluminium alloy pistons, namely, plain and split skirt ones, are illustrated in Fig. 49. The former illustration refers to a piston ground oval in shape to have a total clearance of 0·010 to 0·012 in. on the portion beneath the gudgeon pin holes, and 0·002 to 0·0025 in. total clearances on the thrust faces.

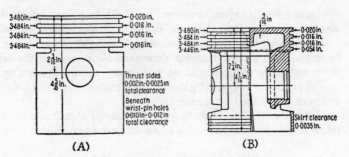

Fig. 49.—Illustrating Aluminium Piston Clearances.
(A) Solid skirt.   (B) Split skirt.

The piston shown in Fig. 49 (*B*) is of a similar pattern to that given in Fig. 44.   It permits close fitting, with skirt clearance of 0·0005 to 0·001 in. per inch piston diameter.   The diagonal slit is arranged on the thrust face.

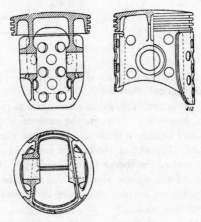

Fig. 50.—The Ricardo Slipper Piston.

The Slipper Piston.—Fig. 50 illustrates a good example of light aluminium piston design, in which the weight has been reduced to an absolute minimum by the removal of all redundant or surplus metal.   It is known as the *Slipper Type Piston* from the fact of the "skirt" portion having the bearing surfaces only on the two piston thrust sides; that on the firing thrust side is of the larger area.   In the case of a 4 in. diameter slipper piston, complete with its rings and gudgeon pins, the weight is only about 1·4 lb., whereas that of the ordinary light cast-iron piston is about 3·3 lb.   The slipper piston has been shown to reduce piston friction

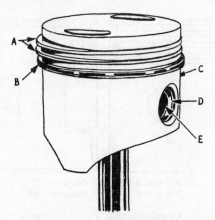

Fig. 51.—Typical American Slipper-type Piston of a Vee-eight Engine. (The recesses in the piston crown are for clearances of the inclined valve heads of the high compression engine.)

by 25 to 35 per cent., giving an increase of power at normal speeds of 5 per cent. and up to 15 per cent. at the highest speeds.   The gudgeon pin in the slipper piston illustrated (Fig. 50) actually bears on the aluminium alloy; there is no bronze brush.   This arrangement works perfectly well, and there is very little wear.

Fig. 51 shows a typical American vee-eight engine aluminium alloy piston of the three-ring pattern.   The cross-section is of oval shape, the piston skirt being about ·009 in. less in diameter, at right angles to the thrust faces.   The piston, which is cut away between the thrust faces is also tapered in length, being smaller in diameter at the top piston land than at the skirt.   The two recesses shown on the crown of the piston are to prevent the fully-open valve heads from touching the piston.

Oval Section Pistons.—In order to insulate, so far as possible, the skirt from the hotter crown portion of the piston, and thus prevent distortion of the former part, it has become the practice to provide slots between the head and the skirt, as shown in

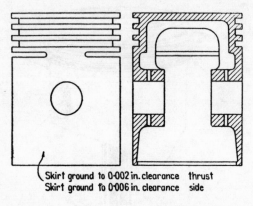

Skirt ground to 0·002 in. clearance   thrust
Skirt ground to 0·006 in. clearance   side

Fig. 52.—An Oval-Section Aluminium Alloy Piston.

Figs. 49 (A) and 52. These slots are always on the piston thrust faces. They also serve to give flexibility to the skirt when the latter is also split.

The heat from the hot piston head in travelling down the un-cut portions near the piston bosses, raises the temperature of these parts of the skirt more than that of the thrust faces. If the section of the skirt were truly circular when cold it would tend to become oval, the greater expansion occurring across the piston boss, or gudgeon pin diameter.

To obviate this undesirable result, it is now usual to make the piston skirt of oval section, when cold, by grinding away a small amount of the metal on the minimum thrust faces, whilst leaving that on the thrust faces.

Fig. 52 illustrates an oval piston of 3 inch diameter, and shows the clearances usually allowed. These pistons are known as "constant clearance" ones, since they become circular in section when hot.

Further Modern Examples of Pistons.—Modern pistons are of the cam-ground and steel insert types but differ somewhat in individual design. Some typical examples of recent pistons are shown in Figs. 53 to 60.

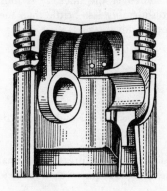

Fig. 53.—The Wellworthy "Welflex" Aluminium Alloy Piston.

The Wellworthy "Welflex" piston (Fig. 53) is also designed on a somewhat similar principle, automatic compensation being provided for heat expansion effects. It has a split skirt and this is supported at its upper ends on most of the trailing side, thus making it much stronger that the usual split skirt design of piston. For this reason there is less risk of "piston rock" owing to the skirt closing; neither do the rings become convex-faced. It is significant that with this type of piston the cold cylinder clearance can be reduced to 0·00075 in. per inch diameter, so that for a 3 in. piston the total clearance is only 0·00225 in.

A recent design of piston, known as the Wellworthy "O.T." (oval-turned) one is similar in principle to the Welflex one but, in addition, is provided with a channel-sectioned scraper ring about half-way down the skirt. This type has been developed, in particular, for oily engines and is recommended as a replacement piston for rebored engines.

Pressed Aluminium Pistons.—A marked improvement in piston manufacture has been that of the *pressed light alloy type*. The results of test have shown that the tensile strengths of pressed aluminium alloy pistons are about 50 per cent. greater than for sand-cast and 25 to 30 per cent. more than for die-cast pistons of the same metal; moreover the ductility is much better so that the pistons are considerably less brittle than the two other types mentioned. Another advantage is that owing to the precision

Fig. 54.—Section of Knurled Piston Skirt.

with which the pistons can be pressed the final dimensions are such that less machining is required than for sand-cast and forged pistons.

Knurled Pistons.—A more recent innovation in aluminium alloy pistons is that of *knurling the piston skirt*, by means of closely-pitched fine serrations—similar to those that might be made with a screw-cutting tool. These serrations (Fig. 54), of from 0·01 in. to 0·015 in. deep, are made by a combined rolling and cutting operation, known as knurling. The breaking up of the otherwise plain smooth skirt into a very large number of tiny rhomboidal areas, separated by fine grooves, greatly improves skirt lubrication and reduces the temperature of the metal, owing to the better oil circulation. Further, in the case of tin-coated and other types of surface-treated pistons the coatings, being better lubricated, last much longer.

Nickel-Iron Piston Inserts.—The wear that occurs in the two upper piston ring grooves of aluminium alloy pistons is due to the high frequency hammering effects of the harder piston rings on the relatively softer piston metal. In order to reduce such wear to a practically negligible extent, hard nickel-iron ring inserts are sometimes employed (Fig. 55). These are located in the mould and the aluminium alloy cast around them. Since the nickel-iron ring carrier has about the same expansion coefficient as the piston metal there is no tendency to loosen during service. The result of tests made originally in Germany, showed clearly that the life of the insert piston was increased by at least twice that of the plain aluminium alloy one. In one instance, the measured wear between the gray-iron piston rings and the cast-iron insert grooves was just under 0·004 in. after 64,000 miles. The pistons with new rings were then replaced in the

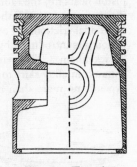

Fig. 55.—Two-ring Nickel-iron Piston Insert.

cylinders but without any machining work having been done on them and they gave a further long period of efficient service. Wear of the cylinder was so small that there was no measurable increase in the oil consumption and new liners were fitted only

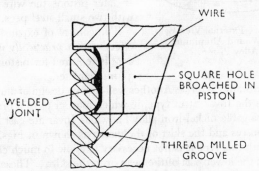

Fig. 56.—Wire-wound Piston Skirt Method.

after 120,000 miles. The piston ring insert is of particular value in high-speed Diesel engines.

Wire-wound Pistons.—The split-skirt piston, of oval-ground form, is fitted so that, when cold, the skirt portion is tight enough on the thrust axis to prevent piston slap. When, however, the piston heats up, the radial expansion of the skirt portion increases the pressure against the cylinder walls and thus gives rise to additional friction. To overcome this drawback the wire-wound piston was introduced by The Automotive Engineering Company, Twickenham. The wire spiral band is arranged around the aluminium alloy skirt to resist the expansion, and since this band expands no more than the bore it does not increase pressure on the cylinder walls as the engine warms up.

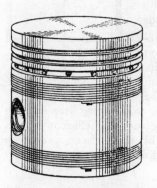

The band, which is shown in sectional detail in Fig. 56, is arranged both above and below the gudgeon pin, the piston being ground after the wire is in place. The wire in the original pistons was held in place by spot-welding the last two turns to two pieces of similar wire inserted in a hole broached in the skirt. In later pistons the wire is fixed with two small steel pegs.

Fig. 57.—External View of Wire-Wound Aluminium Alloy Piston.

The control of expansion is so effective that *practically no clearance* is required for pistons up to $3\frac{1}{2}$ in. diameter.

Invar Strut Piston.—Another well-known design of aluminium piston is the Invar Strut type, in which four strips of a practically non-expansible nickel-iron alloy known as Invar are cast into the piston bosses and the skirt of the piston as shown in Figs. 58 and 59. With this type it is not necessary to allow so much clearance either to the lower part of the head or to the skirt. These pistons are made of a strong aluminium alloy called Birmalite, which has a tensile strength of 16 to 19 tons per sq. in. and is therefore as strong as wrought iron.

Another design of compensating piston, known as the Nelson Bohnalite Autothermic one, shown in Fig. 60, employs a pair of low carbon steel plates P, with their ends anchored to the skirt

but not to the gudgeon pin bosses; the skirt, bosses and head
are a one-piece casting which relieves the steel inserts of

Fig. 58.—The Invar Strut Piston.

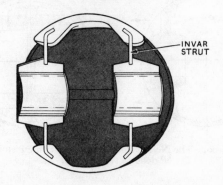

Fig. 59.—Plan View of Cut-Away Piston
to show the Invar Struts.

thrust. Since the aluminium alloy when heated expands more
than the steel the two metals form a kind of bi-metallic element
which tends to bend outwards as shown by the arrows, with

the result that the expansion of the skirt is decreased in the direction perpendicular to the axis of the gudgeon pin and is increased in the direction parallel to the pin. In this way the

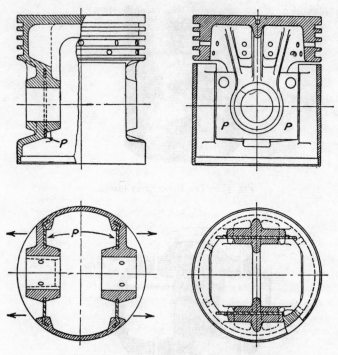

Fig. 60.—The Nelson Bohnalite Autothermic Piston.
(*P* denotes steel plates).

expansion of the piston is effectively controlled and by suitably apportioning the dimensions of the steel members practically any degree of expansion can be obtained. Usually the net expansion of the piston when heated is about the same as that of cast-iron. It is, however, slightly less in the direction perpendicular to the gudgeon pin at both top and bottom of the skirt and slightly greater at the bottom of the skirt in line with the gudgeon pin.   The piston is thus oval when cold and cylindrical when hot, whilst it requires a much smaller cold clearance than a plain piston of the same alloy.   The latter is of high-silicon aluminium and has a lower expansion coefficient than most other commercial aluminium alloys.

Another class of piston employs an aluminium head and piston bosses with a steel skirt, thus combining the excellent heat-conducting properties and lightness of the former material with the good wearing qualities of steel for the skirt.

Coatings for Aluminium Pistons.—The ordinary aluminium alloy pistons are more liable to wear and "scuffing" action than cast iron ones, so that in more recent models the cylindrical surfaces have been given a protective coating of aluminium oxide or tin.

When tin is used as a coating it not only prevents "scuffing", or partial seizure during the running-in period of the engine, but forms a highly polished bearing surface of low frictional resistance.

The *tin-coating* of pistons when new is now more widely used, as it has proved an effective preventative of scuffing and seizure during the running-in stages.   The tin coating required need only be very thin; thus, deposits of 0·0002 in. thickness have given excellent results.   Other surface treatments, including anodizing and Parco-lubrizing are also used for pistons to prevent initial running-in troubles and to give subsequent low friction surfaces.   In some cases the deposited layer is porous, in order to act as an oil absorbent and thus prevent seizure during the early stages of the piston's life.

The *oxide-coating method* results in an extremely hard fine texture surface which resists wear and prevents seizure under normal running-in conditions.   The piston ring slots and the gudgeon pin holes are treated as well as the piston itself.

It is considered advantageous to treat the coated pistons with a solution of *colloidal graphite*, such as Aquedag, in water, allowing the solution to dry off, when the microscopic grains of graphite remain embedded in the pores of the metal and reduce the frictional coefficient appreciably.

The surface of a colloidal graphite impregnated new piston

is of smooth black appearance and the graphite remains during the whole of the new (or rebored) engine running-in period.

Piston Rings.—These are now made of cast-iron, although spring steel ones are sometimes used. The principal requirement of a piston ring is that when compressed into the cylinder it shall bear evenly all round, and shall not have too large a gap. If a concentric type of ring is sawn through at one place and inserted into the cylinder it will not expand to a cylindrical shape, but will bear on the portions near the cut end A and on the diametrically opposite part of the ring D as shown in Fig. 61. There will be no contact with the cylinder on either side at C and B. The type of ring illustrated is therefore unsuitable for the purpose of petrol engines, and must be modified either by reducing the radial thickness near the ends A or by suitably shaping the ring when free so that when it closes within the cylinder it is perfectly circular.

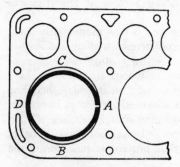

Fig. 61.—Effect of inserting a Plain Concentric Ring in Cylinder Barrel. (Brico)

A method which is sometimes employed for attaining the latter result is to turn the rings to slightly larger diameter than the cylinder bore and then to cut out a small portion leaving a parallel-side gap. The ring is then closed by pressure and in this condition is clamped between circular discs of rather smaller outside diameter and the exterior surface is then ground down to exactly the same diameter as the cylinder bore.

In order to make this circular ring provide *an equal radial pressure* against the cylinder wall, all around its periphery, it is usually hammered inside, the hammer blows being greatest opposite the gap, and decreasing in intensity to nothing at the gap. Another type of ring is the *Eccentric* one: this, if properly made, and hammered, gives a uniform wall pressure, so that leakage of the gases past the piston is minimized.

The disadvantage of the eccentric ring is that it is difficult to ensure that there is not an excessive pressure outwards at the two joint ends—for the examination of rings taken from engines that have been in service for appreciable periods always shows that the greatest wear occurs at these places. It has also the drawback of providing more space behind the ring for carbon deposit to accumulate. For these reasons the eccentric ring has been superseded by the parallel or concentric type.

Wall Pressure of Piston Rings.—In order to prevent the escape of gases past the rings it is necessary to arrange for the latter to exert a certain radial pressure outwards against the cylinder walls. This pressure must only be sufficient for the purpose during the normal working conditions for excessive pressures reduce engine power and also give rise to greater ring and cylinder wear.

Piston Ring Information and Data.—Modern piston rings are of the narrow axial thickness type, of good quality alloy cast iron having elasticity properties, or flexibility, so that they can readily be sprung over the piston without breakage or permanent set. The rings are now always supplied by specialist piston ring firms and are of a high standard. The type preferred is the diagonal-cut one with uniform or graduated radial pressure produced by peening followed by peripheral grinding with the ends clamped together. It has been shown that if the correct ring gap measurement is given there is no more tendency for gas leakage past this kind of joint than the step-cut or any other one. The gap clearance is sometimes taken as being $\frac{1}{400}$ of the cylinder diameter; so that a 3 in. ring would have $\frac{3}{400} = \cdot008$ in.; a 4 in. one, 0·010 in., and so on. The principal reason for *using narrow axial thickness rings* is that in the event of gas leakage into the back space of the slot, there is less radial pressure on the cylinder wall than for wider rings. Other advantages include lighter weight, greater flexibility and less wear in the piston grooves owing to their lower inertia force.

In regard to the radial thickness, this is usually made to $\frac{1}{30}$ of the diameter with a tolerance of plus $\frac{8}{1000}$in. and a free gap equal to 3 to 4 times the thickness; these dimensions correspond to a minimum wall pressure with alloy iron rings of 9 lb. per sq. in. sandcast and 10·5 lb. per sq. in. for centrifugally-cast rings.

For higher wall pressures the ratio of radial thickness and cylinder diameter should be increased. Thus a ratio of $\frac{1}{25}$ will give radial pressures of 14 to 17 lb. per sq. in., for a gap of $\frac{1}{400}$ the cylinder diameter, with cast iron having a modulus of elasticity of 14,000,000 lbs. per sq. in.

The free gap, i.e., the distance between the ends of the ring before placing in the cylinder should be about $3\frac{1}{2}$ times the radial thickness. When the ring is stretched to go over the piston in order to insert it in the ring grooves of the latter the gap opens usually to about 8 times the thickness, but with the free gap previously mentioned the material does not become unduly stressed. The clearances of piston rings in cylinders is dealt with later in this chapter.

Typical Piston Ring.—A typical example of an automobile piston ring illustrating the usual dimensions is that of a 75 mm. (2·953 in.) one which has a maximum width of ·0938 in. and minimum width ·0928 in.; radial thickness of ·098 to ·106 in. and nominal free gap of ·34 in. Such a ring will give a radial pressure of 9 to 13 lb. per sq. in. according to the width and thickness and would be suitable for use in aluminium alloy pistons.

Vauxhall Piston Ring.—A special type of piston ring developed by the Vauxhall engineers enables a liberal amount of oil to be delivered to the cylinder walls, in order to reduce cylinder bore wear, whilst preventing escape of this oil into the combustion chamber; thus adequate lubrication with minimum oil consumption has been the aim in view.

Instead of using the conventional pattern piston ring, shown on the right, in Fig. 62, the Vauxhall ring exerts a pressure on the cylinder walls graduating from maximum values at the two ends of the ring and at the place diametrically opposite, to minimum values between these two positions, as shown by the lengths of the radial arrows. This method, it is claimed, *prevents piston ring vibration* and gives better sealing of the combustion chamber charge and gases; at the same time oil leakage past the rings is reduced to a minimum.

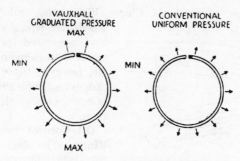

Fig. 62.—Piston Ring used in Vauxhall Engines.

Compression Ring Shapes.—With the increased compressions of modern engines, new types of piston rings have been developed to give better compression and combustion pressure sealing and at the same time to provide adequate oil supply to the cylinder

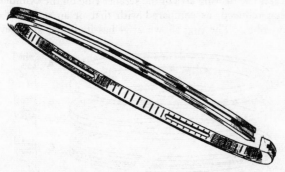

Fig. 63.—A Typical Slotted Oil-Control Piston Ring.

walls. There are over twenty different ring sections in commercial engines, so that it is possible only to mention the more important typical examples, in this section.

Compound Rings.—There are certain compound piston rings in which the gases are given long and tortuous leakage paths.

These rings, of which the Clupet, Wellworthy, and McQuay-Norris, are typical examples, have been used for replacements on worn engines. They are, however, liable to great carbon formation and a gumming tendency, than plain rings.

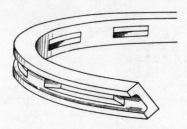

Fig. 64.—Showing Construction of the Slotted and Grooved Oil Scraper Ring.

Oil Control or Scraper Rings.—The object of these rings is to reduce excessive oil consumption by scraping off the surplus oil on the cylinder walls *on the down stroke of the piston*; they have little or nothing to do with the maintenance of compression.

Fig. 65.—*A*, *B* and *C*—illustrates three typical oil scraper rings, the sections shown being taken through the piston wall. In all cases the bearing area of the scraper ring on the cylinder wall has been reduced, as compared with that of an ordinary compression ring. The ring shown at *A* has its upper side bevelled

Fig. 65.—Oil Scraper Piston Rings.

away; ring *B* has its lower portion turned away so as to remain clear of the cylinder wall, whilst ring *C* has a central groove turned away and slots made at intervals through the bottom of this groove, similar to the ring shown in Fig. 64.

Owing to their reduced bearing areas these rings give a greater pressure on the cylinder walls and, therefore, exert a satisfactory scraping action.

It is important to observe two points when fitting scraper rings, viz.: (1) Oil holes must be provided through the piston wall, as shown in the right hand sectional views (Fig. 65), to conduct away the surplus oil scraped off the cylinder walls. (2) The rings must be fitted with their scraping edges downwards, as shown, in order to scrape the surplus oil away on the down stroke of the piston.

There are two alternative methods of fitting oil control rings. These are illustrated in Fig. 66. The method shown at *A* leaves the piston skirt quite clear, but the oil scraped off the cylinder

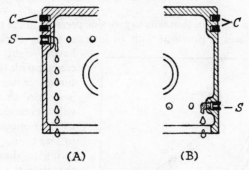

(A)   (B)

Fig. 66.—Methods of Fitting Scraper Rings. *C*—Compression Rings. *S*—Scraper Rings.

walls is hotter than for the arrangement shown at *B*; the latter method gives a slightly heavier piston but there is a reduced tendency for the oil to gum and carbonise in the scraper ring groove.

Piston Rings for Worn Cylinders.—If a standard pattern ring is fitted to a cylinder that has worn oval or is tapered as a result of wear, it will not fit the cylinder correctly and leakage of gas and lubricating oil will occur. In order to overcome this disadvantage piston rings of a flexible nature, having an internal steel expanding device are frequently employed. Owing to the combination of flexibility and internal expansion thus obtained the ring will adapt itself to the inequalities of bore and produce a better gas-tight fit than is possible with a plain ring. A typical example of such a ring is the Wellworthy "Simplex" one illustrated in Fig. 67.

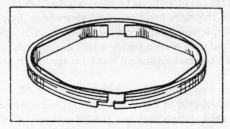

Fig. 67.—The Wellworthy "Simplex" Ring.

It has a number of thin slots to give side pressure in the piston ring grooves and is provided with an internal spring steel expander

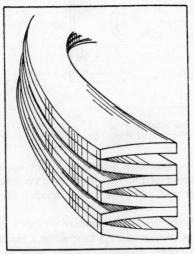

Fig. 68—The Cords Flexible Piston Ring.

having eight points of contact with the interior surface of the ring. It is claimed that this ring will compensate for cylinder inequalities up to ·015 in. and that owing to the design of the slots the space behind the ring becomes filled with oil, thus providing adequate lubrication for the ring; all surplus oil is scraped by the ring through holes drilled below its groove so that it is returned to the crankcase.

Fig. 68 illustrates the Cords compound piston ring made up of single sheet-metal annular members, welded at the common junctions. This flexible ring acts as a compression and scraper ring. It has sufficient lateral spring to seal the groove and to compensate for wear in the latter; moreover, it will compensate for any irregularity in the cylinder bore. Usually these rings are fitted in the two lower piston grooves and a plain compression ring in the top groove.

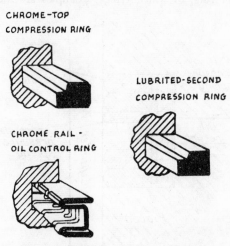

Fig. 69.—Typical set of Piston Rings for American engine (A.M.C. Rambler).

American Piston Ring Set.—The standard American practice in modern engines is to use two compression and one oil control ring. Usually, the upper compression ring is of the chromium-plated stepped or tapered section kind. The lower compression ring is of similar design but has its bearing surface phosphate-coated. The oil control ring is usually of the three-element type with top and bottom chromium-plated flat rings, having rounded edges and an inner serrated steel expander ring. Fig. 69 illustrates a typical ring set, in which the compression rings are of the stepped kind with the second ring surface-treated or "lubrited".

Recent Oil Control Rings.—Some notable advances have been made in the design of *oil-control piston* rings in recent years, with the object of obtaining increased flexibility, both in an axial and radial sense. This type of ring, however, is intended for re-placing the usual oil control ring when the cylinder shows signs of uneven wear such that, although it reduces oil consumption and, to some extent increases the power output, it does not actually necessitate re-boring of the barrels. The principle of this type of replacement ring, of which typical examples are the Simplex,

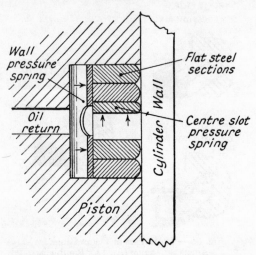

Fig. 70.—The Wellworthy Duaflex Piston Ring.

Duaflex, Cords, Transcoseal, Muskegon Type XSS, Perfect-Circle, etc., is to use very flexible alloy steel or cast-iron rings of the slotted oil control type and to employ an internal spring steel expander—usually of polygonal form, although helical springs are sometimes used—which forces the former ring to conform with the section of the cylinder at all times. In most of the rings mentioned is is also arranged that there shall be flexibility of the ring in the piston slot, also, in order to prevent oil leakage behind the ring.

The multiple flexible steel type piston ring has been popular in the U.S.A. as a replacement one for worn cylinders. In this connection it may be noted that there are two all-steel piston rings available in this country. One is the Wellworthy expander type, comprising a number of thin steel rings, about five being used in a typical application to fill the piston ring slot. These rings have no special side flexibility properties but use a steel polygonal expander ring as in the Simplex model. They act mainly as compression rings but with a certain amount of oil control action. The other type has an identical set of thin steel rings possessing radial flexibility but without the use of an expander. They are used for piston slots that are not deep enough for the expander type or cannot conveniently be machined to size.

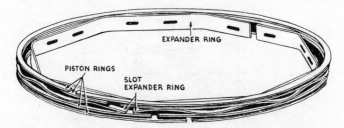

Fig. 71.—The Wellworthy Duaflex Piston Ring for worn cylinders.

Figs. 70 and 71 illustrate the Wellworthy Duaflex, a more recent example of flexible oil-control ring, having advantages over the two types mentioned

It consists of a number of narrow flat-section alloy steel rings with their cylinder wall edges rounded off. In a typical example there would be four of these rings. Between each set of two rings there is arranged a crimped expander ring of spring steel which forces the rings to maintain contact with the sides of the slot. In order to ensure cylinder wall contact all round, the four rings are pressed against the wall by means of a steel expander of octagonal shape—similar to the type used in the Simplex ring; since the thin unit rings are very flexible, good contact is made against the cylinder wall. The rounded edges of the rings ensure immediate bedding-in so that the walls are not injured. This type of ring tends to wipe the surplus oil from the walls, instead of scraping it, as in the case of the ordinary oil control ring. Tests made with these rings show that a marked reduction in oil consumption is obtainable.

The American Perfect Circle Type 86 oil-control ring has a long coil spring of circular periphery, with its ends together and placed behind the flexible alloy cast iron oil-control ring; this gives uniform wall pressure.

Tapered or Wedge Rings.—A later type of piston ring, known as the *Taper* or *Wedge ring*, has a tapered section of about 10° angle. It can be shown that with this shape of ring there is less friction on the piston slot faces and the ring maintains better contact with the cylinder wall than for the usual rectangular-

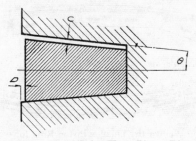

Fig. 72.—The Wedge-Type Piston Ring.

section rings. It has also been found to be much less prone to *piston-ring flutter* and *blow-by*. To ensure that there is no increased groove wear, the wedge faces of the ring must be hone- or lap-finished. This type is particularly suitable for high gas pressure engines, including high compression-ratio petrol and Diesel ones. Referring to Fig. 72 the angle $\theta = 5°$; the recommended side clearance $C = 0.001$ in. to 0.002 in. and distance D of the ring below piston land, 0.002 in. to 0.006 in.

It is usual to fit the taper ring between the compression and oil control ring, in car engines and to use two taper rings between the compression and oil control ring in certain commercial engines.

Tapered Face and Stepped Rings.—The top ring is almost invariably of straight-face rectangular section, to give full surface contact with the cylinder face. It has been found advantageous, in modern engines, to use a second compression ring having a tapered face, as shown at (A) in Fig. 73. This design provides

for a high surface pressure at the lower edge, so that the ring
has also an oil scraping action.   Sometimes, however, the tapered
face ring has an inner bevel, as at (*B*), with the object of increasing
the upper pressure on the ring and providing a certain amount of
twist or torsion, as depicted at (*B*), so as to give a higher bearing
pressure on the lower edge of the ring.

The stepped-section compression ring, shown at (*C*) acts in a
similar manner and provides a high wall pressure combined with an
oil scraping action; this ring is used in the second groove from the
top of the piston.

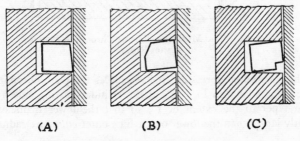

(A)   (B)   (C)

Fig. 73—Tapered Face and tilted Rings.

Chromium-Plated Rings.—A later innovation in piston rings
of the compression type is to *chromium plate* the area that bears
on the cylinder wall.   In the Muskegon ring (Fig. 74) there is an
inner layer of 0·0025 in. to 0·0035 in. of extremely hard chromium
and on this another layer of 0·0015 in. to 0·0025 in. of *porous*
chromium, as depicted in Fig. 74; the total thickness is 0·004 in.
to 0·006 in.   It will be observed that the bearing portion of the
ring is radiused.   This type of ring gives a much longer useful
cylinder and also ring life, reduced friction and freedom from
scuffing action; in some instances the chromium-plated ring
gives from 3 to 5 times the wear resistance of the uncoated ring.
Chromium-plated rings should *not be used in chromium-plated
cylinders*, however.   It is necessary to use the ring, in the top
piston ring groove.

More recently there has been an extension of the chromium

plated rings to include the tapered compression ring and oil-control rings for work cylinders.

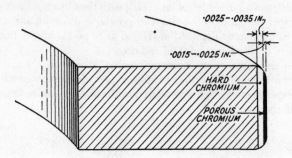

Fig. 74.—The Muskegon Porous Chromium-Plated Piston Ring.

The Vacrom compression ring resembles the plain chromium-plated ring shown in Fig. 74 but the cylinder-bearing side has a slight taper to the extent of 1°, so that the lower diameter is slightly larger than the lower one. The outer edges are radiused

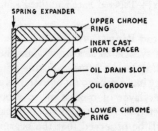

Fig. 75.—The Hepocrom Four-element Oil Control Ring for Worn Cylinders.

and the tapered and radius surfaces are chromium-plated. When used in new or rebored cylinders and properly bedded in, this ring, it is claimed, will have a useful life of at least 100,000 miles.

The Hepocrom four-element oil control ring shown in Fig. 75 has an upper and lower flat-type of ring, with rounded edges,

with their piston slot bearing faces chromium-plated. Between
the two flat rings there is a cast-iron spacer having oil drainage
slots; this ring is not in contact with the cylinder wall but acts only
as a separator and locator of the flat rings. A flexible steel ring,
located at the back of these rings maintains them in contact with
the cylinder walls. The bottom of the piston ring slot is drilled
for oil drainage purposes.

Delayed-Action Ring.—*The Hepolite delayed action* oil-
control ring is somewhat similar in general design to the ring

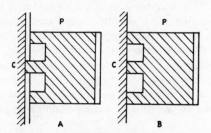

Fig. 76.—The Wellworthy Delayed-
action Oil Central Ring.

shown at *C*, Fig. 65, but, instead of having a truly cylindrical
periphery, it is of polygonal form, i.e. has a large number of small
flats all around its circumference, which allow free passage of
oil to the upper rings. During the engine operation the "corners"
gradually wear away until, finally, the periphery becomes truly
cylindrical and the ring then exerts its full scraping efficiency.
This type of ring is *additional* to the ordinary oil-control ring or
rings, so that an extra piston slot must be provided; it should not
be used in place of the usual oil-control ring.

It is claimed that this delayed-action ring will have an effective
life, as an oil scraper ring for two to three times the road mileage
of the ordinary type of ring.

Another type of delayed-action ring is the Wellworthy ring, the
principle of which is shown in Fig. 76. It consists of a three-
ridge type slotted oil control ring in which the centre ridge is
raised by about 0·002 in. above the outer ridges. After the ring
has completed 10,000 to 15,000 miles the raised portion has worn

down to the same diameter as the two outer ridges, so that the ring then acts as a new oil scraper ring and thus doubles the period before inspection of the ring becomes necessary. In practice, this period is about 60,000 miles.

Piston Ring Clearances.—The piston ring must have the correct clearance in its piston slots to enable it to slide freely, but without compression loss behind the ring. It is usual, in the case of pistons of the sizes used in car engines to allow a side clearance of the compression ring in its slot, between ·002 in. and ·004 in., when new. This clearance is usually measured with a feeler gauge.

The piston ring must also be a good compression fit in the cylinder and an allowance must be made for thermal expansion, so that the distance between the ends of the ring, i.e. the gap, should be sufficient to prevent the ends touching when hot. It is now usual to allow a piston ring gap, in the cylinder, and when cold of about 0·003 in. to 0·005 in. per inch diameter of piston; thus a 3-inch piston ring would have a compression ring gap of ·009 in. to 0·15 in.

The oil control rings usually have rather larger gaps, although in many British engines the same clearances as for the compression rings are used. A typical American vee-eight engine, with pistons of 4-in. diameter, employs ring gaps of ·013 in. to ·023 in. for the compression rings and ·015 in. to ·045 in. for the oil control rings. It is usual for American engines to employ wider gaps for the oil control rings than for the compression rings.

The Valves.—Although many designs of rotary, piston, cuff, sleeve and other types of valve have been tried in the past, present-day practice is confined to the poppet valve examples. The poppet, or mushroom-head valve, is made of high-tensile alloy steel, since it is subjected to severe impacts, tension and high temperatures. Whilst the inlet valve is kept fairly cool by the passage of the wet mixture the exhaust valve is exposed to the intense heat of the out-flowing exhaust gases, so that whereas in water-cooled engines the average temperature of the head on the inlet valve seldom exceeds 250° C. to 275° C., that of the exhaust valve is generally about 700° C. to 760° C. The material for the inlet valve need not therefore be so strong at its working temperatures as the exhaust valve. The latter must also be made of a suitable steel to withstand the hot corrosive effects of the exhaust gases under its head and around the upper part of the stem. When fuels containing tetra-ethyl lead are used the corrosive action is accentuated.

Valve Materials.—Various alloy steels are used for the inlet valve, including plain nickel ones, nickel-chromium and chrome molybdenum.

The steels employed for exhaust valves include nickel chrome high tensile steels, stainless steels, silicon-chrome steel, cobalt-chrome steel, high speed steel, high nickel-chrome (known as *austenitic* steel) and tungsten steels. Of these steels those possessing the necessary tensile strengths and hardnesses at the operating temperatures are the cobalt-chrome, silicon-chrome and the high nickel-chrome *austenitic* steels. A typical cobalt-chrome steel has a tensile strength of 45 tons per sq. in. cold, and 25 tons per sq. in. at 700° C. Similar properties are given by certain silicone-chrome steels. A typical high nickel-chrome steel, gives a cold tensile strength of about 43 tons per sq. in.; 34 tons per sq. in. at 700° C. and 24 tons per sq. in. at 800° C. All of these steels possess good anti-scaling properties.

Typical examples of commercial alloy steels and alloys which are used for exhaust valves of high-performance engines, include the following: Silchrome 1, XB Steel, Nimonic 75, Nimonic 80 and in the U.S.A., Inconel M (or T.P.M.), Sil 10, XCR.

A group of valve steels, often used for aircraft engines, is the "Silchrome" one, having from 1 to 4 per cent. silicon and 8 to 24 per cent. chromium, with 1 to 4 per cent. nickel and sometimes up to 3 per cent. molybdenum, the rest being pure iron.

The Silchrome, cobalt-chrome and tungsten steels are not so good in their corrosion resistance properties to the action of leaded fuels as the high nickel-chromium austenitic steels. The latter, however, have a higher thermal expansion coefficient, so that a greater allowance must be made in the valve clearances when such steels are employed.

With austenitic steels it is necessary to provide the valve stem with a hardened tip to resist impact effects; it is now usual to provide a welded button of the extremely hard alloy Stellite on the valve stem end for this purpose.

Metal-coated Valve Heads.—With the higher compression ratios and outputs of modern engines, cylinder and valve head temperatures have increased. Moreover, with the necessary use of premium-grade leaded fuels for such engines, the valve heads are exposed to oxidation or corrosion effects which, if allowed to continue over a period, cause scaling effects, causing pre-ignition and other detrimental results.

To obviate these troubles modern valve heads of the previously

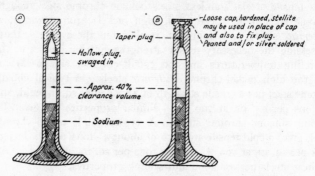

Fig. 77.—Sodium-filled Hollow Stem Exhaust Valves.

mentioned alloy steels are now being treated with highly corrosion-resistant materials, such as an alloy consisting of 80 per cent. nickel and 20 per cent. chromium. A typical process, known as the Eaton Chronicote, employs this coating in an economical manner. The alloy is sprayed on to the valve head and then treated by a thermal process, which gives a bond with the valve metal itself.

Another method is to *coat the head with aluminium*, by dipping it into the molten metal, or by spraying with powdered aluminium and then heating in a flux bath. The aluminium-coated valve heads have been shown to double the useful life of the untreated valves.

Sodium-cooled Valves.—Although employed to a limited extent only on automobile engines this type of valve is standard on aircraft engines. It is made with a hollow stem and head and part of the interior space is filled either with the metal sodium or a mixture of salts, such as potassium nitrate and lithium nitrate

Sodium melts at 97·5° C. and the latter mixture at 130° C. Sodium boils at 880° C. In each case the molten salt and its vapour in the hollow exhaust valve stem conduct the heat from the valve head to the cooler parts of the stem, thus cooling the head. Fig. 77 illustrates typical sodium-cooled exhaust valves as used on aircraft engines.

Valve Seatings.—Although the ordinary valve seatings of cast-iron cylinders give satisfactory service, a greatly increased

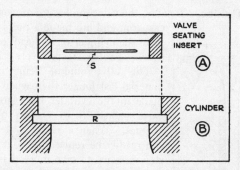

Fig. 78.—Showing one method of inserting Hard Valve Seating into Cylinder. The insert ring shown in (*A*) has a pair of slots *S*, which spring into the seating groove *R*, shown in (*B*).

life, before re-seating is necessary, is given by special seatings, which are either screwed or shrunk into position. The materials used for valve seating insert rings include aluminium bronze (used in aluminium alloy cylinders and heads) and alloy steels. The best of these steels appears to be the one having the requisite strength and hardness combined with a coefficient of expansion equal to that of the cylinder metal. If the latter is of cast iron then a stainless or austenitic (high nickel chrome alloy) steel can be used. When the cylinder is of aluminium alloy it is now usual to employ a nickel-chromium manganese steel having the same expansion coefficient (0·000022) as that of the usual aluminium alloys employed for the cylinder.

In order to increase the hardness of the valve bearing or seating face in many recent engines it is given a coating, by the oxyacetylene process, of Stellite or Brightway (80 per cent. nickel and 20 per cent. chromium). Stellite is an extremely hard synthetic cutting alloy containing cobalt, tungsten and carbon, used also to tip the cutting portions of machining tools in order to provide much higher cutting speeds than for carbon or high speed tool steels. The valve seatings are often shrunk into position by first immersing them in liquid oxygen or in dry ice, or by first heating the cylinder or head, before fitting the valve inserts.

Valve Guides.—Usually, the valve guides are of close-grained cast iron and are made a press or shrink fit in the cylinder or its head, for side or overhead valves, respectively.

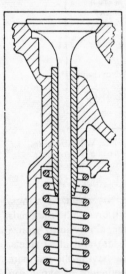

The valve guides, being of harder material last longer than when holes are made in the cylinder block to take the valve stems—as in certain production engines. When a valve guide wears it can readily be replaced and, if the valve stem is worn to give more than ·005 in. clearance, a new valve is fitted.

When the cylinder is made of aluminium alloy it is usual to heat it to a temperature of about 350° C. in order to expand the valve guide hole and then to insert the cold guide. When the cylinder cools down the guide is held very firmly in position; for this purpose the diameter of the guide is made slightly greater than that of the guide hole when both are at the normal air temperature, so as to give what is known to engineers as an *interference fit*.

Fig. 79.—A Powdered Iron Valve Guide.

Another method that is used for aluminium and also cast iron cylinders is to make the guides to an interference fit in their holes and to first cool them down in a mixture of dry ice (solid carbon-dioxide) and alcohol, which has a temperature of about minus 60° Fah. (if liquid oxygen is used the temperature is minus 182°

Fah.), before inserting them in their holes in the cylinder.   When in contact with the cylinder metal the guides expand and become a very secure fit.

Exhaust valve guides are occasionally made of phosphor bronze

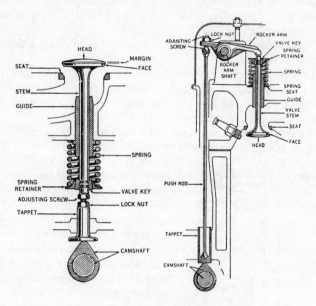

Fig. 80.—Side and Overhead Valves and their Components.
(*Left*) Side valve.   (*Right*) Overhead valve.

as this metal has been found to give less wear and better heat conduction; this type of guide was used on Rolls Royce "Merlin" aircraft engines.   After being shrunk into place the holes are burnished by forcing hardened steel balls through them.

Another material for valve guides is made of powdered pure iron by a die pressing process and has an oil absorption capacity up to 30 per cent. of its volume, according to density.   The guide as die pressed is given a mirror finish and requires no further machining.   It is claimed to extend the life of the valve stem considerably owing to its low coefficient of friction and excellent oil-retaining properties.

Valve Guide Clearances. The valve guides are made a press, or interference fit in the cylinder block sides. Usually the outside diameter of the guide is ·001 to ·0025 in. larger than the hole diameter. The valve guide is inserted from the valve seating side of the block.

Valve stems are usually given a new clearance in their guides of ·0015 in. to ·0025 in.

In some high performance engines the exhaust valves are given a greater clearance. Thus, in a typical example, for a valve stem diameter of ·315 in. the inlet clearance was ·0015–·0025 in. and that of the exhaust, ·0025–·0040 in.

Valve Components and Dimensions.—The components of typical side and overhead valves are shown in Fig. 80*. The

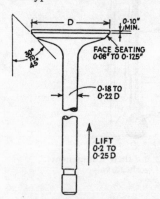

Fig. 81.—Proportions of Typical Valve.

diameter of the head should be as large as possible consistent with the combustion chamber design. It is good practice to make the valve diameter $d = 0.414$ (piston diameter), or alternatively to so choose d that the gas velocity through the port does not exceed 120 feet per second; this consideration leads to the formula:

$$d^3 = \frac{2D^2lN}{86,400} \text{ in.}$$

where D is the piston diameter, l the stroke (both in inches) and N the r.p.m.

The *lift* of the valve in most production engines is from $\frac{1}{4}$ to $\frac{1}{5}$ of the valve head diameter.

The *stem diameter* is from $\frac{1}{5}$ to $\frac{1}{6}$ of the head diameter and the valve stem is made a free sliding fit in the valve guide.

Two alternative valve face and seating angles are used, namely, 30° and 45°. In some modern engines the inlet valve angle is

*Courtesy Black and Decker Ltd.

30° while the exhaust is 45°. The earlier types of valves had relatively wide seating widths, namely, of $\frac{1}{8}$ in. to $\frac{3}{16}$ in. Modern valves, however, have narrower seating widths, namely, from $\frac{1}{61}$ in. to $\frac{1}{4}$ in.

In certain makes of engine, instead of making the valve face and valve seating angles the same, the *valve face angle is made from $\frac{1}{2}$° to 1° smaller* than the *valve seating angle*. The effect of this variation is to cause the new valve to make contact with the valve seating only near the top of the seating. As the valve beds into its seating the width of the contact area increases.

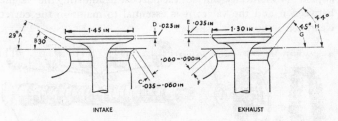

Fig. 82.—Showing the Principal Dimensions of Modern Overhead-type Inlet and Exhaust Valves.

Typical Example of Valve Dimensions.—The principal dimensions and angles of the Vauxhall Victor 92 cu. in. engine are shown in Fig. 82. The inlet (intake) valve angle is 29° while that of its seating is 30°. The exhaust valve face and seating angles are 44° and 45° respectively. The widths of the inlet and exhaust valve seating faces are ·035 to ·060 in. and ·060 in. to ·090 in., respectively. The valve head plain portion thicknesses, for the inlet and exhaust valves are ·025 in. and ·035 in., respectively. The ratio of valve stem to-inlet-valve head is ·240 and for the exhaust valve, ·215.

In some modern engines the *inlet valve is larger in diameter* than the *exhaust valve*; the purpose of this is to obtain the greatest possible amount of charge in the cylinders, at full throttle operation. In Fig. 82 the diameter of the inlet valve is 1·45 in. and that of the exhaust valve, 1·30 in.

Valve Springs.—The *valve spring* must be sufficiently strong to prevent *valve bounce* when the engine is working at high speeds, i.e., it should keep the valve and tappet on the valve cam under these conditions. With the increased compressions and maximum speeds of more recent engines, the valve springs have been made stronger than hitherto. In a typical car engine the valve spring exerts a force of 40-45 lbs. in the valve closed position and 110-120 lbs. in the lifted position. The tappet, or plunger, is interposed between the cam and the valve stem At its lower end it carries a roller (or ball) which obviates the rubbing action of the cam. The top of the tappet is provided with a screw adjuster (with its lock-nut) for the purpose of adjusting the distance between it and the valve; it is essential to maintain the correct clearance amount.

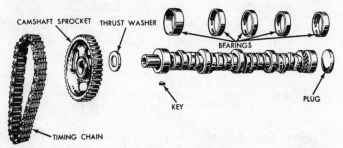

Fig. 83.—Camshaft Assembly for Vee-eight Engine (Ford, U.S.A.).

The Camshaft.—The camshaft is made in forged alloy steel although in the U.S.A. it is, in many cases, e.g. the Ford Company's car engines, made of precision-moulded alloy cast-iron, with the cams hardened by the electric induction process. The camshaft usually has a pair of cams between each pair of its main bearings, e.g. an eight-cam shaft would have five bearings, as indicated in Fig. 83.

In some engines the bearings for the camshaft journals are bored through the cast-iron cylinder block and the journals run directly in these holes. In other instances white metal shell or bush-type bearings are used, so that after long periods of service the bearings can be renewed. Provision is always made for taking any *end thrust* of the camshaft by a thrust washer, or thrust plate at the

driving end of the camshaft.   Any wear, due to thrust is taken up
by fine washers termed "shims".   The usually permissible end
clearance is 0·003 to 0·005 in. and the camshaft bearing clearance,
0·001 to 0·003 in.

In the example shown in Fig. 83 the camshaft is driven by a
relatively wide chain and sprocket from a half-size sprocket on the
crankshaft.   As the ignition distributor is driven at one-half the
crankshaft speed it is usual to provide the camshaft with a helical
gear which engages with an equal helical gear on the distributor
shaft; the latter is generally arranged vertically or at a sloping
angle to the vertical so that the contact breaker and distributor
assembly are readily accessible.

Usually the camshaft gear is located in the central region and
the distributor shaft is extended downwards into the crankcase
sump so as to drive the oil pump.   In the example shown in Fig.
83 the camshaft gear is shown on
the extreme right next to the rear
bearing journal.

Valve Clearances.—The clearance allowed depends upon the
design of the engine, that is, upon
the relative heating of the valve and
cylinder; most manufacturers specify
the appropriate clearance values for
their engines.   In the absence of
such information, it was previously
usual to give side valve engines a
clearance of $\frac{3}{1000}$ in. for the inlet and
$\frac{5}{1000}$ in. for the exhaust, the adjust-

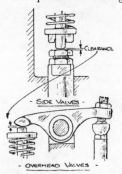

- SIDE VALVES -

- OVERHEAD VALVES -

Fig. 84.—Showing Clearances
of Side and Overhead Valves.

ment being made whilst the engine is hot.   Overhead cam shaft
operated valves had clearances of $\frac{3}{1000}$ to $\frac{4}{1000}$ in.

Large Clearance Valves.—In order to avoid the necessity
of frequent attention to tappet clearances and any possibility
of burnt-out valves due to insufficient clearance or stretch,
it is now the practice, in the case of side and overhead valve
engines to employ special designs of valve cams, which will
enable the engines to run quietly with valve tappet clearances of
the order $\frac{15}{1000}$ to $\frac{25}{1000}$ inch.

Automatic Tappet Adjusters.—In order to obviate regular adjustments of valve tappet clearances, some manufacturers have fitted automatic adjustment devices, thus ensuring maximum engine efficiency at all times. In the U.S.A., hydraulic tappet adjusters are now standard on most engines.

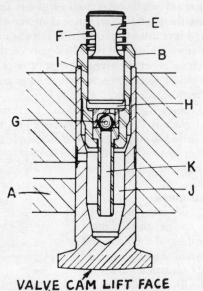

VALVE CAM LIFT FACE

Fig. 85.—The Zero-Lash Hydraulic Valve Lifter, which maintains the Valve Stem clearance, automatically.

The method usually adopted is to utilize the oil pressure of the engine lubricating system to hold a plunger member of the tappet against the valve stem.

There have been several types of hydraulic tappet adjusters, known also as *hydraulic lifters*, of which the Eaton Zero-Lash (external spring) and General Motors (internal spring) designs are the most widely used.

Fig. 85 illustrates the principle of the *Zero-Lash hydraulic lifter* which consists basically of a cylinder *B*, plunger *E*, ball check valve G and light spring F. Oil from the engine lubricating system is fed through the tappet guide just above A to a supply chamber J in the tappet body, whence it can feed into the tube K

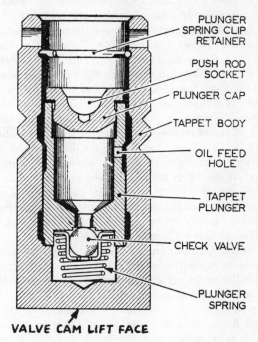

VALVE CAM LIFT FACE

Fig. 86.—The General Motors Hydraulic Valve Lifter.

and past the check valve G into the space H between the bottom of the cylinder B and plunger E. During the closed position of the valve, i.e., when the tappet is on the circular or non-lifting part of the valve cam, the spring F lifts the plunger E to make contact with the valve stem. When the cam begins to lift the tappet, pressure is increased in the space H, forcing the valve G on to its seating. the further cam lifting action on the tappet increases the compression chamber H pressure so that the whole assembly acts as a solid member, lifting the valve stem. Any initial air bubbles in the compression chamber oil leak out through the clearance I, between the plunger and cylinder, creating a certain noise, until the air is fully exhausted. Compensations for wear of the tappet lifting faces or valve stem end, are made by allowing a slight leakage of oil, under load, between the plunger and cylinder.

The *General Motors hydraulic lifter*, illustrated in Fig. 86, is more compact, due to its internal control spring. As before, the light spring holds the plunger against the valve stem or push rod end. and the space between the bottom of the tappet body recess and lower end of the plunger is filled with oil, under pressure. As the valve cam lifts the complete tappet the check valve closes and the much increased oil pressure causes the whole valve lifter unit to act as a "solid" tappet, but, during the lifting period there is a slight leakage of pressure oil past the plunger for compensation purposes. It may be mentioned that the hydraulic-lifter can also be fitted to the *upper end* of an overhead *valve push rod*, to adjust the clearance between the rod end and the valve actuating rocker arm. Also, a flat disc can be used instead of a ball check valve, as in the Chicago Screw Company's lifter.

The Valve Components.—The complete valve assembly includes the valve itself, valve collar or spring cap, the cotters that hold the collar to the valve stem under the pressure of the valve spring, the valve spring, and oil seal.

Fig 87 shows a typical overhead valve assembly, the various numbered items being described in the caption below. The removeable valve guide is also included in this illustration.

Notes on Valve Springs.—In many earlier designs of automobile engine the valve springs fractured and caused engine breakdown. This was due to the unsatisfactory grade of steel and the heat-treatment employed; to some extent, also, it was due to the design of the spring.

In later engines the trouble has been cured by the use of better steels for the springs and by the use, in certain cases, of *compound springs*, i.e., one spring within another; occasionally, triple valve springs are employed. The object of compound springs is to avoid the surging or wave travel effects when the engine is running at higher speeds, for with a single spring it sometimes happens that the natural frequency of the spring's vibration coincides with the normal operating frequency of the valve and resonance effects then occur which give rise to surging effects; by using compound springs of different natural frequencies these effects are obviated.

Owing also to the liability of springs to fail by *fatigue effects*,

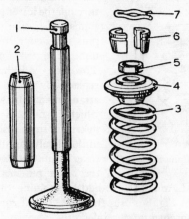

Fig. 87.—The B.M.C. Overhead Valve Components, dismantled.

1. Valve.   2. Valve guide.   3. Valve spring.   4.   Valve   spring   cap. 5. Oil seal.   6. Split conical cotters. 7. Circlip.

through initial surface defects it is now the practice to employ ground steel wire for making springs or to *shot-blast* the surfaces after coiling so as to give a work-hardening effect.

The steels employed for valve springs include a hard-drawn carbon steel with 0·7 to 0·8 per cent. carbon and 1·0 per cent. manganese (maximum). The carbon spring steels when oil hardened and tempered give 80 to 90 tons per sq. in. tensile strength. Another spring material is a steel having 0·4 to 0·5 per cent. carbon and 1·0 to 1·5 per cent. chromium with 0·15 per cent. of vanadium; when suitably heat-treated it has a tensile strength of 90 to 115 tons per sq. in.

Silicon-chrome and silicon-manganese steel is also used for springs.

Valve Spring Retainers.—Several methods have been used for retaining the lower end of the valve spring, i.e., for transmitting the downward spring pressure to the valve stem.

Previously, round or rectangular-section holes were machined

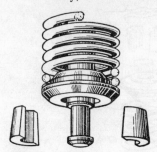

through the lower part of the valve stems and similarly-shaped cotter pins inserted through the holes, under the valve spring collars. These methods have since been replaced by the split conical cotter arrangement shown in Fig. 88, which is not only more economical in first cost, but gives a uniform valve spring pressure, instead of only at two places, when the cotter pin method was used.

Fig. 88.—Split Conical Cotter Method of Spring Retainer.

The favoured American method for holding the valve spring collar, shown in Fig. 89, uses the split conical cotter, but with a single internal groove for the inlet valve stem and three grooves

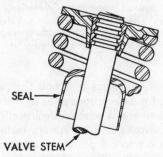

SEAL

VALVE STEM

Fig. 89.—Typical American Overhead-type Exhaust Valve Spring Collar and Oil Seal.

for the exhaust valve stem. Sometimes a single rectangular groove is used. With the rotating valve device (Fig. 93) a different design of split cotter is used.

Simple Valve Spring Collar.—A novel method of holding the valve spring collar without the aid of split cones, cotters or other devices, is used on the Standard Vanguard engine overhead valves. The valve collar (Fig. 90) has two holes, the outer one H being large enough to allow the stem of the valve to pass through; the other h is of smaller diameter. The valve stem is recessed, as shown, and after the collar is placed over the stem, through the larger hole, it is moved sideways to the central position, the spring being compressed. Upon releasing the spring pressure the smaller hole part of the collar is forced against the tapered portion T of the valve stem, where it is held centrally and securely.

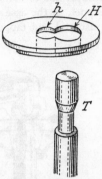

Fig. 90.—Method of Holding Valve Collar without the use of cotters on Vanguard Engine.

Reducing Overhead Valve Noise.— With the increasing adoption of the push-rod and rocker-arm type of overhead valve mechanism the problem of reduction of noise in operation has arisen. It is well known that overhead valve gear is apt to be much noisier than side valve gear. By the use of suitably designed cylinder head covers for enclosing the valves and a plentiful supply of oil the noise has to some extent been reduced. In the case of the Austin overhead engines, oil-cushioned push-rods have been used and there is also a felt-insulated double cover for the rocker gear. The use of the hydraulic automatic valve stem clearance adjusters eliminates noisy operation.

Referring to Fig. 91 there is a constant oil cushion between the push-rod upper end and the hollow cup-rod above; this ensures that the cup portion is always in contact with the spherical end of the rocker arm. When it is necessary to check the valve clearance, i.e., between the other end of the rocker and the end of the valve stem, it is necessary to break the oil seal by pressing down with a screwdriver on to the adjusting screw. The oil seal is then broken, as shown at B.

Valve Bounce.—When a valve is operated, the valve cam lifts

the valve off its seating in a positive manner, compressing the
valve spring to a greater extent than when the valve is on its
seating. Valve springs are made sufficiently strong, when thus
compressed to maintain the valve operating components, e.g. the
tappet, in the case of the side valve engine and the rocker arm,
push rod and tappet, with overhead valves, in contact with the

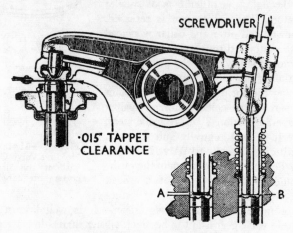

Fig. 91.—Oil Silencing Cushion Device,
on Austin Overhead Valves.

valve stem and cam face. It is the valve spring which returns
the lifted valve on to its seating, but if the valve spring is too weak
the engine speed too high there will be a lag in the valve's motion,
so that there will be an increased clearance in its operating mechanism, such that eventually the latter will take up this clearance, with
a characteristic noise, or clatter known as *valve bounce*. At higher
engine speeds than the maximum for which valve operation is
intended the weight of the valve when lifted by the cam causes the
valve to lift higher than its normal position. This inertia effect
results in the increased clearance or lack of contact of the valve
mechanism tappet with the return or decelerating face of the
operating cam, which then gives rise to the noisy "valve bounce"

operation. Valve bounce can cause valve spring fracture, a broken valve or—in the case of modern high compression engines, having smaller clearances between the piston and cylinder head—to the valve head striking the piston crown. Valve springs in engines which have given long service should always be checked for correct compression loading, with either a new spring or a special spring tester provided for this purpose.

Desmodromic Valve Operation.—The possibility of valve bounce occurring with the orthodox valve gear can be prevented by operating both the opening and closing periods of the valve, positively, by means of a special type of actuating gear. Perhaps the simplest method is to fit the valve stem end portion with an extension giving two parallel faces, marked O and C, in Fig. 92, and to arrange the operating cam between these faces. It will be seen that the valve is opened, closed and held closed during the closing period by means of a single cam and without the usual valve spring and collar. In practice two cams, arranged side by side are used, namely one to open and the other to close the valve. The advantages of this type of valve operating gear are, as fol-

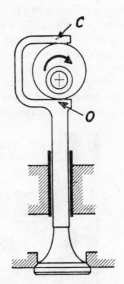

Fig. 92.—Illustrating the Principle of the Desmodromic Valve Method.

lows: (1) Valve operation is always positive and independent of valve spring return, or deceleration. (2) The valve gear can be made lighter in weight. (3) A wider range of valve opening and closing rates is possible. (4) Higher valve operating speeds can be employed. (5) Smaller valve opening periods can be used.

Against these advantages must be offset (1) the greater complication of the overhead type desmodromic valve gear. (2) The greater accuracy required for the cams. (3) The continual maintenance of the correct working clearances. (4) Adequate lubrication of the rubbing surfaces.

The Mercedes-Benz engine, type 300 SLR, used one cam to open the valve and a separate cam, by its side, to close the valve. The closing operation was performed by a kind of elbow lever, one arm of which contacted the closing cam face, while the other arm was forked at its outer end. The forked portion straddled the valve stem (which had a pair of parallel flats machined for the fork to clear) and contacted the lower face of a mushroom-type end of the valve. The inlet and exhaust valve maximum lifts were 0·570 in. and 0·512 in., respectively. The maximum output of the eight-cylinder in-line engine was 300 B.H.P. at 7,450 r.p.m.

Rotating Valves.—A more recent innovation for valves is to provide means for rotating them during the engine operation. In this connection experience has clearly shown that the valve seatings keep cleaner and more true for much longer periods; there is a much longer valve life; the valve stems do not tend to gum up in their guides and the rotary movement tends to maintain a film of oil on the valve stems and their guides. Two notable examples of rotating valves are the Eaton and Thompson types, of American origin. In the case of the Eaton valve unit (Fig. 93) the valve is quite free to turn in either direction during the greater part of its lifting and closing period, due to the design of the valve end, valve collar and the fitments shown; this type has been in commercial service on various types of engine with satisfactory results.

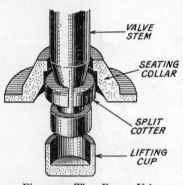

VALVE STEM

SEATING COLLAR

SPLIT COTTER

LIFTING CUP

Fig. 93.—The Eaton Valve Rotating Device.

The Thompson valve unit operates on the spring-loaded ball and inclined plane principle. Fig. 94 (*A*) shows the valve in the closed position with the load applied at (1) and (2) to the retaining cap. When the valve is lifted the valve spring pressure deflects the spring washer, thus causing the load to be transferred from (2) to the spring washer. This, in turn, causes the ball to roll down the incline, as shown at (*B*) and produces a rotation of the valve stem relatively to the spring seating collar.

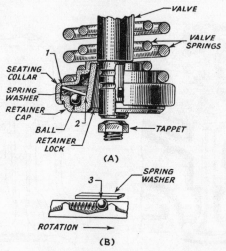

Fig. 94.—The Thompson Valve Rotator.

Overhead Valve Oil Seals.—If there is any appreciable wear or clearance between the overhead valve stem and its guide, lubricating oil will leak into the cylinder and cause carbon formation and, possibly, sparking plug fouling trouble. To prevent this, it is usual to fit a felt washer oil seal on top of the valve guide. This washer is held in place by a thin metal collar upon which one end of the valve spring bears. Alternatively, a synthetic rubber washer can be used. A simple oil deflector, known as the Hycar, is shown in Fig. 95 (A). Made of synthetic rubber it is pushed down the valve stem, inside the coils of the valve spring. The rubber is oil and heat proof, and will withstand the usual engine valve stem temperatures, indefinitely, as lengthy road tests have demonstrated.

Overhead Valve Operation.—Overhead valves are operated either (*a*) by push-rods and rocker arms or (*b*) by an overhead cam-shaft and rockers. An example of the former method is shown in Fig. 80. The fulcrum for the rocker arm is mounted on the cylinder head, whilst the valve end of the arm is usually

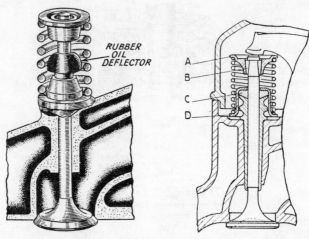

RUBBER
OIL
DEFLECTOR

A
B
C
D

Fig. 95.—(A) The Hycar Oil Seal.   (B) Alternative Type.
A—Valve collar.   *B*—Split cone cotters.   *C*—Rubber oil seal.
D—Seal retainer.

rounded, or provided with a ball or roller, and bears direct on the valve stem end. An example of the ball-ended rocker arm is shown in Fig. 91. The push-rod is now invariably provided with ball-and-cup ends, to minimize the friction constraint in the working parts. Sometimes the cup is in the rod end, and sometimes in the rocker or tappet. Means are provided for adjusting the valve stem clearance, usually, by means of a screw, with a locking nut. Another example of this method of adjustment is shown in the lower diagram of Fig. 84. This method is a convenient one, since the valve rockers are readily accessible.

The tappet rods are often made hollow, for lightness. The camshaft can be kept in a low position, for this type of overhead valve engine, so that a similar location and short chain drive can be employed, as in the case of the side-valve engine. An advantage of this arrangement is that cost can be kept down, while the effects of chain wear and stretch—as compared with the chain-driven overhead camshaft engine—are minimized. For these

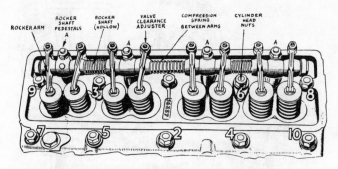

Fig. 96.—Overhead Valve Rocker Arms and Shaft.
(Standard Vanguard).

reasons most of the production overhead valve engines are of the push-rod and rocker arm type.

In regard to the rocker arms these are all located on a long tubular member which is held by brackets, with removable caps, on the cylinder head. The rocker arms are located and have their bearings on a long hollow shaft which is held securely in position by pedestals or brackets which are integral with the cylinder head (Fig. 96). There are usually four of these pedestals (A) and a pair of rocker arms located by compression springs between the pedestals, in the case of four-cylinder engines. The end-located arms are located by end collars. The locations of the valves, reading from left to right are: Inlet valves, 2–3–6–7. Exhaust valves 1–4–5–8. The cylinder head holding-down stud ends and nuts are indicated by the numbers alongside them and the order of tightening the nuts is 1, 2, 3, 4 etc.

Overhead Camshaft Engine.—The overhead camshaft method of operating the valves is employed upon the more expensive engines; it requires a greater number of working parts, but is usually quieter, better lubricated and more compact. When the valves are operated directly from the cams, as shown in Figs 97 and 98, by means of hollow tappets an appreciable

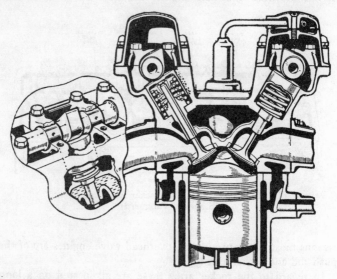

Fig. 97.—Valve Gear of High Output 2½-litre Engine.

amount of weight is saved, compared to that of the usual tappet, push-rod and rocker arm method. The valves can thus be operated at much higher speeds, due to the lower inertia effects and more positive action of the cams on their tappets. It is for these reasons that this type of valve operation is favoured in racing car and similar high output engines. The more popular method is to employ a single overhead shaft having all the cams cut on it in their appropriate positions; Figs. 97, and 98 show typical examples of single overhead camshafts. In the case of high output and competition engines, two camshafts are often used, one for the inlet valve, and one for the exhaust, this method enables the valves to be inclined to a greater degree, and leaves a convenient "trough" in the cylinder head for the sparking plugs.

An example of the twin camshaft overhead valve high output engine is the 2½ litre example, illustrated in Fig. 97. This arrangement is particularly suited to high speed, high efficiency engines,

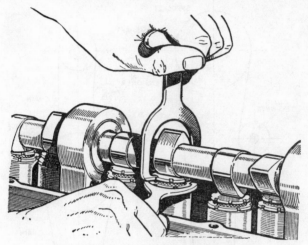

Fig. 98.—Method of Adjusting the Valve
Clearance on Wolseley Engine.

since it enables the weights of the reciprocating parts to be minimized, thus reducing power absorption and vibration effects; further, it enables the valve seatings to be adequately cooled, by the circulated water in the cylinder head.

Referring to Fig. 97, it will be seen that the valves are operated directly, i.e., without push rods or rockers, from the camshaft cams, so that the number of operating parts is reduced to a minimum. The same arrangement was employed in the Morris and Wolseley overhead camshaft engines.

The method of adjusting the valve stem clearance in the direct-operating type of valve is illustrated, in the case of the Wolseley engines (Fig. 98). The tappet which operates the valve is located between the cam and valve stem end, and has a serrated head adjusting screw which is used to alter the length of the tappet, in order to vary the valve stem clearance. A serrated head locking nut is used to secure the adjusting head in position. The special spanners shown are used for the head and locking nut

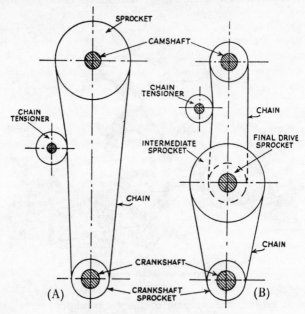

Fig. 99.—Two Alternative Chain Drive
Methods for Overhead Camshaft.

serrations; the distance between each serration on the head represents ·001 in. and the correct valve stem clearance when the engine is hot, is ·015 in.

The method of driving the Wolseley engine camshaft, consisted of a vertical shaft which was driven from the front end of the crankshaft, through a pair of spiral gears. At the upper end of this shaft there was another pair of split spiral gears to drive the camshaft at one-half engine speed. The vertical shaft was continued below the crankshaft so that it drove the gear-wheel type oil pump.

When *a single overhead camshaft* is employed to operate the inclined type inlet and exhaust valves in a cylinder head, the valve actuating mechanism is liable to become rather complicated, since it is then necessary to use short push rods and rocker arms to operate the valves, when the relative inclinations of the valves to the cylinder axis is more than about 30°. The inertia effect of these rockers and push rods is liable to become appreciable at the higher engine speeds.

In the case of the 90° vee-type eight-cylinder engine, which is popular in the U.S.A., a single rigidly-designed and mounted camshaft is located above the crankshaft in the vee between the cylinders. The camshaft is driven by a multi-link silent chain direct from the crankshaft, so that the drive is both compact and simple. All of the valves are operated from the camshaft's cams. When the side-valve engine is employed (as shown in Fig. 182), the left and right cylinder block valves face each other and are operated directly by tappets.

When overhead valves are used it is usual to employ push rods and rocker arms. In this case the push rods receive their movement from tappets which bear directly on the cams of the single camshaft.

Overhead Valve Camshaft Drives.—Owing to the relatively great distance between the crankshaft and camshaft in overhead camshaft engines the ordinary chain drive used for side-valve engines has to be modified. If a direct chain drive is employed the chain will tend to stretch and the non-driving side will "flap" sideways. The effect of chain link wear and stretch is more serious in this case, since the chain "lengthens" on this account and this results in valve timing alteration. Thus, the general effect is that of rotating the camshaft backwards by a small amount, so that opening and closing points of the valves occur later; this results in a loss of power.

To obviate chain "flap" due to stretch, spring-loaded sprocket or flat spring tensioning devices are fitted on the non-driving or "slack" sides of the chains; these devices are automatic in their action, so that no hand adjustments are needed. (Fig. 99 (*A*)).

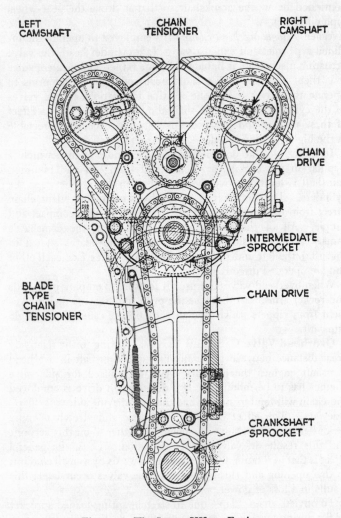

Fig. 100.—The Jaguar XK120 Engine
Camshaft Driving Arrangement.

Another method, which obviates the use of long chains is to employ an intermediate double sprocket, with two shorter chain drives. (Fig. 99 (*B*)).

The method of driving the two overhead camshafts of the Jaguar XK.120 engine is illustrated in Fig. 100*. In this case the

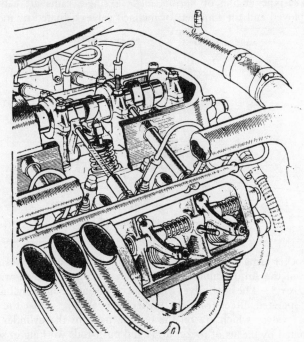

Fig. 101.—Auto Union Engine Valve Operating Mechanism.

drive is taken from the engine chain sprocket to an intermediate sprocket, keyed to the hub of a second sprocket; the latter with its separate chain, drives the two sprockets of the left and right camshafts.

The first chain drive has a blade-type spring tensioner on its non-driving side, while the second chain drive as a sprocket

*Courtesy *The Automobile Engineer*.

chain tensioning device located above the intermediate sprocket. With this arrangement both camshafts rotate in the same (clockwise) direction.

Single Camshaft Operating Four Sets of Valves.—In order to obviate the use of two separate camshafts in the case of a Vee-type engine of *narrow angle* a single camshaft can be employed and the four sets of inclined valves actuated by means

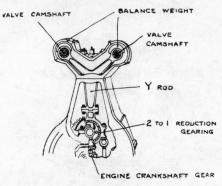

Fig. 102.—A Novel Method of Driving
the Camshafts.

of rocker arms and levers.   An example of such a mechanism is that of the Auto Union sixteen cylinder racing engine shown in Fig. 101.*   The inlet valves which are on the inside of the cylinders are operated direct from the rocker arms shown, whilst the exhaust valves which are on the outside parts of the cylinder are actuated by means of rocker arms and push-rods working in suitable guides.

The disadvantages of this valve gear are its complexity and therefore additional cost of manufacture and the increase in the inertia of the high speed moving parts.

Two Camshaft Drive.—A unique method of driving the two camshafts that was used on one particular commercial engine consists in providing a gear-wheel above the crankshaft, driven off it at one-half speed.   This gear-wheel, together with each of

*Courtesy *The Motor.*

the camshaft ends, has a small crank; each crank has the same "throw". The Y-shaped member shown in Fig. 102 has roller-bearings at the end of its three arms, which engage with the three cranks mentioned; the rotation of the 2 to 1 crank causes each of

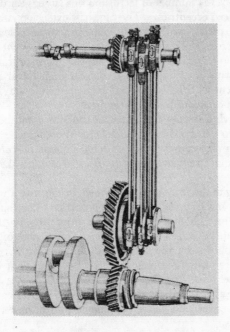

Fig. 103.—Triple Eccentric Drive for Camshaft used on a Bentley Engine.

the overhead camshaft cranks to turn in a similar manner, just as as two pairs of connecting-rods and cranks would do. The advantages of this method of operations are that it avoids the backlash and wear of the usual trains of gearing, and gives a direct, quiet and inexpensive method of operating the camshafts. Actually there is an eccentric and strap (similar to steam engine practice) at the 2 to 1 shaft, and not a crank.

A somewhat similar method, following this "link-plate" idea, had previously been employed by Ricardo in the case of the Tylor-Ricardo engine. An ingenious overhead camshaft drive that was used on the 8-litre Bentley car is shown in Fig. 103. In this arrangement the crankshaft has a gear meshing with another gear, of twice the number of teeth, and this larger gear drives three equally spaced eccentrics. The camshaft extension also has three similar eccentrics, which are driven from those below by light connecting rods. This method gives a positive, silent drive that is free from backlash effects. The balancing of this type of camshaft drive is an important factor, for high speed engines.

The recent N.S.U. small car engine, mentioned later, also uses a somewhat similar eccentric drive.

The Valve Cams.—The camshaft contains a number of cams, namely, two cams for each cylinder of the engine. The *shape of each cam* determines (1) The total amount of *valve opening* (2) The manner in which the *valve is lifted* and (3), the *maximum lift*.

Referring to the three diagrams shown in Fig. 104 it will be seen that each cam consists of a circular portion—corresponding to the closed or "no lift" period and a projecting portion which gives the lifting action to the valve. The period of the valve opening is shown by the angular portion marked *P*. Since the cam operates at one-half engine speed the equivalent crankshaft angle to the valve opening period will be 2*P*. The maximum lift of the valve is indicated in the diagram and is equal to the difference between the outer cam radius and the base (or no-lift) circle radius.

As the subjects of valve timing diagrams and valve lifts have already been dealt with, it is here proposed to deal with the cam shapes and their effects on valve operation.

The shape shown at (*A*) gives a quick lift to the valve, since immediately after the flat end of the tappet rests on the flat part of the cam, the curved peak portion lifts the valve quickly, but maintains the valve fully opened for a very short period; the use of large area valves of limited maximum lift is associated with this shape of rapid acceleration lifting and closing cam.

The *sides of the cam can be curved*, instead of flat, as shown at (*B*). This shape gives a more gradual opening (or acceleration) and therefore a greater effective opening and closing action than for (A), while retaining the same total opening period.

The shape shown at (C) gives a very quick lift and fall to the valve, with a longer period of maximum opening. This type of cam is used for high output, e.g., racing engines, but is apt to be noisy in action. It will be observed that it is necessary to use a roller-ended tappet for this shape of cam has a marked influence

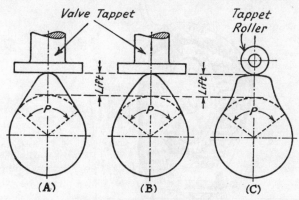

Fig. 104.—Valve Cam Shapes.

on the power output of the engine; the best shape is that which gives the quickest (practicable) opening and closing, consisten' with the maximum allowable period of opening. The sides of the cam must not be too steep, however, or there will be too much noise at opening and closing, due to impact, too much side force on the tappet or rocker arm, and the tappet will tend to "jump" or leave the cam during the opening period, thus necessitating the use of stronger valve springs to keep it in contact.

Number of Valves.—For engines up to 3 or $3\frac{1}{2}$ ins. diameter of cylinder, it is the common rule to employ one inlet and one exhaust valve per cylinder. For larger sizes, high speed, and racing engines, two exhaust and one inlet valve (or two inlets) per cylinder are sometimes employed. It is considered better practice to use two valves of smaller diameter than one valve of larger diameter, although the operating mechanism is more complicated.

The Camshaft Drive.—Since the camshaft must actuate each pair of valves of any cylinder once every two revolutions of the main crankshaft it must be driven at one-half engine speed by gearing down from the crankshaft.

In earlier types of engine it was usual to employ a pair of gear-wheels—sometimes an intermediate wheel was used to avoid

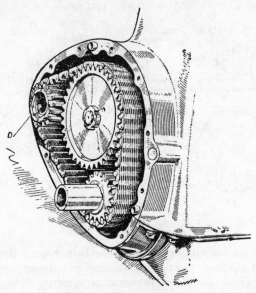

Fig. 105.—Illustrating the American-type Silent Chain Driven Camshaft Arrangement.

having the camshaft too near to the crankshaft. The camshaft wheel had twice the number of teeth of the engine shaft pinion to give it the required half-speed ratio. Unless, however, special herringbone or spiral gear-wheels or one gear-wheel of laminated resin are used this form of camshaft drive is apt to become noisy in service. The later practice with many side-type and push-rod operated overhead valves is to employ a silent chain drive for the camshaft as shown in Fig. 105. The lower shaft is the engine crankshaft extension. The silent chain-wheel sprocket on this drives the larger chain-wheel attached to the camshaft above at one-half engine speed. The sprocket marked *D* on the left drives the lighting dynamo at engine speed.

The inverted-vee type of silent chain has more recently been replaced by the roller chain in this country, since the latter is lighter, less bulky and rather more efficient, mechanically.

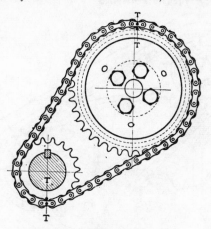

Fig. 106.—Roller Chain Drive for
Camshaft (B.M.C. Engine).

Fig. 106 shows a typical production car engine roller chain drive for the camshaft. In this example the distance between the crankshaft and camshaft axes is kept small, so that no chain tensioner is required.

The diagram shows also the method used for ensuring the correct valve timing, should the chain be removed and afterwards replaced. The chain has two bright links, at *T*, *T*, while the sprocket wheels are marked with similar letters opposite the two teeth with which the bright links should engage for correct timing.

The more recent examples of car and commercial vehicle engines employ the double or triple roller chains, according to the size of engine. These chains may be regarded as two, or three, roller chains placed side by side, but with common links for the inside members.

Chain Tensioners.—To permit the quiet operation of roller-type chains and also to take up any chain wear or stretch effects it is usual to fit some form of chain tensioning device, on the normally slack side of the chain. In some instances, e.g., the

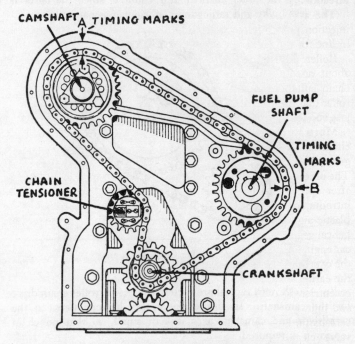

Fig. 107.—The Perkin's engine Chain Drive to the Camshaft, showing Automatic Chain Tensioner.

more recent B.M.C. engines, a synthetic rubber pad is held by spring pressure against the slack, i.e. left-hand side of the chain, as viewed from the radiator end.

In the Rover engines hydraulically-operated chain tensioners are used.

The Perkins and several other makes of engine use an idle chain sprocket having within its hub a spring and ratchet device which pushes the idler sprocket towards the chain and holds it there when the tension is correct. This is known as the Reynolds tensioner and its location is shown in Fig. 107, namely, on the slacker side of the chain, since the engine rotates in a clockwise direction.

The arrows A and B, denote the manufacturer's valve and injection pump timing marks. These pairs of arrows should be in line for the correct timings.

Roller chains operate silently and only require replacement about once every 50,000 miles or so. A typical double-roller chain suitable for a $1\frac{1}{2}$ to 2 litre engine has a pitch of $\frac{3}{8}$ in., roller diameter of $\frac{1}{4}$ in., and total width of 1 in.; the breaking load is 4,000 lbs.

More Recent Camshaft Drives.—As distinct from the roller chain and sprockets drive for the camshaft, the earlier all-steel gear wheel drive is still employed in certain automobile engines. The objections previously made to the noisiness of the gear wheels after appreciable periods of usage have been overcome by the introduction of intermediate gear wheels made from laminated plastic and similar synthetic materials. Further, apart from the helical gears and shaft drives to overhead camshafts, a recent small car engine, namely, the N.S.U. Prinz employs a method of driving the overhead camshaft that uses a similar principle to that used on the earlier Bentley car, shown in Fig. 103. This arrangement of eccentrics on the engine-driven lower half-speed shaft (Fig. 108) and light connecting rods which actuate similar eccentrics on the camshaft gives a quiet and positive drive which maintains constant valve clearances, regardless of engine temperature changes. Further, it is found that lighter valve springs can be employed, yet another advantage is the extremely low rate of wear in the drive members, so that no changes in valve timing are experienced.

A recent departure from the conventional steel chain drive for the camshaft, illustrated in Fig. 109 is a toothed reinforced rubber belt which employs a pair of steel gear wheels on the crankshaft and camshaft, respectively. Made by the North British Rubber Company, for the German 993 c.c. Goggomobil, miniature car engine, this belt is made from helically-wound high tension steel wire which is embedded by moulding into the artificial (neoprene) rubber belt which, as stated, has moulded gear teeth on its inner surface. This gives a quiet and positive drive to the camshaft.

The roller chain and sprocket method of driving the camshaft is now practically universal for side valve and overhead valve

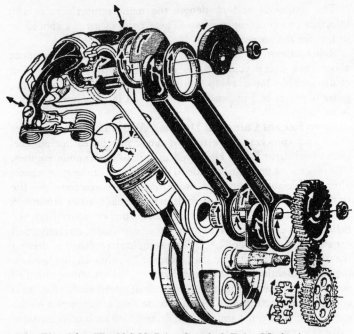

Fig. 108.—The N.S.U.-Prinz Camshaft Drive Mechanism.

engines of the push-rod and rocker arm type. The duplex pattern roller chain with some form of automatic tensioner is favoured. The tensioner is generally of the spring blade kind (as shown in Fig. 100) and gives satisfactory results.

A method of tensioning and silencing the timing chain drive to the camshaft used in the more recent Austin engines employs an oil-resistant synthetic rubber ring between the two sets of teeth of the larger (camshaft) sprocket wheel (Fig. 110). The timing chain makes contact with the rubber which projects above the

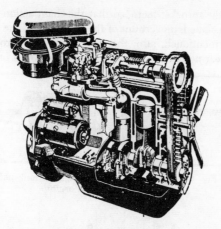

Fig. 109.—Toothed Reinforced Rubber Belt Drive for
the Overhead Camshaft.

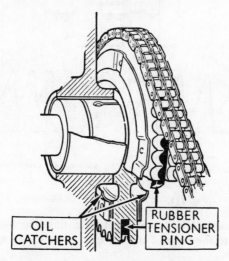

OIL
CATCHERS

RUBBER
TENSIONER
RING

Fig. 110.—The Austin Roller Chain Sprocket
Tensioning and Silencing Device.

bottom of the sprocket teeth, so that the chain action is quietened and also tensioned, on account of the resiliency of the rubber ring. It may be mentioned, also, that the overflow oil from the front camshaft bearing is collected in a hollow steel ring on either side

Fig. 111.—A Four-Cylinder Engine Crankshaft made from a Drop Forging.

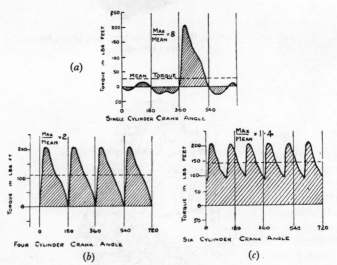

Fig. 112.—Crankshaft Torque Diagrams.

of the sprocket wheel and is fed from this ring by centrifugal action on to the teeth of the sprocket and also the side links of the chain.

The Crankshaft.—The crankshaft and connecting-rod (or rods) convert the reciprocating motion of the piston (or pistons) into one of rotation. It is made very stiff, since it is subjected to severe and varying twisting and bending stresses due to the combustion pressures, and also to the "inertia" effects of the reciprocating parts; the latter effects are the forces due to the acceleration and deceleration of the piston and connecting-rod in their strokes.

The twisting or turning action on the *crankshaft*, which is generally spoken of as the *Torque*, is constantly changing; this

Fig. 113.—A Two-Cylinder Opposed Engine Built-up Crank-shaft, with Connecting-rods and Slipper Pistons.

fact necessitates a stronger shaft than for a steady motion. The manner in which the torque varies in the case of a single cylinder engine is shown in Fig. 112 (*a*). It will be observed that the firing stroke gives the greatest torque. The dotted line shows the value of the average torque. In this case the greatest torque is no less than eight times the mean value. In order to reduce this great variation of torque, which puts severe stresses on the driving and driven members of the automobile, it is usual to employ a number of smaller cylinders, so that the maximum values of the torque are correspondingly lower, and occur more frequently. Fig. 112 (*b*) and (*c*) illustrates the torque values in the case of a four and a six-cylinder engine; it will be noticed from this that the greater the number of cylinders the smoother becomes the torque curve. Interpreted practically, this means a smoother running engine, with a lighter crankshaft, and incidentally lower minimum top-gear speeds. The single cylinder engine is, relatively speaking, very jerky in its running, on account of the wide torque variations, and it is necessary to provide a relatively heavy fly-wheel to reduce the speed and torque variations.

For four-cylinder engines, the crankshaft is usually made in one piece by a forging process, known as "drop-forging"; in this process a powerful steam or pneumatic hammer is provided with a pair of dies, having internal impressions corresponding to the required shape of the crankshaft. The almost white hot steel is

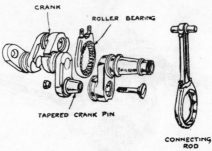

Fig. 114.—Example of a Built-up Crankshaft.

pressed into these dies, and assumes the desired shape. The journals and crank-pins only are machined.

In some earlier engines the crankshaft was planed, milled and turned from a solid slab of steel; this was a more expensive process, however. Another method, which has been developed, is to build up the crankshaft from a number of components, as shown in Fig. 114. This method, although more complex, enables roller bearings to be employed for the big ends of the connecting-rods.

Main bearings.—In the single and twin cylinder engines only two main bearings are used for the crankshaft. Also, in the smaller four cylinder engines of 500 to 600 c.c., a relatively stiff crankshaft has been used with only two main bearings with four-cylinder engines of 700 to 1500 c.c. three main bearings have become the accepted practice, but in certain recent instances, e.g. the Ford $1\frac{1}{2}$ litre Classic engine, five main bearings are used. Thus, with a main bearing on either side of each crank this engine runs very smoothly.

The 90° vee-eight engines have five main bearings and four double-width crank-pins, upon each of which an opposite pair of connecting-rods works.

Fig. 116 illustrates a vee-eight crankshaft with its main bearings, flywheel, timing chain and sprocket, and crankshaft vibration damper. This is of the five main bearing type, the upper half shell bearings being housed in machined seatings in the cylinder block. The lower half-shell bearings are held in place by the

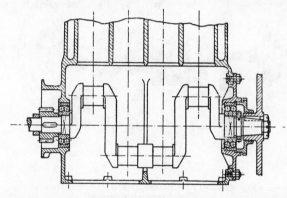

Fig. 115.—A Two-bearing Four-Throw Crankshaft.

bearing caps, shown below; long bearing cap bolts are used to secure these caps.

With six-cylinder engines, while the cheaper type of engine often employs one main bearing for each pair of crank-pins, i.e., four main bearings in all, the better practice is to use a main bearing on either side of each crank-pin, making seven main bearings in all.

The crankshaft in addition to converting the reciprocating movements of the pistons into rotary motion, must also drive the camshaft and certain accessories, indirectly, such as the oil pump, dynamo, ignition distributor unit, water-pump and cooling air fan. Usually, the drive for these components is taken from the front end of the crankshaft; in addition the starting handle dog, when provided, is fitted to this end. The other end of the crankshaft carries a flange to which the fly-wheel is secured; the clutch drive is taken at this end.

The crankshaft, with its full set of components for a typical B.M.C. car engine of the four-cylinder vertical class is shown in Fig. 117.

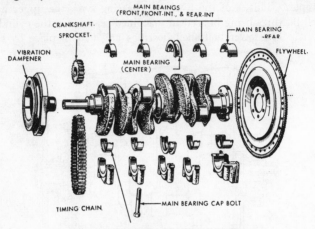

Fig. 116.—The Crankshaft and its Components
(Ford Mercury Engine).

Crankshaft Materials.—Inexpensive and slow-speed engine crankshafts were frequently made of a medium carbon steel known as "40 ton steel," unhardened. The majority of modern crankshafts are made of alloy steels, including 3 per cent. nickel-steel, chrome-vanadium and nickel-chrome. After machining to approximate dimensions the crankshaft is hardened, and the bearing surfaces finally ground to size. In the hardened condition these alloy steels have tensile breaking stresses of from 50 to 90 tons per sq. in.

More recently high grade crankshafts have been made of alloy steels containing molybdenum; typical steels of this class include the nickel-chromium-molybdenum and somewhat similar steels with the addition of a small proportion of vanadium. When suitably heat-treated such steels give tensile strengths of 65 to 75 tons per sq. in., combined with high fatigue resistance.

Another class of high tensile steel that has been used for

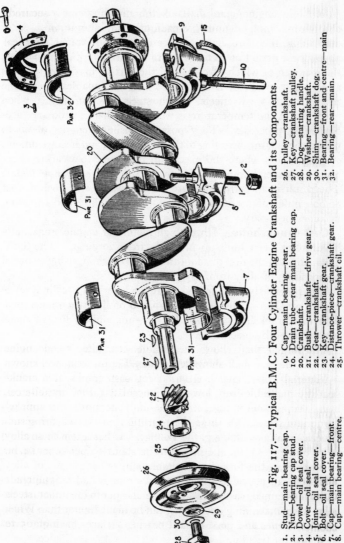

Fig. 117.—Typical B.M.C. Four Cylinder Engine Crankshaft and its Components.

1. Stud—main bearing cap.
2. Nut—bearing cap stud.
3. Dowel—oil seal cover.
4. Cover—oil seal.
5. Joint—oil seal cover.
6. Bolt—oil seal cover.
7. Cap—main bearing—front.
8. Cap—main bearing—centre.
9. Cap—main bearing—rear.
10. Drain tube—rear main bearing cap.
20. Crankshaft.
21. Bush—crankshaft—drive gear.
22. Gear—crankshaft.
23. Key—crankshaft gear.
24. Distance-piece—crankshaft gear.
25. Thrower—crankshaft oil.
26. Pulley—crankshaft.
27. Key—crankshaft pulley.
28. Dog—starting handle.
29. Washer—crankshaft.
30. Shim—crankshaft dog.
31. Bearing—front and centre—main.
32. Bearing—rear—main.

automobile engine crankshafts is the nitriding one containing chromium, molybdenum, vanadium and aluminium. After machining, the crankshafts are heated to about 500° C. and exposed for a period of 40 to 90 hours to a stream of ammonia gas, the nitrogen of which takes part in the surface hardening process and finally produces a glass hard surface capable of extreme resistance to wear effects. Such steels do not distort when hardened as the temperature is well below that of ordinary case-hardening processes. They possess tensile strengths of 60 to 90 tons per sq. in. according to the composition and heat treatment.

In instances where the maximum resistance to wearing is required, the crankpin and main *journals are chromium plated.* The result is an extremely hard surface which will give at least 80,000 miles of useful service before re-grinding and replating becomes necessary.

Cast Crankshafts.—Hitherto all motor car engine crankshafts have either been built-up or made from forgings. The Ford Company was the first to adopt the cast crankshaft, using a special composition of alloy iron.

The previously used forged shaft for a certain 8-cylinder Vee engine weighed 90 lbs. in the rough and 66 lbs. when machined. The cast crankshaft weighed 69 lbs., and only 9 lbs. of metal was removed by machining.

The forged shaft shows measurable wear after 10,000 miles, whereas the cast shaft shows less than $\frac{2}{10000}$ in. wear.

Materials that have been used for cast crankshafts include pearlitic malleable iron, low-alloy inocculated iron, graphite cast iron, copper-chromium, chrome-molybdenum and nickel-chromium irons. Of these the pearlitic iron is used for certain American crankshafts, e.g. the Pontiac. It has a tensile strength of 50 tons per sq. in. and is said to be about 15 per cent cheaper to make than the forged steel crankshaft.

In general, provided the necessary strength and fatigue resistance are obtained, cast crankshafts are cheaper to manufacture, require fewer machining operations, can be made lighter than forged alloy steel ones and possess hard bearing surfaces having a lower coefficient of friction.

A typical chrome-molybdenum iron crankshaft was fitted to a

four-cylinder 75 H.P. commercial vehicle engine running at a maximum speed of 2,500 r.p.m. which covered a distance of 40,000 miles in 2½ years. At the end of this period the wear on the main crankshaft journal averaged 0·001 in. to 0·002 in.; that on the crankpins was found to be negligible. The big-end bearings were made of R.R. 56 light aluminium alloy and had a running clearance of 0·003 in.

Eliminating Crankshaft Vibrations.—The petrol engine crankshaft is subject to two different sources of vibration, that may cause the engine, itself, to vibrate seriously, namely, (1) *Torsional* and (2) *Engine Unbalance* vibrations.

(1) Torsional vibrations are caused by the irregular turning efforts or torques on the crankshaft, due to the firing strokes of the different cylinders. These variable torques tend to make the crankshaft vibrate, to and fro, about its own axis of rotation in the manner of a coil spring—as distinct from any bending action. When the period of these vibrations coincides with the natural period of vibration of the crankshaft this resonance effect is liable to make the vibrations increase to such an extent that the crankshaft is severely stressed above its designed values; many cases of crankshaft fracture have been due to this effect. In practice, the selection of working speeds away from the resonant speeds and the use of vibration dampers fixed to the end of the crankshaft are the means used to eliminate or minimize torsional vibration effects.

One successful type of vibration damper is shown in Fig. 172, while a neater and more compact damper is illustrated in Fig. 118. It consists of a metal flywheel attached by a rubber mounting ring to the fan pulley drive at the front end of the crankshaft. The rubber member damps out the vibrations and has been shown to be most effective for automobile engines on account of its high energy absorption per unit weight. Referring to Fig. 118, this Metalastik vibration damper consists of a steel flywheel F to which is attached, by an efficient rubber-to-metal bonding process, a rubber ring of angle-section; this is also bonded to a steel pressing D, attached to the fan-drive pulley P, the latter being keyed to the end of the crankshaft C. The damping action takes place between the plate D—which may be regarded as being in a state of axial vibration—and the flywheel

F, through the medium of the rubber unit *E*. The inertia effect of the rubber-insulated flywheel is an important factor in the final vibration damping out result.

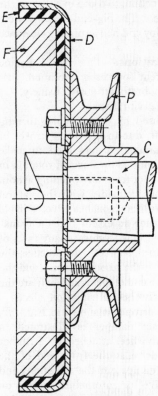

Fig. 118.—Rubber Ring-type Vibration Damper.

(2) Engine unbalance vibrations may be due to two separate causes, namely, (A) Lack of static balance of the crankshaft unit or (B) Lack of balance of the reciprocating forces due to the pistons and connecting-rods.

(A) If the centre of gravity of the crankshaft and flywheel does not coincide with the axis of the crankshaft the latter will be out of balance and will cause centrifugal vibrations, which will increase in intensity as the engine speed is increased. By suitable removal of metal on the heavier part of the flywheel, balance can be restored.

(B) The balancing of the effects due to the various reciprocating forces is liable to be a complicated process, but in most examples the values and positions of the balancing weights—which are arranged as projecting webs on the crankshaft on the opposite side to the crankpin—can be calculated accurately for all standard engines. In practice, the car manufacturer uses a special dynamic balancing machine in which the crankshaft is rotated and the positions and amounts of the balancing weights—or the weights to be removed by drilling the webs—are shown by scale readings.*

*Fuller particulars are given in *Automobile and Aircraft Engines*. Vol. I. Mechanics of Petrol and Diesel Engines. A. W. Judge. (Messrs. Pitmans Ltd., London, W.C.2.).

The relatively smooth running of most modern engines is due to the use of crankshaft counterbalance weights, vibration dampers and the individual balancing of crankshafts on balancing machines.

Connecting-Rods.—These rods are subjected to heavy loadings of a variable nature. To resist such forces, the connecting-rods should be designed as struts subject to side bending. The tubular sections have been used but the H-sections are now invariably used; these give the greatest strengths in compression and bending. Connecting-rods are usually made in one operation as drop-forgings; they are afterwards machined, and the bearings fitted. The small-end bearing, when the gudgeon pin is fixed in the piston bosses, takes the form a of phosphor-bronze bush, which is pressed into the small-end of the rod. An oil hole above, or two holes below, and oil grooves are arranged for the purposes of lubrication. The gudgeon pin is sometimes secured to the small end and rocks in the piston bosses. Another method is to bush the small end, and to allow the gudgeon pin to "float" (i.e., to be free to rotate) in it and also the piston boss bearings; in this case, means, e.g. circlips, are provided to prevent end-movement of the gudgeon pin; otherwise it would score the cylinder. The big-end bearing is very important. The common practice is to employ split white-metal-lined shell bearings, bolted together with two bolts, as depicted in Fig. 119, which shows the connecting-rod unit of an Austin engine, in dismantled view. In this modern practice, the gudgeon pin (7) is clamped to the split small end of the rod by the pinch bolt, shown at (9); a spring washer locks the bolt in position. The gudgeon pin has a groove on one side for locating it in relation to the pinch bolt. In regard to the big-end bearing (11) this consists of two white-metal-lined half-shells with steel backs. Each half bearing has a small projection at the split face, which locates in a corresponding depression in the connecting-rod and cap—seen on the left, near the split machined faces, the object of these projections is to locate and also prevent the bearings from rotating in the connecting-rod bore. The cap is secured rigidly to the connecting-rod by special bolts which screw into holes in the rod end. A small hole is drilled through the upper bearing shell and crankpin at (10) to allow a jet of oil to be emitted when the two holes coincide; this is arranged to

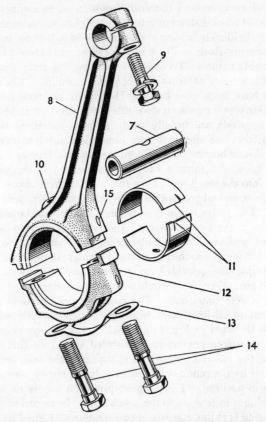

Fig. 119.—B.M.C. Engine Connecting-rod Components.

7—Gudgeon pin. 8—Connecting-rod. 9—Pinch screw and washer. 10—Cylinder wall lubrication jet. 11—Connecting-rod bearings. 12—Connecting-rod cap. 13—Lock washer. 14—Set screws. 15—Mark on rod and cap.

occur when the connecting-rod is nearing the upper end of its movement, so that the oil jet will lubricate the cylinder walls.

Previously, to the half-shell bearing method, the connecting-rod and its cap were lined with white-metal the two halves of the big-end bearing clamped together with a number of thin metal packing pieces known as shims, at the joint; when any wear occurred, one or more of these thin strips was removed from each joint to reduce the diameter of the bearing; the white-metal could then be scraped away by hand until a good fit on the crank-pin was obtained.

In regard to the split shell-type of bearing it *is not possible to take up the effects of bearing wear,* by filing or machining the split faces, as with solid white-metal-lined bearings. When appreciable wear occurs, new under-size shell bearings are fitted, to suit the existing or reground crankshaft journal diameters.

Although it has become the normal practice to split the connecting-rod big end bearing at right angles to the length or axis of the rod, as shown in Fig. 120, some of the modern connecting-rods now split at an angle which is usually 45°, as shown in Fig. 119. This arrangement possesses certain advantages in regard to accessibility and is especially suited to the use of the white-metal lined steel shells used. The bearing cap sometimes has serrated faces which mate with corresponding ones on the connecting-rod so that any shearing action due to the piston loads is taken off the bolts holding the cap to the rod.

The use of the inclined type of big-end joint results in a more compact crankcase and enables the piston and rod to be withdrawn upwards through the cylinder barrel. Usually, with the right-angle connecting-rod joint the big-end part of the connecting-rod is too large to pass through the cylinder barrel.

Crankshaft Bearing Clearances.—The clearances of most British car engine big-end bearings, for journals of about 2 to 2·75 in. diameter, lie between the limits of 0·0005 in. and 0·0025 in.; the average values are ·0015 in. to ·005 in. (new).

The connecting-rod must be given a certain amount of side-play to allow for piston and connecting-rod alignment in the cylinders. This side- or end-clearance ranges from ·006 in. to ·012 in.

The clearances for big-end bearings of American vee-eight engines usually lies between ·0005 in. to ·0025 in. for the journals and ·005 in to ·009 in. for side-play. Fig. 120 illustrates typical clearances for a modern Studebaker engine.

The main crankshaft ournal clearances usually lie between

·0005 in. and ·003 in. (new). The side clearances or *end float* is generally ·004 in to ·008 in. Provision is made on the side of one of the main bearings to adjust this end float, by means of thrust liners.

Lead-bronze bearings require more clearance than the white metal lined shell-bearings, for which the above data is given.

Big-End Cooling.—Fig. 121 shows a modern big-end design of light yet very strong construction. Special features of this

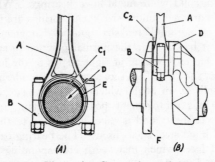

Fig. 120.—Illustrating Connecting-rod Clearances.

A—Connecting-rod. *B*—Bearing cap. *C1*—Bearing clearance (0·0005–0·0025 in.). *C2*—Side clearance (0·005–0·009 in.). *D*—Crankshaft. *E*—Big-end bearing. *F*—Counterweight.

arrangement include the connecting-rod cap which is provided with fins for assisting the cooling of the bearing; light cap securing bolts with bevelled heads and reduced section, namely, to that of the bottom diameter of the nut thread and H-section rod with central oil hole in the enlarged web to lead oil to the small-end bearing.

Big-End Bearing Lubrication.—The lubrication of the big-end bearing is important. The crankshaft is drilled in such a way that oil under a pressure of 30 to 60 lb. per sq. in. from the oil pump is delivered to the centre of the shaft and then passes out through holes in the main and big-end crank journals to the insides of the bearings. Oil grooves radiating from the oil supply holes are cut in the white-metal, and the latter is recessed near the joints.

Fuller information on this subject is given in Chapter VIII. Mention has already been made of the use of ball bearings (Fig. 115) or roller-bearings for the big-end (Fig. 114). These are quite satisfactory, if properly designed, and their use enables a compact and narrow crank-pin bearing to be obtained; these bearings give a rather lower frictional resistance, also.

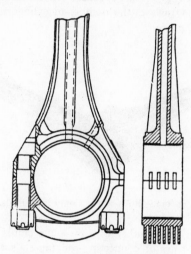

Fig. 121.—Connecting-Rod with Ribbed Cap.

Connecting-Rod Proportions.—Connecting-rods in automobile engines are usually made equal in length to from $1\frac{3}{4}$ to $2\frac{1}{2}$ times the stroke of the piston. Where H-sections are used the approximate proportions are: *Depth* from $\frac{1}{3}$ to $\frac{1}{4}$ the cylinder bore, *Width* from $\frac{1}{7}$ to $\frac{1}{10}$ the cylinder bore, and *Web Thickness* from $\frac{1}{25}$ to $\frac{1}{52}$ the cylinder bore. The diameter of the gudgeon pin should be from 0·20 to 0·25 times the piston diameter and its length according to the design used from 0·9 to 0·95 times the latter.

Connecting-Rod Materials.—Connecting-rods are usually made in 3 per cent. nickel steel, nickel-chromium and chrome-vanadium steels, heat-treated. Aluminium alloy connecting-rods are also used, and have given satisfaction. The alloys used are forged Duralumin, and "Y" alloys, which enable the connecting-rods to be made about 30 to 40 per cent. lighter than the drop forged steel ones of equal strength.

Connecting-Rods of Vee-type Engines.—In connection with the Vee-type of engine the crankshaft has one-half the number of crank pins or journals as the number of cylinders; each pair of

Fig. 122.—Rolls Royce Connecting-Rods.

cylinders facing one another uses the same crank journal. The corresponding pairs of connecting-rod big-ends therefore work on the same journal. Two alternative arrangements for the rods are as follows, namely: (1) Similar designs of rod working side by side on the journal as in the modern vee-eight engines, and (2) one forked rod and one central rod with its bearing between the forked portion. Another method used in the case of the twelve-cylinder Maybach engine is to have a master connecting-rod running on roller bearings and a link rod working on a pin attached to one side of the master rod for the opposite cylinder. The link rod has a phosphor-bronze bush type of bearing; this arrangement is somewhat similar to that of the connecting-rods of radial engines. With this type of articulated rod, the stroke of the articulated rod cylinder is different to that of the other cylinder, so that suitable compensation must be made to the compression ratio.

In the earlier Rolls Royce Vee-type engines a forked rod is used for one cylinder unit and a plain rod for the facing cylinder rod. The forked rod works on the crank journal and the plain rod on the big-end portion of the forked rod; in each case the bearing surfaces are of white-metal and of equal bearing area. Fig. 122 illustrates the connecting-rod arrangement of the engine in question. The material used for the rods is $3\frac{1}{2}$ per cent. nickel steel, machined all over and heat-treated.

The forked rod carries a nickel steel bearing block lined inside and out with a special lead-bronze alloy. The bearing block which is divided across its bore bears directly on the crank-pin and is secured to the forked rod by four bolts, whilst the outer lining of the block forms the bearing for the plain rod.

Recent Bearing Improvements.—Although white-metal is a satisfactory material for the main and big-end bearings of petrol engines, provided the bearing pressures are not excessive and that oil filters or cleaners are fitted, when high compressions are used, as in high output petrol and compression-ignition engines, this material does not always give good service.

In the latter type engines, some trouble was experienced owing to cracking and breaking down of the white-metal, so that it became necessary to seek stronger bearing metals having low frictional qualities. Many of the recent designs of engine are now fitted with *lead-bronze* main and big-end bearings. Lead-bronze is an alloy of copper, lead and tin having a distinct bronze colour; it is very much stronger in compression than white metal, and has a comparable friction coefficient. As it is a more expensive material to employ, some makers only use it for the more highly loaded halves of the bearings, viz., the upper half of the big-end and lower half of the main bearings; white-metal is employed for the other halves. Lead-bronze bearings can be run for mileages of 80,000 to 100,000 before replacement or attention is necessary; white-metal under similar circumstances will not last longer than about 25,000 to 30,000 miles.

Another strong bearing alloy is one containing *cadmium and nickel* with small proportions of *copper, magnesium, silver*, etc. This is about twice as strong as the best white-metal.

In later engines the white-metal linings of steel shell bearings have been reduced to about 0·003 in. to 0·005 in. in thickness and it is claimed that the increase in bearing fatigue strength has resulted in the bearing life being prolonged to four or five times that previously obtained. The bearing metal is bonded to the steel shell by means of a special porous matrix.

Indium-Coated Bearings.—The best combination *for maximum life*, consists of an hardened alloy steel (or nitrided steel) journal, used in combination with a coated lead-bronze-lined steel shell bearing. The coating consists of a thin layer of lead, namely, of about ·001 in., which is then "flashed" or infused with the element, *indium.* Such bearings are immune from corrosion attack due to the lubricating oil products; further there is a much reduced tendency to journal scoring.

The Main Bearings.—The split steel-backed shell type of bearing is used for the main crankshaft bearings, in modern engines and, as with big-end shell bearings, it is not possible to take up the effects of wear by machining the split faces and then scraping in the bearings. Usually, the crank journals after much service are found to have worn slightly oval and (or) tapered, by about ·003 in., and must therefore be reground slightly undersize and new undersize main bearings fitted.

Preventing Oil Leakages.—If the recommended bearing clearances are exceeded *oil leakages* will occur, and the oil pressure may fall below the normal value. In this connection special provision is always made to prevent oil leakage from the two end main bearings into the timing gear and the clutch casings, respectively. At the front end bearing a felt ring oil seal is usually employed, while at the rear end the oil that escapes towards the clutch casing is caught by a sharp-edged *oil thrower ring* which throws the oil outwards by centrifugal action, whence it is collected in a groove, formed in the main bearing and returned through a passage, or pipe to the oil sump below.

Fig. 123 (*A*) illustrates the principle of this method, the web of the rear crankshaft journal being shown on the left and the flywheel flange on the right. It is usual to prevent any seepage of oil from the outer wall of the thrower chamber along the crankshaft by means of a felt ring, as shown at (*A*), but an alternative method is to use a screw thread on the right of the thrower ring to return the oil that tends to leak along the shaft, back into the oil thrower chamber as depicted at (B), in Fig. 123.

Referring back to the B.M.C. crankshaft unit, shown in Fig. 117, it will be seen that, instead of fitting a felt oil leakage preventing washer or insert at the timing case end of the crankshaft, an oil thrower ring (25) is used. There is also another oil thrower

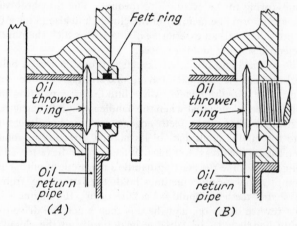

Fig. 123.—Methods of Preventing Oil Leakage from Rear Main Bearing.

ring at the rear end of the crankshaft and also gaskets for sealing the oil sump and bearing end joint.

Other Components and Accessories.—*The fly-wheel* is provided primarily for the purpose of storing up part of the energy of the firing or combustion period, and utilizing this stored up energy to carry the crankshaft, rods and pistons over the three idle strokes; the fly-wheel thus tends to re-distribute the energy received, so as to give more uniform running. The essentials of a fly-wheel are a relatively heavy rim, situated as far away from the axis as possible; this is equivalent to stating that the "moment of inertia" of the fly-wheel should be as great as possible (see Fig.

116). Practical considerations fix the size and weight of the fly-wheel; thus for a motor-cycle engine the diameter should not exceed from 3 to 4 times the piston bore, and the weight about 1 to 2 lb. per cubic inch of cylinder capacity. For four-cylinder engines, owing to the more even torque and to the overlapping of the cycles, the fly-wheel can be made much lighter in proportion, namely from $\frac{1}{2}$ to 1 lb. per cubic inch capacity. There is hardly any theoretical need for a fly-wheel in a six or eight-cylinder engine, owing to the evenness of torque. The fly-wheel serves, however to form one member of the clutch, and also in most cases it is provided with an exterior gear ring with which the pinion of the electric starting motor engages.

The forward end of the crankshaft is provided with a pair (or more) of teeth or dogs, each having one straight and one sloping tooth. *The starting handle*, when fitted, has a pair (or more) of similar teeth, such that when the handle is put into engagement, the straight sides of the teeth engage, and the handle can then rotate the engine crankshaft in its normal direction of rotation. When the engine fires the crankshaft teeth move the faster, and the sloping sides force the starting handle's teeth out of engagement. A compression spring sometimes holds the latter away from the crankshaft's teeth. Should a backfire occur, the engine rotates in the reverse direction, and there is then a positive drive to the starting handle. If the latter is held rigidly in the closed fist, it is very easy for the wrist to be broken. The handle should therefore be held on the under-side only, when pulling up, and without the fingers encircling it. Patent safety starting handles, having pawls and ratchet devices, have been marketed; these enable the risks of backfire to be avoided.

Engine Mountings.—The engine mounting to the frame of the chassis has to provide for the following factors, namely, (1) The engine dead weight. (2) The unbalanced vibrations due to uncorrected out-of-balance of the engine. (3) The engine torque reaction. (4) Torsional vibration effects at resonant speeds.

The torque reaction is the torque, equal to, but opposite in direction to the engine torque, i.e., the torque, tending to cause the engine to rotate in the reverse direction, which the engine mounting must cater for.

Hitherto, it was the practice to bolt the engine unit solidly to the chassis frame at three or four places. The "three-point" mounting consists of a front central fixing and two side ones; this method enables the engine to accommodate itself to small frame movements, without undue stresses on the engine casing. The modern method of mounting the engine dispenses with solid fixings and employs springs or rubber blocks between the engine bearers and the frame. In this way any vibrations due to misfiring or lack of correct engine balance, etc., are not transmitted to the frame and bodywork. From the car occupants' viewpoint the flexible mounting method is both quieter and free from transmitted vibrations.

Rubber can be used in tension, shear or compression, in flexible mountings for the engine-gearbox unit, but the former method is not used since in the event of small surface injuries, e.g., cuts, the rubber would rapidly lose its strength properties. In engine mountings rubber bonded to metal is employed in shear or compression and the bond is such that it gives a direct tensile strength of 800 lb. to 1,400 lb. per sq. in. Whilst in compression loadings up to 10,000 lb. per sq. in. can be employed. There is a wide range of alternative rubber-bonded metal mountings for automobile engines of the shear, compression and combined shear-compression type, the usual method being to ensure that the engine metal side does not move sufficiently to make contact with the chassis mounting side; in some cases engine movement is limited by the use of stop units on the chassis-fixed part of the mounting. Synthetic rubbers, on account of their marked resistance to the effects of heat, sunlight, oil, petrol and air influences, are preferred to natural rubbers.

Arrangement of the Mountings.—There are three principal arrangements for engine flexible mountings, namely, (1) A single mounting at the front, centrally under the engine; usually a full compression mount is used. At the rear of the engine-transmission unit two mountings are used; these are often of the combined shear-compression kind. (2) Two mountings at the front end and a large central compression-type mounting, centrally under the transmission unit. (3) Four mountings arranged in symmetrical pairs at the front and rear, respectively.

Usually, the arrangements (2) and (3) are employed for the vee-eight engines. Thus, for the more recent Buick vee-eight

engines, there is an inclined shear-compression mounting on either side of the engine at the front and a large compression unit supported on the chassis frame cross-member and arranged under the rear end of the automatic transmission casing. Synthetic rubber units are employed for these mountings. A similar arrangement is used for the later Ford, e.g. the Galaxie vee-eight engine. The Chevrolet vee-eight engine, in a recent model, is supported at the front by two strut or compression-type engine mountings and, at the rear end by two shear-type rubber mountings, under the torque converter housing.

Fig. 124.—One of a Pair of Flexible Mountings for the Front End of the Mercury Vee-eight Engine Unit.

Fig. 124 illustrates one of the front pair of flexible mountings for the Mercury vee-eight engine. The upper flange of this shear-compression unit is bolted to the engine bearer and the lower metal member is bolted to the cross-frame of the chassis. The single rear-end mounting (Fig. 125) is installed under the lower end of the transmission casing, the large moulded rubber unit being secured between the upper bracket—which is bolted to the underside of the transmission casing—and the retainer steel member which is bolted to the chassis frame cross-member.

The Floating Power Method of Engine Mounting.—Introduced by the Chrysler Company, of America, the "floating power" method consists in mounting the engine and gearbox unit on rubber blocks positioned in respect to the centre-of-gravity of the unit in the manner depicted in Fig. 126. There is a relatively large rubber mounting at the front end of A of the engine and another at C near the rear end. In addition there is a pair of rubber compression members B, one pair on either side of the engine axis (only one pair being shown in Fig. 124). These members limit the rotational movements about the axis passing through the

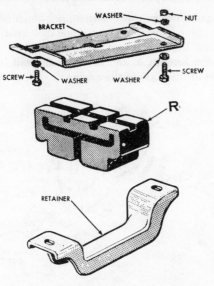

Fig. 125.—Components of the Single Rear Engine Unit Mounting of the Mercury Vee-eight Engine.

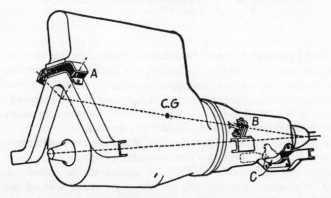

Fig. 126.—The Floating Power Method of Flexible Engine Mounting.

C.G. as indicated by the upper dotted line, which is the normal rocking axis. The supports are located and their elastic properties selected so that the unbalanced engine vibrations, weight and torque are taken account of and the general result is that of an elastic support at the C.G. of the engine.

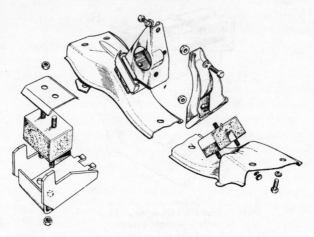

Fig. 127.—(*left*) The Single Front and (*right*) Twin Rear Mountings of the Vauxhall F-Series Four-cylinder Engine Unit.

The Vauxhall Engine Mountings.—Previously, the Vauxhall engines had four flexible mountings, but in the more recent 92 cu. in. F-Series engines the engine is mounted on a single front compression-type rubber unit and at the rear, by two shear-type units in mountings bolted on one side to the transmission casing and on the other to a chassis frame cross-member.

Fig. 127 shows, on the left, the front mounting components and, above, on the right the two rear mounting components; also, the frame cross-member to which these mountings are bolted. It will be observed that the rear left and right brackets which are attached to the transmission casing are of different shape to conform to the shape of the casing.

Notes on Mountings.—The type of rubber used for engine mountings is of a special kind having a marked resistance to ageing (or perishing) whilst being proof against the action of petrol and oil. More recently, synthetic rubbers such as Buna and Neoprene have been employed as these materials offer increased advantages in connection with their resistance to the effects of sunlight, ozone, oil, petrol and grease and can be made to give excellent strength properties.

It is important to note that where an engine is mounted flexibly, *the exhaust pipe* should also be given two or more *flexible supports* on the frame; otherwise it will restrain the engine's movements and may, itself, be damaged. The engine controls and the fuel pipes must also be designed to allow for engine movements. Another point is the possible electrical insulation of the engine by the rubber blocks. As the engine forms the earth return of the ignition system, it must be connected electrically with the chassis frame by a flexible copper or brass strip.

The Exhaust System.—The exhaust gases leave the cylinders of a modern automobile engine, soon after each of the exhaust valves open, at a pressure of about 55 to 80 lbs. per sq. in., and at a temperature of between 600° C. and 800° C., with a high velocity, namely, about 150 feet per second in the case of a high speed engine. At each discharge, i.e., on every exhaust stroke, a compression, or sound wave, is sent out, giving the characteristic exhaust note of the engine. It is necessary in public interests to reduce this noise considerably, and for this purpose the exhaust gases are expanded into a silencing chamber, where the strong sudden discharges are broken up into a more or less continuous one, emitting little noise. The best design of silencer enables the exhaust gases to cool down and expand into the atmosphere continuously and to emerge as a more or less uninterrupted stream. Usually a number of baffles are fixed in the silencing chamber, but careful proportioning of these is necessary in order to avoid back pressure and overheating of the engine. Fig. 128 illustrates some examples of automobile engine silencers that have been used and shows how the gases are broken up before emergence. The internal diameter of the exhaust pipe should not be less than one-quarter to one-third of the cylinder diameter, whilst the ordinary design of silencer should have a capacity of about 14 cubic inches for every B.H.P., at the maximum value. It should not weigh more than about 0·1 lb. per B.H.P. (maximum) developed by the engine.

A well-designed silencer will not reduce the power of the engine appreciably; the results of tests show that very good silencing can be obtained with a loss of power of about 2 to 3 per cent.

An indirect effect of back pressure is to leave more of the hot exhaust gases in the cylinder when the exhaust valve closes. This

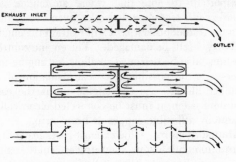

Fig. 128.—Types of Exhaust Silencer.

not only increases the exhaust valve temperature—and therefore enhances the tendency to detonation—but lowers the volumetric (or charge) efficiency by heating the incoming charge that follows.

Multiple Exhaust Pipes.—When a multi-cylinder engine is fitted with a single exhaust manifold with one exhaust pipe outlet, the effect of successive and rapid discharges of gases from the various cylinders is to increase the mean pressures in the manifold and to create a series of pressure pulsations that may lead to power loss. Further, the exhaust valves temperatures are higher, than when separate exhaust pipes are used for each cylinder or pairs of cylinders. It is therefore necessary, when tuning engines for improved outputs, to modify the exhaust systems as indicated. Thus, a four-cylinder vertical engine would have four separate exhaust pipes leading into a common pipe of larger diameter and thence to the silencer.

In many American higher-powered vee-eight engine cars, two separate exhaust systems, as shown in Fig. 129 are employed, while in other instances, notably for the lower-powered eight-cylinder engines, a single system, usually consisting of two exhaust pipes leading to a common pipe which leads to a silencer (muffler) and then through a longer pipe to a resonator at the rear end of the car.

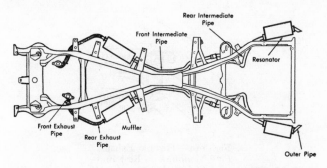

Fig. 129.—Exhaust System of the Cadillac Vee-eight engine, showing the two separate exhaust pipes, silencer (muffler) and resonator units.

Exhaust Pipe Mountings.—In further reference to the method of mounting exhaust pipes, these mountings are usually in the form of suspension members consisting of an angle bracket bolted to the chassis frame and a hanging, or dependent member of a reinforced artificial rubber strip—sometimes known as a "tyre carcase" strip—which is attached to the chassis angle bracket and to a clip around the exhaust pipe, as shown in Fig. 130. The rubber strip is usually in tension, but in some American cars, is in shear. In some instances, a rubber block is used between the upper angle bracket and the chassis frame, to assist in insulating the frame from exhaust pipe vibrations.

Results of Silencer Researches.—The results of researches on petrol engines show that the exhaust noise which originates from the fluctuating pressures at the exit is due to a combination

of notes of various frequencies, i.e., the sound wave is of complicated form made up of fundamental notes and overtones, etc.
For practical purposes the exhaust noise may conveniently be
regarded as consisting of two main parts, namely, a *low-to-medium
frequency band* of about 50 to 600 cycles per second and, dis-
regarding an intermediate frequency band from 600 to about
3,500 cycles per second, of little noise, *a high-frequency band* of
about 3,000 to 10,000 cycles per second.

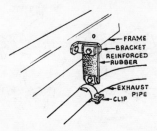

Fig. 130.—Flexible Exhaust
Pipe Mounting on Rear End
of Chassis Frame.

The lower frequency part appears to be due to a resonance
effect between the varying capacity of the cylinder and exhaust
system as the piston moves in the exhaust stroke, and the area
through the exhaust valve as it opens and closes. This conjecture
has actually been confirmed by tests made upon a petrol engine
driven by an electric motor at the same speeds as when running
under its own power, when the low-pitch noises were shown
to be almost as great as when the engine was working normally.

The high-pitch part of the noise is due primarily to the release
of high-pressure gas—at pressures of 55 to 80 lbs. per sq. in.—
through the exhaust ports and silencing system. Alterations of
engine speed whilst affecting the intensity of the total noise do not
materially influence the pitches of the two parts of the noise.

In studying the problem of effectively silencing the exhaust
noise it is necessary to consider the means to deal with each of the
two main frequency bands. Unfortunately any silencing system

designed to suppress or damp down the high-frequency band will not reduce the low-frequency notes appreciably.

Thus it has been ascertained that a silencer of the *absorption type* consisting of a straight-through perforated tube in an outer casing filled with absorbent material, such as glass, silk or bundled fine wires, will damp down and absorb most of the high-frequency notes; in this case the peaks of the high-frequency pressure waves pass out through the perforations into the absorbent material and are thereby reduced in magnitude, but may return after some delay out of phase with other peaks; in effect, these waves are smoothed

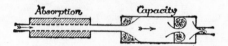

Fig. 131.—Principle of Ideal Silencer.

down so as to give a more or less continuous pressure condition associated with low intensity noise at exit.

This type of silencer is, however, ineffective for dealing with the lower frequency band, for which the best type of silencer is the *capacity type*, i.e., a plain silencer of relatively large capacity and big changes of sections; its action is based upon the absorption of sound waves in the turbulent areas of each change of section.

By combining the two types of silencer it is possible to deal with both the higher and lower frequency bands and thus to obtain the best silencing of the complete exhaust noise. Fig. 131 illustrates, diagrammatically, how this can be effected, whilst the practical interpretation of these principles to composite silencers is shown in Fig. 132.

One type of silencer is designed so that the exhaust gases are divided into two parts, each following a path through the silencer of a different length. It is so arranged that the two paths differ in length by half a wave-length of the sound vibration. Thus one wave will arrive at its maximum positive vibration whilst the other will be at its negative maximum, so that the two cancel out and, theoretically, the vibrations should be silenced out.

It is, of course, only possible to arrange for this cancellation of sound waves at one particular note frequency, viz., that of the normal engine speed; at most other speeds only partial cancellation occurs.

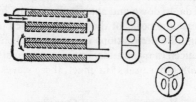

Fig. 132.—Typical Composite Reverse Flow Silencers.

Fig. 133 (*A*) and (*B*) show the Quincke and Herschell silencers respectively. In the former instance there is an enclosed tube of one-quarter the wave-length to be cancelled; as the sound wave in this part is reflected at the end, it has a half-wave effect when

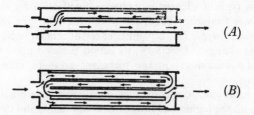

Fig. 133.—The Sound Wave Cancellation Type of Silencer.

meeting the direct wave. The latter diagram shows a divided path in which the upper one is half a wave-length longer than the straight-through lower one.

Resonant-type Silencers.—While based upon similar principles to those previously referred to, namely, to the higher and lower sound frequencies and their modification, the resonant silencer depends upon the use of modified resonators, known as Helmholtz chambers. The larger the resonant chamber the more it will suppress the low frequency noises, while the smaller its size the better the higher frequency noises will be reduced. It is usual to combine both types of resonator in a single compound silencer.   In regard to the position of the resonator, for maximum supression results it should be located at or near to a pressure antinode, i.e. a position in the exhaust pipe where the particular range of frequency notes is at maximum pressure.

The silencer used in many General Motors vee-eight exhaust systems are of oval section and about 30 in. long.   The silencer is

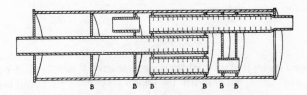

Fig. 134.—Silencer used in some General Motors cars fitted with Vee-eight engines.   B denotes baffles and tube supports.

of all-welded construction and contains three separate tuned resonance chambers which are of the reverse flow kind with diffusers to suppress the periodic pressure vibrations or waves in the exhaust system.   This type of silencer must be designed for its particular engine and exhaust pipe size and diameter; and for reducing the exhaust noise effectively under full-load conditions.

A type of silencer that was used on the racing track, is illustrated in Fig. 135.   In this case it was laid down by the R.A.C. that the exhaust gases must be led into an expansion chamber having a capacity of not less than six times the volume swept by one cylinder of the engine, the diameter D (Fig. 135), if circular, or the equivalent dimensions, if of any other form, being not less than one-fourth of the length, the tail pipe shall have an internal diameter of not more than half the equivalent diameter of the silencer, and shall be so arranged that the gases from the pipe cannot impinge upon the road, and the pipe shall extend beyond the rear axle.

The Burgess Silencer.—This design of absorption silencer represents a scientific attempt to silence the exhaust by storing some of the gas at high pressure and returning it at low pressure in a more or less continuous stream. The silencer consists of a

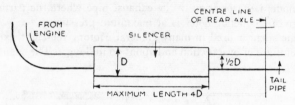

Fig. 135.—Racing Car Silencer.

central perforated tube, which is unobstructed throughout its whole length. It is surrounded by another cylinder of sheet metal, the intervening space being filled with a special sound absorbing material. The theory of this silencer is that during the high

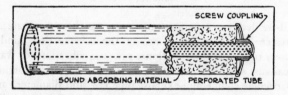

Fig. 136.—The Burgess Absorption Type Silencer.

pressure fluctuations the gases pass through the perforations where they are stored in the sound absorbing material; in effect the high pressures are reduced by the damping action of the latter material. The fluctuations of pressure in the exhaust gases, which are the cause of the exhaust noise are appreciably reduced in the gases flowing out of the central tube, so that the noise must also be reduced in intensity.

Notes on Exhaust Systems.—The results of much experimental work on various types of engines have shown that exhaust noise depends upon a number of factors, including the number of cylinders, or firing intervals, the valve timing, length and diameter of exhaust pipe, the cylinders (total) capacity, position of silencer along exhaust pipe and, of course, the design of silencer. The maximum noise occurs under full throttle operation.

The loss of power due to a good silencer does not depend upon the average pressure in the exhaust system, but upon the pressure fluctuations, so that for a full investigation indicator diagrams should be taken at several positions in the exhaust system.

With a well-designed exhaust pipe-silencer system, it is possible to obtain rather more B.H.P. than with an open exhaust, i.e., short pipe and no silencer.

The single cylinder engine is notoriously the most difficult to silence effectively, but the results of tests have indicated that it is possible to reduce the exhaust noise by special silencers, rather larger in dimensions than certain standard models, and without any appreciable loss in power output.

With certain absorbent material silencers, there may be a risk of partial blockage of the material by soot or even oily deposits, which would reduce the efficiency of the silencer.

Silencers made from sheet steel are liable to cold corrosion, which leads eventually to holes in the casing; this has been known to occur after as little as 15,000 miles of road service.

The results of tests made by L. E. Muller, in the U.S.A., showed the following periods of service at which failure occured:

Plain steel, 14·5 months. Zinc-plated steel, 20 months. Terne (tin) plate, 22 months. Zinc-coated or galvanized steel, 36 months. Aluminium-coated steel, 36 months.

The measurement of noise, by special acoustical apparatus, enables the various noise intensities to be measured on a scale using decibels as units. It is thus possible to assess or compare noise effects, but so far,* no standard scales or values of automobile noises, from the viewpoint of annoyance to the public has been agreed upon.

The Ignition System.—Reference has already been made to the fact that the compressed air-fuel charge is ignited at the correct moment by a high tension spark. It is not possible in

* 1962.

the present limited space to give more than a very brief outline of the principles of the ignition systems used on car engines.

Hitherto, the most popular method of obtaining the high tension spark was by means of the magneto. This device consists of a low-tension current generator, operating upon a similar principle to that of the dynamo. This generator, in addition to its low-tension or *primary circuit* also embodies a *secondary circuit* consisting of a very large number of fine wire windings around the primary.

The primary circuit has a circuit-interrupting device, known as a *contact breaker*. Now, it is a well-known fact that if a current is flowing in the primary circuit of an electrical transformer, and this current is suddenly interrupted by breaking the circuit, it

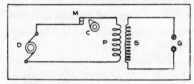

Fig 137.—Principle of Magneto.

will induce a high voltage current in the secondary circuit. Thus, if the latter circuit has several thousands turns of wire and the primary only a few turns, the voltage of the current induced in the secondary circuit will be of the order of thousands of volts if that in the primary be very low, viz., from 4 to 8 volts.

Referring to Fig. 137, which illustrates the principle of the magneto. D is a dynamo or, current generating device, alternating current being taken from it to circulate around a primary circuit P having a primary coil, consisting of a few turns of thick insulated wire, and a circuit make-and-break device M; the latter has two contacts M, one of which is fixed and the other opened by the engine-driven rotating cam C. As the latter rotates, it breaks the circuit once every revolution, a spring returning the contact after the cam has passed its opening position.

The secondary circuit S has several thousand turns of fine insulated wire wound around a central soft iron core, the primary usually being wound over the secondary. The latter's circuit contains a spark gap G, such that when the contacts M are separated the high tension current in the secondary circuit leaps across this gap. This spark gap corresponds to the sparking plug's points.

The magneto, which operates on this principle, uses permanent magnets of horse-shoe form to create the required magnetic field in which the primary coil (armature) rotates. The armature, in the case of a single cylinder four-cycle engine runs at one-half engine speed and, therefore, gives one spark every two revolutions.

The four-cylinder magneto also runs at engine speed, but it has a *high tension distributor*, giving two sparks per revolution of the engine; the six-cylinder magneto has a six-contact distributor with rotor arm running at one-half engine speed, so as to give three sparks per revolution of the engine.

The magneto is not used on modern car engines, but is still employed on special commercial vehicles, tractors, some motor cycles and on stationary petrol power units.

Camshaft Magnetos.—The earlier magnetos, such as the Lucas and Bosch, had fixed permanent magnets and rotating armatures containing the primary and secondary coils as described previously.

More recent models employ stationary coils and rotating magnets of special shape. The contact-breaker is operated by a rotating cam, as in coil-ignition systems. The rotating magnet magneto is less liable to failure, since the coils do not rotate and can be kept free from oil.

The camshaft magneto has a rotating magnet and, it is claimed, possesses the advantages of both the coil and magneto. Examples of camshaft magnetos are the Lucas, Bosch and Scintilla (Vertex).

One advantage of the modern magneto is that it is possible to move both the contact-breaker and armature together for engine timing purposes, and thus to obtain the maximum spark at both the fully advanced and retarded positions; this cannot be done with the rotating armature type magneto.

The camshaft magneto is usually made so that it is exchangeable with the coil-ignition unit, thus employing the same shaft drive.

Coil Ignition.—Instead of employing a separate electrical generator, *viz.*, the permanent magnet and rotating armature device described, it is possible to use the current direct from the lighting battery, so that the primary circuit (Fig. 139) now contains the battery, a switch for the purpose of cutting out the ignition when stopping the engine, the primary circuit and contact-breaker.

The secondary circuit has the secondary coil, a distributor arm which rotates around a number of equally-spaced contacts and the sparking points, or plugs.

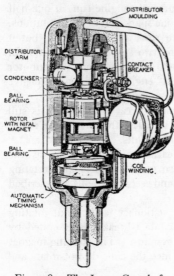

Fig. 138.—The Lucas Camshaft Magneto with Stationary Coils.

It will be noticed that, instead of employing separate insulated circuits, as in Fig. 137, a good deal of wiring has been dispensed with by connecting one side of the battery and primary, one end of the secondary coil and one side of the sparking plugs to a common electrical conductor, known as the "earth" of the system; this is usually the metal frame of the chassis and the engine. The system in question is termed the *Single Pole* ignition one. In modern British systems the positive pole of the battery is earthed. This *positive earthing* method has certain advantages over the previous negative earthing method, including much longer lives for the sparking points, contact breaker and distributor rotor. Most recent American systems have negative earths.

Advantages of Coil Ignition.—Apart from being cheaper than the magneto, coil-ignition gives a practically uniform intensity of spark, whereas the usual design of rotating armature magneto gives a more intense spark at the higher speeds so that, unless suitable precautions are taken the sparking plug points are apt to wear quickly. The coil-ignition system gives a better spark at engine starting speeds than the magneto, but there is a tendency for the spark to fall off in intensity at high engine speeds, so that for racing engines magnetos are usually employed.

One drawback of the coil system is that it depends upon the state of the battery, so that if the car has been standing for a period of several weeks and the battery has run down, it is usually impossible to start the engine—even by hand-cranking.

Some Coil Ignition Details.—With the coil ignition system, the contact breaker and distributor form a single unit mounted on a convenient part of the engine.

Fig. 140 shows the complete components of the Lucas distributor and contact breaker unit, in "exploded" form. The engine drives the vertical shaft (T) at one-half engine speed, usually from a helical gear pair on the camshaft. The shaft drive is taken

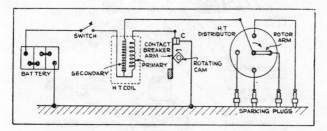

Fig. 139.—Illustrating the Principle of the Coil Ignition System.

through a mechanism known as the *automatic timing control*, which operates on the centrifugal principle to advance the ignition timing, i.e., to make the sparks occur earlier in the engine, as the engine speed increases. Just above this timing control unit is the contact breaker unit, the contact breaker of which is operated by a cam (R) which, for a 4-cylinder engine has four lobes. In Fig. 141 the contact breaker unit is shown separately, with the centre shaft cam and rotor drive shown in the centre of the unit. The two contact points are indicated at (1); the operating four-lobe cam at (2); contact rocking lever pivot at (5) and the condenser, which prevents sparking at the contacts is shown at (4). The two screws (3) secure the fixed contact member and are for adjusting the opening gap of the contacts.

Referring to Fig. 140 the slotted ring on the top of the shaft cam (R) drives the rotor arm (C) which is of bakelite with a projecting metal spark conductor plate. The distributor cover has

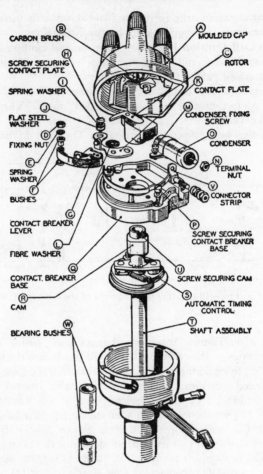

CARBON BRUSH — B

SCREW SECURING CONTACT PLATE — H

SPRING WASHER — I

FLAT STEEL WASHER — J

FIXING NUT — D

SPRING WASHER — E

BUSHES — F

CONTACT BREAKER LEVER — G

FIBRE WASHER — L

CONTACT BREAKER BASE — Q

CAM — R

BEARING BUSHES — W

MOULDED CAP — A

ROTOR — C

CONTACT PLATE — K

CONDENSER FIXING SCREW — M

CONDENSER — O

TERMINAL NUT — N

CONNECTOR STRIP — V

SCREW SECURING CONTACT BREAKER BASE — P

SCREW SECURING CAM — U

AUTOMATIC TIMING CONTROL — S

SHAFT ASSEMBLY — T

Fig. 140.—The Lucas Four-cylinder Engine
Coil-ignition Unit, in Exploded View.

four equally-spaced contacts, to each of which the rotor delivers
high voltage current, in turn; the sparking plug cables are connected to these contacts.   The distributor cover also has a central

high tension cable connector, to convey high tension current from the H.T. coil to the carbon brush (*B*) which bears on the centre of the metal part of the rotor arm.

The contact-breaker formerly used platinum or platinum-iridium contacts, but nowadays tungsten--a cheaper metal—is employed, as it is equally effective in resisting corrosion.

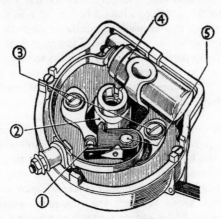

Fig. 141.—The Contact Breaker Unit.

The contact-breaker points are only opened by a very small amount, viz., from $\frac{14}{1000}$ to $\frac{16}{1000}$ inch by the cam; the latter has as many lobes as the engine has cylinders, when it is driven at one-half engine speed, i.e., off the engine's camshaft.

The *high tension coil* is a separate unit mounted in any convenient position under the engine bonnet. It contains only the primary and secondary coil and a laminated iron core; the primary is generally arranged inside the secondary coil. The high tension coil is cylindrical in shape, and the central terminal is the high tension one.

The primary coil usually has a few hundred turns of enamelled copper wire of about 1 to 2 ohms total resistance; the current is from 1 to 2 amperes and voltage from 6 to 10.

The secondary coil has from 10,000 to 15,000 turns of very

fine insulated wire—usually about $\frac{5}{1000}$ inch diameter and enamelled. It gives a spark of several thousands volts but of low current intensity. The layers of wire are also insulated with varnished silk.

In regard to the *H.T. Distributor*, instead of allowing a carbon brush to rub over an insulated disc having brass contacts inserted at equal intervals around the periphery as was previously the case, it is now the practice to have a bronze arm, as shown at (*C*) in Fig. 140, which rotates inside an insulated ring having equally-spaced brass inserts. The outer end of the rotor does not actually touch these brass parts, but passes very close to them, so that the spark leaps across the very small gap as the motor passes each brass insert. This does away with the frictional resistance wipe-contact method. The system described is known as the *Jump Spark* one.

New Ignition Systems.—While the ordinary coil ignition system has proved fully satisfactory for automobile engines operating at the usual maximum speeds of 4,500 to 5,500 r.p.m., it has been found that at higher speeds, the very short time intervals available for the contact-breaker to open and close the primary circuit tends to limit the maximum possible speed. Thus, for a six-cylinder engine running at 5,000 r.p.m. the total period per spark is only $\frac{1}{250}$ sec., i.e. ·004 sec. and the limiting speed about 7,500 r.p.m., corresponding to a period of $\frac{1}{750}$ sec., or ·0026 sec. At the intermediate speeds, between those mentioned the high speed opening of the contact-breaker arm tends to fling it away from its fixed contact. This trouble has to some extent been overcome by lighter arms, stronger control springs and different ignition cam profiles.

However, fouling of the sparking plugs due to the use of premium (leaded) fuels, to contact-breaker points pitting, inaccuracy of ignition timing at the higher speeds, e.g. 7,000 to 12,000 r.p.m., and the necessity to use higher sparking voltages, due to the higher compression and speed factors, has led to the development of entirely new ignition systems*, which, briefly, are summarized, as follows:

* Detailed information is given in Motor Manual, Vol. 6, *Modern Electrical Equipment for Automobiles.*

(1) *Low Voltage, Surface Discharge Plug Systems.*—This system requires an operating voltage of only 3,000 to 4,000 volts, a condenser in the H.T. circuit and a current rectifier (to charge the condenser). Timing the ignition sparks is done by the distributor instead of the usual contact-breaker. It is necessary to use a special kind of sparking plug, known as a *surface discharge plug.* This plug is unaffected by soot, carbon or moisture, on the plug insulator.

(2) *Transistorized Systems.*—Use is here made of the properties of the very small transistorized elements to amplify the voltage and provide a very clean cut-off for circuit make-and-break purposes; further the element will only allow current to flow in one direction, as in ordinary current rectifiers. It enables the ordinary contact-breaker to be dispensed with, and much higher sparking speeds to be attained in practice. In the Lucas system an electromagnetic pick-up is used, together with pole pieces fixed to an engine-driven disc. In the Delco-Remy system a distributor shaft type drive is used with a multi-lobed iron timer which rotates between the poles of a permanent magnet unit.

(3) *The Piezo-Electric Systems.*—These systems make use of the property of special piezo-electric crystals when suddenly compressed to develop currents of relatively high voltages, of the order of 14,000 to 18,000 volts. The timing of the spark is by means of an adjustable cam-operated timing switch. The latest elements are capable of producing voltages up to 30,000 and to have useful lives exceeding 1,000 hours, or about 30,000 miles of road service.

Sparking Plugs.—The high voltage current from the ignition apparatus is conveyed to the inside of the cylinder by means of the sparking plug. This consists of an outer metal body or "shell" which is screwed into the wall of the combustion chamber—usually, although not always, over the cooler inlet valve—and an inner or a central electrode of nickel or a platinum tip, as in the Lodge plugs. This electrode is insulated from the outer shell by means of ceramic or, more recently, aluminium-oxide insulators. The high tension cable is attached to a terminal on top of the central electrode, and the spark occurs at the end between the electrode and the metal shell across a spark

gap measuring 0·016 to 0·020 in. for magneto ignition and
0·020 to 0·025 in. for coil ignition. Many more recent engines,
including American ones, use wide gaps, namely, from 0·030 to
0·038 in. Special high voltage high tension coils are employed
with wide gap plugs.

The sparking "points" of the plug should *be level with or
slightly below the combustion chamber* walls.

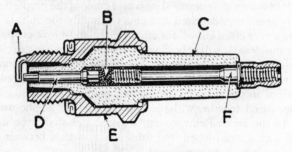

Fig. 142.—Typical Modern Sparking Plug.
A—Earth electrode. *B*—Conducting glass seal.
C—Aluminium-oxide insulator. *D*—Insulated electrode.
E—Metal shell. *F*—Cable connector.

Fig. 142 shows the construction of a modern sparking plug,
having a 14 mm. diameter threaded portion. It has an aluminium-
oxide insulator *C* which is glazed all along, except the end
surrounding the central electrode *D*, where the outer surface is
left unglazed to discourage the formation of lead salts from
premium fuels. The sparking plug cable connector shown to
the right of the central metal electrode *F*, is of the push-on
type. There is a conducting glass seal *B* connecting the two
central electrodes *D* and *F*. The insulator is sealed to the inside
of the metal body *E*, usually by a cement on the left-hand joint and
a special ring-type member on the right-hand side, which is
compressed by spinning over the metal of the body. All modern
plugs are of the single electrode type as shown at *A*. The
sparking electrodes are made of nickel or nickel alloy. The
platinum-electrode plug has a much longer useful life, however.

Hot and Cold Plugs.—For each design of engine there is a certain range of temperatures for the exposed surfaces in the cylinder, at which the plug will operate satisfactorily and without forming carbon deposits. For engines which normally operate hotter than usual, plugs having shorter heat flow paths, as at *B*, in Fig. 143 are used; these are known as *Cold Plugs*. For engines that run cooler than normal ones, *Hot Plugs*, as shown at

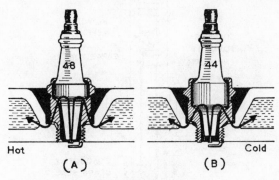

Fig. 143.—Illustrating suitable Sparking Plugs for Cool- and Hot-running engines.

A, in Fig. 143 are used. In these, longer heat flow paths, i.e. longer exposed insulator surfaces are employed.

Low-Voltage Non-fouling Plug.—A new commercial plug, designed to operate with existing coil-ignition systems, having from two to three times the useful life of ordinary plugs and which will not be affected by carbon deposits on the exposed insultaor surface, has an inbuilt condenser (capacitor) with a spark gap formed by end plates (Fig. 144) and two end gaps, namely, a surface and an air gap of 0·020 in. and 0·010 in., respectively. As shown in the end view, there are four earthed electrodes that do not need any adjustment during the life of the plug namely, 25,000 to 30,000 miles.

Timing the Ignition.—If the combustion of the compressed charge were instantaneous, when the ignition spark occurred, then the piston would have to be on the top dead centre of its compression stroke at the moment of sparking. Actually, there is always a time interval, or lag, between the point of sparking

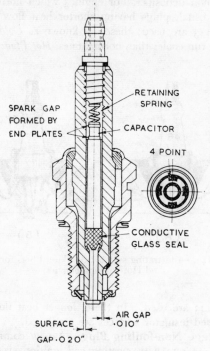

SPARK GAP
FORMED BY
END PLATES

RETAINING
SPRING

CAPACITOR

4 POINT

CONDUCTIVE
GLASS SEAL

AIR GAP
·010"

SURFACE

GAP ·0 2 0"

Fig. 144.—The Lodge Non-fouling low-
voltage Sparking Plug.

and the attainment of maximum pressure in the cylinder, so that the spark must be arranged to occur a short interval before the piston reaches its top centre.

This is termed the *angle of advance*, and its amount varies in different engines according to the engine speed, ratio of bore-to-stroke, compression ratio, nature of fuel used, etc. It is usual to provide for an advance of 25° to 35° for normal car engines and rather more for high speed and performance engines.

Previously, the driver was provided with a hand control for moving the contact-breaker in relation to the rotating cam, or the cam in relation to the rotating contact-breaker according to whether coil or magneto ignition was used. As already referred to, it is now the custom to embody an automatic control

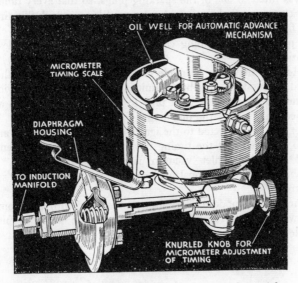

Fig. 145.—Combined Centrifugal and Vacuum Control for Ignition Timing, showing, also, Hand Setting Control.

in the ignition unit. This, for coil ignition, consists of a *centrifugal device* which operates the contact-breaker and advances it as the engine speed increases. At starting, the device is always left in the fully retarded position.

The latter position is usually arranged so that when the piston is on its top dead centre (compression), the contact-breaker points are just opening and the spark is, therefore, occurring.

In some engines the spark is timed so that at low engine speeds it occurs at a few degrees before top dead centre.

The flywheels of practically all engines are marked with lines on their rims which, when opposite a fixed pointer on the crankcase show the top dead centre of No. 1 (front) cylinder piston.

In some cases the flywheel rim has a bright steel ball embedded in it, which comes opposite a fixed pointer to show the top dead centre. For ignition timing purposes a timing light or flashing unit is fitted to one of the sparking plugs, so that every time the spark occurs a flash is given by a special lamp, which illuminates the flywheel and shows the position of the steel ball relatively to the pointer. The timing can then be adjusted to any desired position.

Mention may also be made of the *vacuum timing unit* that is fitted to modern distributor units. This diaphragm device is connected on one side by a pipe to the inlet manifold, so as to subject this side of the diaphragm to the inlet "vacuum". The other side is connected to the ignition advance mechanism and it is so arranged that the timing is varied according to the engine load to which the inlet "vacuum" is inversely proportional. This provides for an additional load control, to the centrifugal speed control.

Finally, the modern distributor unit is fitted with a micrometer screw adjustment for independent setting of the timing to suit the grade of fuel used. (Fig. 145).

CHAPTER III

TYPES OF AUTOMOBILE ENGINE

It is now proposed to consider the various types of automobile engines in use. These range from the small motor-cycle "light-weight" engine of about 1 B.H.P. up to the large high-power car engines of 8 and 12 cylinders and from 150 to 400 B.H.P. output. Automobile engines include, in addition to the poppet valve type described in the preceding chapters, the "valveless" or sleeve-valve types, two-stroke engines, and oil or Diesel engines. Special types of engines, such as the "swash-plate," link-plate, radial and rotary engines, have been suggested and in some cases actually used in automobiles; these have usually been of an experimental nature, however. More recently, the supercharged petrol-type engine, developed primarily for aircraft purposes, has been used successfully on racing motor-cars and some commercial cars.

The ordinary four-stroke poppet-valve engine is by far the more commonly used on automobiles, and, is available in the single, twin, four, six, eight and twelve cylinder designs, for various horse-powers within the stated range.

The Single Cylinder Engine.—This is now mostly employed on motor-cycles, although it has been used on certain small three-wheel cars, e.g., the Bond and Isetta. Experience has shown that for motor vehicles the maximum size of a single cylinder engine should be from 600 to 700 c.c. Above this size, very heavy fly-wheels are necessary, and the vibratory effects due to lack of balance of the reciprocating parts may become marked. The average capacity of the so-called 3½ H.P. engine is about 500 c.c.; a single cylinder engine of 86 mm. bore by 88 mm. stroke gives a capacity of about 517 c.c. The well designed high compression car engine of this size would give from 25 to 30 B.H.P. at 5,000 to 5,500 r.p.m.

An example of a single cylinder four-cycle engine, as used on the

Isetta car, namely, the German B.M.W. engine shown in Fig. 146 is of the overhead inclined valve type, cooled by an air stream directed from an engine crankshaft fan at the front end. The cooling air is directed by metal shrouding around the cylinder and heat fins.

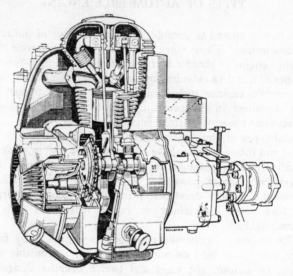

Fig. 146.—The B.M.W. single-cylinder Isetta car engine.

The engine shown is the 300 c.c. model, with a bore of 72 mm. (2·83 in.) and stroke 73 mm. (2·87 in.) giving a cylinder capacity of 298 c.c. (18·61 cu. in.).

With a 7:1 compression ratio this engine developed 13 B.H.P. at 5,200 r.p.m. The Isetta car had a maximum road speed of 53 m.p.h. and fuel consumption (average) of 80 m.p.g.

It is not possible to balance the reciprocating parts of the single cylinder engine very satisfactorily; thus the piston, gudgeon pin and the upper half of the connecting-rod have a reciprocating motion which it is only possible to balance effectively by introducing another reciprocating mass moving oppositely; in the horizontal opposed twin engine this balance arrangement exists. The compromise usually adopted in the case of the single cylinder engine is to supply a counter-weight in the fly-wheels, on the opposite side to the crank-pin, to balance the weights of the crank-pin and the lower part of the connecting-rod Fig. 147 (A). The unbalanced effect in this case is that of a rocking action (or transverse couple) which tends to vibrate the engine sideways in the plane of the flywheel.

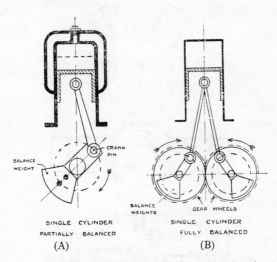

SINGLE CYLINDER
PARTIALLY BALANCED
(A)

SINGLE CYLINDER
FULLY BALANCED
(B)

Fig. 147.—(A.) The Single-Cylinder engine (partly balanced). (B.) Method of fully balancing the engine.

Another alternative is to balance one-half the weight of the reciprocating parts, assumed to be on the crank-pin, in addition to the other parts previously mentioned; this reduces the transverse rocking effect to about one-half of its previous value. In the early Lanchester engine almost perfect balance was obtained as illustrated in Fig. 147 (B).

An idea of the magnitude of the unbalanced reciprocating force will be obtained when it is mentioned that in the case of a single-cylinder engine, with a piston weighing 2 lb., the stroke being

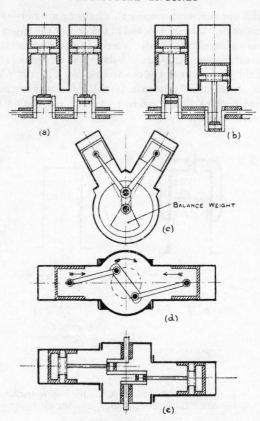

Fig. 148.—Showing the different arrangements
for the Two-cylinder Engine.

4 inches, the greatest out-of-balance force is 907 lb. at 2,000 r.p.m., and occurs at the beginning of the stroke. Its value falls progressively to zero when near the centre of the stroke, and then increases again to a somewhat lower value, viz., 680 lbs. (usually about three-quarters of the former), and opposite in direction.

In addition to this unbalanced effect, the single cylinder engine gives only one power stroke every two revolutions, so that as stated earlier a relatively heavy fly-wheel is necessary in order to store up sufficient energy to carry the engine over its three idle strokes.

The Two-Cylinder Engine.—The cylinders are duplicated in the case of motor-cycle and light-car engines, not only for the purpose of obtaining more power, but also in order to provide a better balance, and more even torque. Twin-cylinder engines for motor-cycles range from about 500 to 1,000 c.c. capacity. The two-cylinder engine for three-wheelers and small cars ranges in size from about 500 up to about 800 c.c.

There are four possible arrangements for the two-cylinder engine as shown in outline in Fig. 148. In (a) both cranks are in line and the pistons move up and down together. The engine fires at 360° intervals, but the reciprocating parts are unbalanced; in effect this arrangement is equivalent to two single-cylinder engines side by side. The arrangement in (b) gives a better (although still

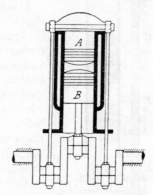

Fig. 149.—The Gobron Brille Engine. The pistons are shown at *A* and *B*.

only a partial) balance to the reciprocating forces, but a sideways, or transverse rocking couple is introduced due to the lines of action of the reciprocating parts not being coincident. This type of engine has unequal firing intervals, namely at 0°, 180°, 720°, 900°, etc.

A method of balancing the reciprocating parts of an engine, originally employed in the early Gobron-Brille motor car engine as shown in Fig. 149 and later, applied in the German Junkers compression-ignition aircraft and motor vehicle engines, is shown diagrammatically in Fig. 150.

Although both of these types have a common cylinder of

approximately twice the length of the normal engine cylinder they may conveniently be regarded as two one-cylinder engines with a common combustion chamber, giving twice the power output of a single cylinder one. The chief point of difference, however, is that the number of power impulses is one-half that of a twin cylinder engine having separate cylinders, unless—as is the case with the Junkers engine the two-cycle principle is employed.

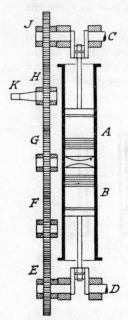

Fig. 150.–The Geared Type Junkers Engine. *A* and *B*—Opposed pistons. *C* and *D*— Upper and lower *H* and *J*—Gear crankshafts. *E*, *F*, *G*, Gear wheels. *K*— Output shaft. The Power.

The arrangement shown in Fig. 149 consists of two pistons *A* and *B* which work in opposite directions, moving together during the compression and exhaust strokes and away during the suction and firing strokes. The combustion chamber is formed in the space between the piston crowns when in their nearest positions so that the valves would be arranged in the sides of the central portion unless use were made of ports in the cylinder walls near their outer ends as in two-cycle practice.

The upper piston *A* has a crosshead arrangement above carrying trunnions on which the upper ends of a pair of connecting-rods rock; the lower ends of these rods work on cranks on either side of the crank for the piston *B*. With this arrangement, and also that shown in Fig. 150 the reciprocating forces due to the pistons could be in almost perfect balance, although there are certain small unbalanced effects due to the fact of the connecting-rods being of different lengths.

The disadvantage of this type of engine lies in its somewhat excessive height and greater weight, but by careful design the latter can be reduced almost to the value of a two-cylinder unit engine of conventional design.

In the Junkers engine (Fig. 150) which, as previously mentioned works on the two-cycle principle the two pistons A and B move symmetrically in a common cylinder of about twice the usual length.   The combustion chamber is formed in the space between the pistons when they are closest together.   Each piston has its own connecting rod and operates its own crankshaft.

The two shafts rotate in the same direction, and they are coupled together by the gears shown at E, F, G, H, J.   Power was taken off from the shaft K.   The two crankshafts are shown at C and D.

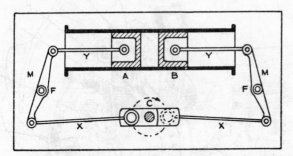

Fig. 151.—Arrangement of the Sulzer Engine.

Another method of using opposed pistons is shown in Fig. 151. This arrangement illustrates the Sulzer two-cycle engine used for motor vehicles.   The opposed pistons A and B are connected by the rods Y to rocking levers M, having fixed fulcrum bearings at F.   The other ends of these levers are coupled by the connecting-rods X to the crank-pins of the crankshaft C; these cranks are 180 degrees apart.   This type of engine is used in certain modern Diesel engines.

Typical Two-cylinder Engines.—Typical side-by-side two cylinder engines of the kind shown at (a), in Fig. 148 include the Fiat 500 c.c. and N.S.U. "Prinz" 600 c.c.   By using two smaller cylinders instead of one larger cylinder the parts can be made smaller and the overall height reduced.   Further, the unbalanced reciprocating forces are appreciably less than for the larger single cylinder engine of equal capacity.   The cylinders, although giving a wider engine, can also be cooled more effectively. Fig. 152 illustrates the Fiat 500 c.c. Model 101 rear-type, two-cylinder air-cooled engine in situ and showing the *cooling arrangements*. The metal shroud, normally fitted around the cylinder unit has been removed to show the finned cylinder barrels and head.

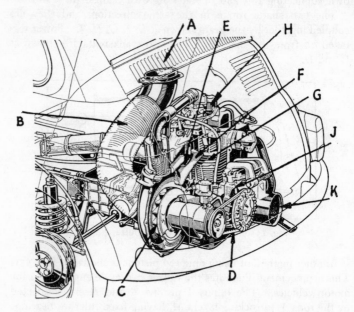

Fig. 152.—The Fiat 500 c.c. Twin-Cylinder engine, as installed in rear end of car, showing also the Air Cooling system.

The cylinders in the previous 500 c.c. engines were arranged, horizontally, on the left-hand side, but in the engine shown, they are vertical. The engine has a crankshaft pulley at its rear end which, by belt *D*, drives a smaller pulley on the dynamo and cooling fan shaft *C*. Air from louvres just below the rear window of the car is drawn down a flexible duct *A* and *B* of large bore into the fan and thence discharged by a smaller horizontal duct and shrouding *E* and *F* past the cylinder barrels *G* and head.

Air for the carburettor intake is also taken through an air filter from the cooling air duct, while at the exit end K of the air stream, hot air leaving the cylinders is taken through a hand-operated valve J to a duct leading to the front of the car for heating and windscreen demisting purposes. A thermostatically-operated air valve in the exit duct controls the cooling air temperature. The Fiat 110 engine has a bore of 66 mm. (2·60 in.) and stroke of 70 mm. (2·75 in.) giving a cylinder capacity of 479 c.c. (29·2 cu. in). It has a compression ratio of 7:1 and has a maximum output of 16·5 B.H.P. at 4,000 r.p.m. The valve timing is, as follows: the inlet valve opens at 9° before T.D.C. and closes at 70° after B.D.C. The exhaust valve opens at 50° before B.D.C. and closes at 19° after T.D.C.

Vee-Type Engine.—This is still popular on high-powered motor-cycles, and has also been used on three-wheeler cars. It is economical to make, since the same type of fly-wheel and common crank-pin as in the case of the single-cylinder engine can be used, and its external shape enables it to be fitted readily into the frame or chassis. The valve gear can also be made compact, and the crankcase is practically that of the single-cylinder engine. It will be observed that the big-ends of both connecting-rods work on the same crank-pin, but in one or two cases one rod has been hinged to the other or "master" rod near to the crankpin, so as to obtain a larger area of bearing surface. The angle between the centre lines of the cylinders varies in the different makes, and ranges from 40° to 90°: the 60° angle is the more common.

In the case of the 40° twin engine the firing intervals reckoned from the crank position as zero at the rear cylinders are 0°, 400°, 720°, 1120°, and so on. For the 90° twin these intervals are 0°, 450°, 720°, 1170°, and so on; these intervals are less regular than in the preceding case. On the other hand the 90° twin engine can be balanced much better; thus the maximum unbalanced force is only about one-third that of the 60° twin, and the amount of vibration only about one-twelfth.

The much better balance of the 90° twin engine is usually accepted as outweighing the drawback of its less regular firing intervals compared with twin engines of smaller cylinder angles.

The vee-twin engine is balanced, or partly so, in a similar manner to the single engine, namely, by means of a common counter-weight as shown in (c) Fig. 148.

The Horizontally-Opposed Engine.—With this arrangement for the cylinders and crankshaft, as shown at (d) and (e), Fig. 148, the reciprocating forces are in perfect balance, since the pistons

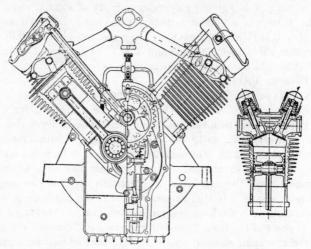

Fig. 153.—Two-Cylinder Vee-Type Engine.
(Right-hand view shows one cylinder in section.)

and connecting-rods move outwards and inwards identically. If it were possible to have the cylinders, shown at (e) in the same line then the complete engine would be in perfect balance—assuming the crankpins and part of the big-ends are dynamically balanced. However, with the cranks at 180°, as shown at (e), Fig. 148, the lines of action of the reciprocating forces will be separated by a distance which we call X inches. There will therefore always be a rocking couple, or torque equal to the value of the reciprocating forces of the two piston assemblies, here denoted by F lb. so that the magnitude of the rocking-couple will be FX lbs. ft. This couple will tend to rock the engine, as shown at (*e*) first clockwise and then anti-clockwise about its centre, as the pistons first move inwards, and then move outwards. As viewed from above in (*e*) the engine will have a variable rocking couple (since *F* varies during each revolution) about a vertical axis. This small out-of-balance effect is eliminated in the case of the opposed-four-cylinder engine described later in this chapter.

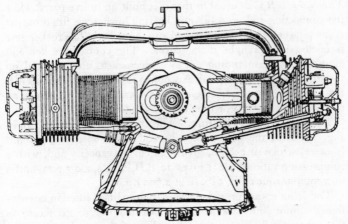

Fig. 154.—The Citroen Two-cylinder opposed engine.

A special feature of the horizontally-opposed engine is that on account of its much better balance than other two-cylinder arrangements it can be run at appreciably higher speeds, and therefore higher power outputs for a given cylinder capacity. It can be designed to give better maintenance accessibility, but it usually requires a wider engine bonnet or hood.

The Citroen Engine.—Typical examples of two-cylinder opposed automobile engines include the 2CV French Citroen and the Dutch D.A.F. engines, shown in Figs. 154 and 155 respectively. The original 2CV engine had a bore of 62 mm. and stroke of 62 mm. with a cylinder capacity of 375 c.c. It developed 9 B.H.P. at 3,500 r.p.m. with a compression ratio of 6·2:1. The later (1962) models are the Bijou, with a bore of 66 mm. and stroke, 62 mm. (425 c.c.), developing 12 B.H.P. at 3,800 r.p.m. with a compression ratio of 7:1 and the larger Ami engine of 74 mm. bore and 70 mm. stroke (602 c.c.), giving 22 B.H.P. at 4,500 r.p.m. with a compression ratio of 7·3:1. The Citroen model illustrated in Fig. 154 has overhead valves operated by push-rods and rockers from a central camshaft below the crankshaft. The engine has an aluminium crankcase, and cast-iron cylinders. The *crankshaft* is unusual in that it is built up in five parts, while the connecting-rods and big-end bearing bushes are fitted on the crank pins before they are pressed into the webs. The rocker gear is enclosed by a light metal cover. The contact-breaker and automatic ignition timing advance mechanism are mounted on the front end of the camshaft. Cooling is obtained by means of a fan which is driven from the end of the crankshaft. There is an oil cooler of the radiator type behind the fan.

The Dutch D.A.F. Engine.—The D.A.F. car engine, shown in Fig. 155 is of the four-cycle type and has a bore of 3 in. and (shorter) stroke of 2·5 in., giving 36 cu. in. capacity and, with a compression ratio of 7·1:1 gives 22 B.H.P. at 4,000 r.p.m. and a maximum torque of 32·5 lb. ft. at 2,600 r.p.m.

The cylinders are provided with cast-iron liners fitted in heavily-finned aluminium jackets. The cast aluminium crankcase is divided vertically across the centre line of the crankshaft. The connecting-rods have lead-bronze big-end bearings. The cam-shaft is located below the crankshaft and is driven from it by a plastic composition gear wheel which provides quiet running. The camshaft has only two cams, each of which operates two horizontally opposed push-rods.

The valves are operated *via* aluminium push-rods and steel-rockers. A special design of valve cotter is employed which imparts a rotating movement to the valve; this rotates evenly on the seat during running and this action minimizes wear of the valve head and seating.

Efficient cooling of the cylinder heads is ensured by mounting the (hot) exhaust valve at the front of the cylinder, thus being directly in the air stream.

The removable aluminium cylinder heads with their deep fins are provided with shrink-fitted valve seats of a special cast-iron alloy. Bronze valve guides are used to ensure good heat dissipation.

All moving parts are pressure-lubricated. The oilways are drilled into the block and thus there is no risk of leakage resulting

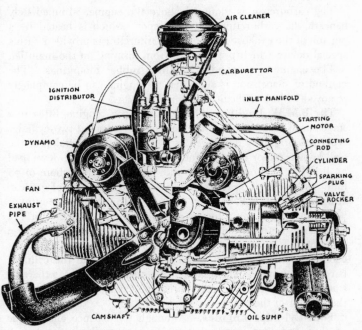

Fig. 155.—The Dutch D.A.F. Two-Cylinder Opposed Engine, as installed at the Front end of the car.

from external damage. The gear-type oil pump is driven from the front of the camshaft.

The distributor driving pinion is mounted on the crankshaft and the distribution driveshaft serves to drive the petrol pump. Effective and reliable cooling is ensured by the fan, which is mounted on the front end of the crankshaft. On this a belt-pulley is also fitted, from which the dynamo—mounted above the engine—is driven.

Above the crankcase is an aluminium case containing the oil filler; this is provided with a large orifice for ease of filling. The case contains a special non-return valve for crankcase ventilation. Fumes passing through the non-return valve are taken to the air filter.

The carburettor is mounted above the engine. Immediately beneath the carburettor is a "hot spot" which is heated by a portion of the exhaust gases. The carburettor is provided with a special device which prevents ice from forming on the main jet.

The engine is suspended on three rubber mountings. The method of suspension is such that the engine can be completely removed without difficulty.

The B.M.W. Engine.—The B.M.W. "700" engine, fitted to a small car of that name is of the horizontally-opposed two-cylinder type, was originally developed as a lower-powered version of the well-known motor-cycle engine which, in racing form developed 35 B.H.P. The more recent "700" car engine has a bore of 78 mm. and stroke of 73 mm., giving a cylinder capacity of 697 c.c. It develops 35 B.H.P. at 5,000 r.p.m., with a compression ratio of 7·5:1 with a maximum torque of 37 lb. ft. and B.M.E.P. of 135 lb. sq. in., both at 3,400 r.p.m. This air-cooled engine has a centrifugal fan mounted at one end of the crankshaft, directing cooling air through sheet metal shrouding, around the cylinder barrels and heads. The quantity of air is controlled by a thermostat and the hot air leaving the engine can be used for heating the interior of the car.

Four-Cylinder Engines.—Although the balance and torque of the opposed two-cylinder engine are excellent, the overall length and the cylinder dimensions limit it to the smaller power units for car use. It is also not so good as regards slow running and low speed torque as the four-cylinder engine.

The overall length, when the engine is set laterally in the chassis, often interferes with the design of the bonnet and the general lines of the car; in the earlier light cars the cylinder heads often projected through the sides of the bonnet, air-scoops for cooling the cylinder heads being arranged in the projections.

The four-cylinder engine is by far the most popular light and medium car engine in present-day use.

The four-cylinder engine arrangement universally adopted is shown in Fig. 156. The two outer and the two inner cranks,

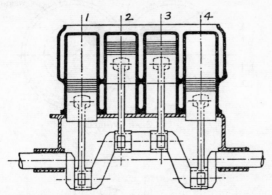

Fig. 156.—The Four-Cylinder Engine, showing a simple Crank Arrangement.

respectively, are parallel, so that each pair of pistons is always in the same relative position. This arrangement enables a firing stroke to be obtained twice every revolution, so that the torque is

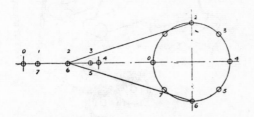

Fig. 157.—Illustrating Unsymmetrical Travel of Piston in Four-Cylinder Engines.

much more uniform than that of the single-cylinder engine. Apart from the more uniform torque, the balance of the four-cylinder engine is also quite good. Although not equal to that of the two-cylinder opposed, it represents an excellent compromise.

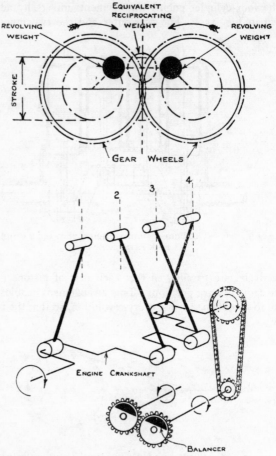

Fig. 158.—Showing how a Reciprocating Force can be Balanced by Two
Equal Revolving Masses rotating in Opposite Directions.

It will be observed that the pairs of pistons always move in opposite
directions so that the reciprocating part forces tend to balance.
They do not quite balance each other, however, for, owing to the
inclination or obliquity of the connecting-rods, the pistons move

rather faster during the first half of the crank-shaft revolution, and slower during the latter. Thus when the crankshaft has moved through 90° from its top centre position, the piston moves through more than one-half of its stroke, as shown in Fig. 157. Hence the two ascending pistons (from the bottom centre) will at first travel slower than the two descending ones, and the forces due to the masses of the reciprocating parts will not balance exactly. If very long connecting-rods could be used these forces would balance; in practice, however, design considerations limit the length of the connecting-rod.

The general result of this want of symmetry in the piston positions is that there is an out-of-balance force which causes vibrations in an up-and-down direction, in the engine shown in Fig. 156, at a frequency equal to twice that of the engine speed. By the employment of light connecting-rods and pistons, the magnitude of this unbalanced force can be reduced to small dimensions.

In the case of certain of the more expensive car engines this *Secondary Force*, as it is termed, is balanced automatically by introducing mechanically another equal and opposite reciprocating force at twice engine speed. In some engines, this is accomplished by driving off the crankshaft, by means of suitable gearing, a pair of weights (as shown in Fig. 158) whose common centre of gravity moves up and down at twice engine speed.

Fig. 159 shows the mechanism that was used in Lanchester car engines to balance the secondary forces.

Ricardo has employed a direct method of balancing the reciprocating forces in the case of a single crank, two-cylinder vertical engine, by operating a pair of balance weights situated in the sump, by means of eccentrics and universally jointed straps. The eccentrics are mounted on the crankshaft, and the balance weight inner members are guided up and down steel rods fixed to the crankcase. The balance weights are cylindrical inside and can rock on separate cylindrical discs, which in turn slide up and down the steel guides. This allows for the sideways tilting movement of the eccentrics. Fig. 160 illustrates the principle and constructional details of this arrangement.

In ordinary unbalanced four-cylinder engines, *the secondary force* is usually about one-eighth to one-twelfth, that is due to one of the piston and connecting-rod reciprocating masses.

It has been mentioned that the more frequent firing intervals

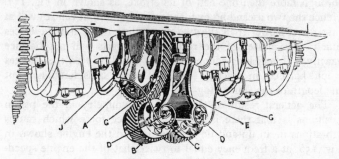

Fig. 159.—The Lanchester Balance Gear.

A—Helical gear on crankshaft. *B* and *C*—Balance wheels moving in opposite directions. *E*—Bracket. *G*—Oil pressure lubrication. *J*—Crankshaft web.

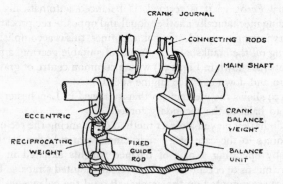

Fig. 160.—The Ricardo Reciprocating Force Balance Gear.

give rise to a better turning effort on the crankshaft; in this respect it is interesting to note that whereas in the case of a certain single-cylinder engine, the greatest torque value (during the firing stroke) was no less than 8·0 times the average value, in the case of a four-cylinder engine of similar capacity the maximum torque value was only 2·0 times the mean. This means that the crankshaft would not be stressed nearly as much, in proportion, and that a much lighter fly-wheel could be fitted.

Opposed-Four Cylinder Engines.—Mention has already been made of this type of engine, with its two pairs of opposed cylinders arranged in horizontal "flat" engine form.

This type of engine may be regarded as equivalent to two opposed cylinder engines of the type shown diagrammatically at (*e*), in Fig. 148 but arranged side by side so that all the reciprocating forces are balanced and also the pairs of oppositely acting rocking couples mentioned earlier. It is thus possible to operate suitably designed engines at relatively high rotational speeds.

The opposed-four engine, while requiring more engine bonnet width, in the larger sizes presents a neat and accessible arrangement, more particularly in the water-cooled versions. When air-cooled, the blower and air ducts tend to complicate the engine in regard to accessibility, valve clearance adjustments, etc. This type of engine has been used over an appreciable period both for the small aircraft engines and also automobile car and racing engines, e.g., the Porsch. An earlier example* of the latter type was the English Butterworth four cylinder engine with a bore of 3·45 in. and stroke of 3·25 in. It had cast-iron barrels with aluminium fins made by the Al-Fin casting method, a cast aluminium crankcase, nitralloy-treated crankshaft mounted on three plain bearings, separate camshafts for each bank of two cylinders, overhead exhaust valves and side-type inlet valves of a special "swing" type such that when fully-opened a larger inlet charge flow area was obtained. Two Amal down-draught carburettors were used.

The engine developed 180 B.H.P. at 6,500 r.p.m. and weighed (dry) 180 lb., i.e. one lb. per B.H.P. The valve timing was as follows: inlet valve opened at 48° before T.D.C. and closed at 68° after B.D.C.; exhaust valve opened at 70° before B.D.C. and closed at 46° after T.D.C.

The Volkswagen Engine.—This is a good example of an air-cooled opposed "flat-four" engine, designed to operate the gearbox and final drive of the transmission, from the rear end of the chassis, thus saving the usual propeller shaft with its universal couplings. The original engine (Fig. 161) has a bore and stroke

* 1951.

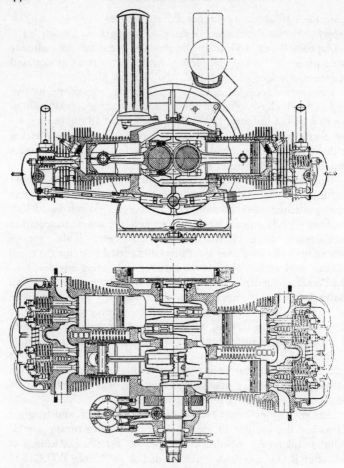

Fig. 161.—The Volkswagen Four-Cylinder Opposed Air-Cooled
Engine, (above) in Front-sectional and (below) Plan-sectional
Views.

of 75 mm. and 64 mm., respectively, being of the shorter stroke
type; the cylinder capacity is 1,131 c.c. The engine has a compression ratio of 5·8 to 1 and develops about 24·5 B.H.P. at

3,300 r.p.m.   The maximum torque value is 49 lbs. ft. at 2,000 r.p.m.

The current Volkswagen engine has a bore and stroke of 77 mm. and 64 mm., respectively, giving a cylinder capacity of 1,192 c.c. It develops 40 B.H.P. at 3,900 r.p.m., with a compression ratio of 7·0:1.   The maximum torque is 65 lb. ft. and maximum B.M.E.P., 101 lb. sq. in., both at 2,400 r.p.m.   There is also a recent 1,500 c.c. class engine which has a compression ratio of 7·8:1 and develops 53 B.H.P. at 4,000 r.p.m.

The engine is of the push-rod and rocker arm operated, overhead-valve type and the push-rods are arranged beneath the cylinders and actuated by a single camshaft.   The engine has a four-throw crankshaft, with three main bearings.   The cylinders are of an alloy cast-iron, being interchangeable; they are fitted with silicon-aluminium alloy heads, cast in pairs. Bronze valve seating inserts, phosphor-bronze valve guides and steel sparking plug insert rings are used with this type of head.

The four-cam camshaft is driven by a magnesium alloy helical gear riveted to the camshaft's steel flange.

Special attention has been given to the cooling of this rear-mounted engine (Fig. 162).   A centrifugal fan, running at 1·75 times engine speed provides the cooling draught.   The fan is enclosed in a semi-circular cowling extending down over the cylinder heads. The interior of the cowling is provided with baffles to direct the flow of air over

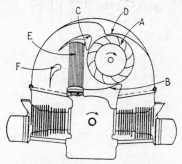

Fig. 162.—Cooling arrangement of Volkswagen air-cooled engine. *A*—centrifugal fan. *B*—oil cooler baffle. *C* and *D*—baffles for L.H. cylinders cooling. (*D* also deflects air to R.H. cylinders.) *E*—oil cooler. *F*—baffle for oil cooler exit air distribution to to L.H. cylinders.

the cooling fins and also through the oil cooler, which is arranged inside the cowling.   Auxiliary blades are fitted on the fan rotor to supply cooling air to the electric generator.   The fan is driven by belt off the crankshaft pulley. The engine is fitted with a Solex down-draught carburettor, having an oil bath type air cleaner.

Square Four Engines.—Instead of arranging the four cylinders in a line as in ordinary car engine practice, they can be located in square formation as seen in plan view. This arrangement is equivalent to a pair of twin cylinder engines connected together.

There are two interesting examples of such engines that have been employed in motor-cycle practice, viz., the Ariel and the Matchless Silver Hawk.

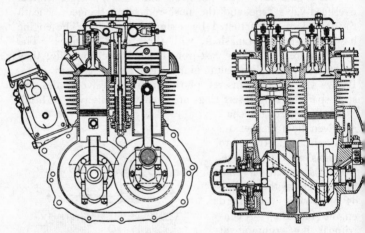

Fig. 163.—Ariel Square Four Engine in side section.

Fig. 164.—The same engine in front section.

The former type is shown in Fig. 163. It will be noticed that it is virtually equivalent to two vertical twin engines, each set having its own crankshaft, the two being geared together. The engine shown has a bore of 2·56 in. and stroke of 2·96 in. giving a cubic capacity of 61 cu. in. (997 c.c.). The four cylinders are in a single casting secured to the top of a vertically split crankcase.

Each two-throw crankshaft is supported in a bronze back babbitt-lined bearing at the timing end and in a large-diameter roller bearing at the driving end, where the coupling gears are housed in a separate compartment. The rear shaft, from which the primary chain drive to the gearbox is taken, is furnished with an additional roller bearing outside the gear wheel. There is one flywheel on each of the crankshafts, the two flywheels being bolted to opposite sides of centrally located flanges on the crank-shafts, thus permitting their rims to overlap. Crankpins and journals are hardened, and connecting rods are made of the R.R. aluminium alloy.

The single inlet manifold is cast integral with the cylinder head casting. Detachable finned exhaust manifolds, each serving two cylinders, are bolted to opposite sides of the head casting. Duralumin push-rods are used. To prevent interference with the free flow of air over the cylinder fins, the combined generator and ignition magneto unit is mounted at an angle.

The engine was capable of running in excess of 6,000 r.p.m.

The other example of four-cylinder engine, the principle of which is shown in Fig. 166, consists of two sets of Vee-twin engines arranged side-by-side with a common two-throw crank-shaft.

Four Cylinder Firing Order.—Numbering the cylinders from left to right in the order 1, 2, 3 and 4, the order in which the spark occurs in the cylinders is 1, 3, 4, 2. There is, however, an alternative order which is occasionally used, namely, 1, 2, 4, 3. If there is any doubt in the case of a particular engine, the firing order can readily be ascertained by turning the crankshaft by hand, slowly, in the direction of normal rotation, and writing down in turn the numbers of the cylinders as their pistons reach the tops of *their compression* strokes. It is easy to distinguish the compression from the exhaust stroke, for both inlet and exhaust valves are closed during the former, and the inlet valve only during the latter—the exhaust valve being lifted. This method applies to any type of engine.

Development Trends.—More recent tendencies in regard to production car engines has been the substitution of the light six-cylinder engines in the 2,000 c.c. class, by four-cylinder engines.

The principle reasons for the adoption of the four-cylinder

engine have been cheapness of manufacture in comparison with the six-cylinder type and the possibility of obtaining smooth performance and ample power for the size of car to which it is fitted. With improvement in engine balancing means and employment of rubber mountings for the engine and gear box unit, the four-cylinder engine is fully satisfactory for its purpose.

By careful design the torque curves of these engines can be made "flat-topped" over an appreciable speed range, to give better slow-running on top gear, but it must be admitted that a six-cylinder engine of equal capacity will run more smoothly at the lower speeds and will show a better acceleration.

A further advantage of the four-cylinder engine is its easier maintenance due to the smaller number of cylinders and their components; this also applies to the ignition system.

Modern Example of Four Cylinder In-line Engine.—This four-cylinder engine is used on the majority of production cars in the 850 to 1,700 c.c. class and, over the past few years its performance has been improved considerably, with the result that the power output, expressed in terms of B.H.P. per litre (1,000 c.c.) of engine capacity has by the use of higher compression ratios and engine speeds, and by much improved engine design and materials increased over the past 12 years from about 30 to 35 to 40 to 45.

Fig. 165 illustrates the Ford Cortina engine in part-sectional view, showing its principal components. It has a bore of 80·96 mm. (3·19 in.) and stroke, 58·17 mm. (2·290 in.), thus giving a stroke-box ratio of 1·4. The cylinder capacity is 1,198 c.c. (73·1 cu. in.). With its compression ratio of 8·7:1, this engine develops 48·5 B.H.P. at 4,800 r.p.m., with a maximum B.M.E.P. of 131 lb. sq. in. and maximum torque of 63·0 lb. ft., both occurring at an engine speed of 2,700 r.p.m. The power output is equivalent to about 41·3 H.P. per litre. As with other modern engines this one has overhead valves operated by push-rods and rocker arms. The crankshaft has three main bearings and both the main and big-end connecting rod bearings are coated with copper-lead alloy. The combustion chambers are all fully-machined to ensure accurate shapes and compressions. The engine weighs 220 lb., complete with all its accessories, giving a weight per H.P. of 4·6 lb. It is of interest to note that the two-door unladen Cortina car weight, namely, 1,716 lb. gives a weight-to-power ratio of 35·5 lb. per H.P.

A more recent innovation in connection with four-cylinder vertical engines is the employment of *five-bearing crankshafts*, which not only reduce the bending stresses but provide a smoother-

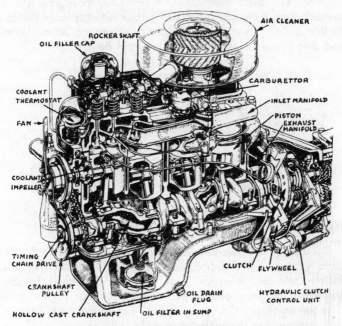

Fig. 165.—The Ford Cortina Four-cylinder Vertical Engine.

running engine. The Ford 1½ litre Classic engine and some Continental smaller engines follow this pattern.

Four Cylinder Vee-Type Engines.—An alternative arrangement for the four cylinders of an engine is that of two sets of vee-cylinders side-by-side, usually with the axes of the cylinders inclined at a relatively small angle to one another. In order to obtain the conventional arrangement of cranks as in the case of the four cylinder vertical engine it is usual to stagger the opposite cylinders by an amount equal to about one-half of the cylinder diameter (Fig. 165). The angle between the cylinder axes varies appreciably in the different commercial engines, but is usually between 14° and 20°. This results in a cylinder block which is not very much wider than that of the vertical type of four-cylinder engine. The layout provides for a very compact engine, which is much shorter fore-and-aft in comparison with the four-cylinder in line engine. A stiffer crankshaft can therefore be employed and side-by-side connecting rods arranged for.

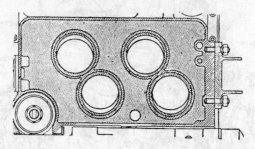

Fig. 166.—Showing Arrangement of Cylinders in the Lancia Four-Cylinder Vee Type Engine.

Balance of the Vee-four Engine.—If the vee-four engine be regarded as a vertical four engine, but with its pairs of cylinders inclined to a central vertical plane it will be understood that the balance of the reciprocating parts of the vee-four engine must depart from that of the vertical four. The difference in balance will depend upon the angle between the cylinder axes.

If the angle is appreciable, e.g. over 40°, the balance of the primary forces becomes more difficult to attain, by the usual crankshaft counterweights method, so that a degree of unbalance will occur, which may cause vibrational effects.

It can be shown that for a typical 60° vee-angle engine, while the direct primary equivalent weight can be counterbalanced, the reverse rotating weight can only be balanced by the provision of another shaft, parallel to the crankshaft, provided with a counterweight; the shaft must rotate in a reverse direction to that of the crankshaft.

This unbalance correction therefore increases the complication and cost of the vee-four engine, if the vee angle is appreciable. In this connection the use of the larger vee-angles, gives a much wider engine than the vee-four.

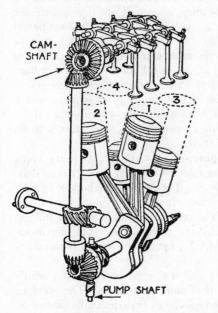

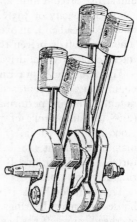

Fig. 167.—Illustrating the arrangement of the Matchless Vee-four Motor-Cycle Engine. The cylinders are shown at 1, 2, 3 and 4.

Fig. 168.—Showing Crankshaft and Pistons of Matchless Four-Cylinder Engine.

Typical Vee-four Engines.—An earlier application of the vee-four principle was that used on one of the Matchless motor-cycle engines, described on the next page.

The Italian Lancia Company has produced a number of different vee-four engined cars, dating from the earlier Lancia Lambda (1921), through the Augusta (1931), Aprilia (1937), Ardea (1939), Appia (1957 on) and Flavia (1961).

The more recent European-manufactured Ford Taunus 12M (Cardinal) vee-four engine and the Russian 746 c.c. vee-four rear-engine, known as the Zaporogets, are other examples.

A motor-cycle example of this type is the Matchless design, illustrated in Figs' 167 and 168, in which the cylinder axes are inclined at 18°. A well-finned air-cooled monobloc cylinder casting is employed. The actual firing angle is 26° as each piston has an inclination of 4° to its connecting rod. Each pair of connecting-rods works on the same crankpin, the two crankpins being arranged at 180° apart. The cylinders each have a bore and stroke of 50·8 mm. and 73·02 mm., respectively, corresponding to a capacity of 593 c.c. The cylinder block is fitted with a one-piece head and, an overhead camshaft for operating the valves. The camshaft is driven through two pairs of level gears with an intermediate vertical drive shaft.

The Lancia Appia Engine.—The more recent (1962) Appia engine (Fig. 169) represents a development of the earlier models, notably in engine performance. The engine has a bore of 68 mm. (2·68 in.) and stroke of 75 mm. (2·95 in.) giving a capacity of 1,090 c.c. (66·5 cu. in.). The compression ratio is 7·8:1 and the maximum output, 48 B.H.P. at 5,000 r.p.m. The maximum torque is 63 lb. ft. and B.M.E.P., 143 lb. per sq. in., both occurring at 3,000 r.p.m.

The cylinder block is of cast-iron and the crankcase, which houses the main bearings, is of aluminium alloy. The previous overhead camshaft has been replaced by twin camshafts located on either side of the crankshaft. The latter has now only two instead of the previously used three main bearings. This later engine also dispenses with the previous wet cylinder lines, thus giving a more rigid cylinder block. The vertical-four type of four-throw crankshaft is used but the front pair of crankpins is offset to one side of their webs; the rear pair are offset to the opposite side. This method is adopted on account of the narrow vee-angle of the cylinder axes, namely 10° 14 min. The pistons have slightly domed crowns and are each fitted with three compression and two oil-control rings; one of the latter rings is below the gudgeon pin. The top compression ring is chromium plated.

The symmetrically-located camshafts are driven from the

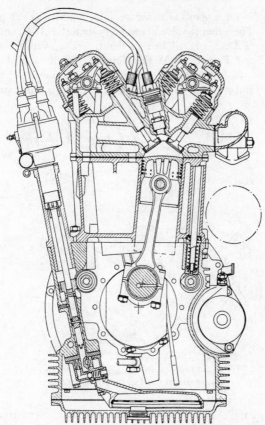

Fig. 169.—The Lancia Appia Vee-four Engine in
Side Sectional View.*

crankshaft by short roller chains, in the manner depicted in Fig.
170.* It will be observed that there is a chain tensioner sprocket
located above the crankshaft sprocket. The inlet valve, of heat-
resistant steel is of 30 mm. (1·18 in.) diameter and the exhaust
valve, of nickel-chrome steel, 27·5 mm. (1·08 in.). The seating
angle is 45° and the valve maximum lift is 7·5 mm. (0·29 in.).

* Courtesy of *The Automobile Engineer*.

The inlet valve opens at 2° before T.D.C. and closes at 40° after B.D.C. The exhaust valve opens at 37° before B.D.C. and closes at 2° after T.D.C. It will be seen that the valve overlap is only 4°.

The Ford Taunus 12M Car Engine.—Built at Cologne, in 1962, this vee-four engine has a 60° vee-angle, a bore of 80 mm. (3·15 in.) and stroke of 58·86 mm. (2·32 in.) so that it is appreciably

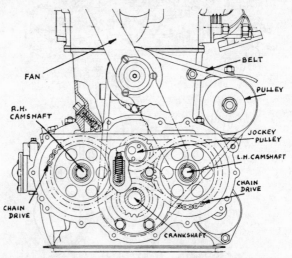

Fig· 170.—Showing the Chain Drive to the Two Camshafts of the Lancia Appia Engine.

"Over-square". The cylinder capacity is 1,183 c.c. (72·16 cu. in.). The compression ratio is 7·8:1 and the engine develops 50 B.H.P. at 5,000 r.p.m. with a maximum torque of 61·5 lb. ft. and B.M.E.P. of 128 lb. per sq. in., both occurring at 2,700 r.p.m.

This engine cannot be balanced properly by means of crank-shaft counterweights in regard to the principal unbalanced primary forces, since, at best, there is an unbalanced reverse rotating force. To overcome this difficulty the manufacturers have fitted a balancing shaft, at a slightly higher level than the crankshaft and near to it, carrying a suitable balance weight; the shaft rotates at crankshaft speed but in a reverse direction. The driving gear on the balance shaft is made of phenolic resin, to reduce gear noise. The engine has a single camshaft mounted in a high position between the vee-angle of the cylinder axes. The pairs of pistons are arranged to reach their top dead centre marks, simultaneously, thus giving equal firing intervals. A special feature of this engine is the relatively wide water jacket spaces around the cylinder barrels and head.

The Six-Cylinder Engine.—A better balance and a more uniform torque can be obtained by using six instead of four cylinders. This latter arrangement is now employed upon the higher powered, and also upon the more, recent cars of moderate powers. A much smoother running, more flexible and quieter engine is thus obtained, but at the expense of extra cost and complication. The six-cylinder engine car will run more slowly on top gear than the four, and will accelerate more rapidly when the throttle is opened; the engine vibration will be less. As regards torque, the ratio of the maximum to the mean torque in the case of the six-cylinder engine in the example considered in Fig. 112 is only 1·4, as against 2·0 for the four-cylinder, and 8·0 for the single-cylinder engine. The six-cylinder engine therefore requires only a light fly-wheel.

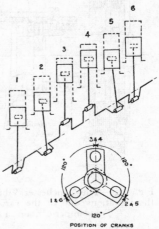

Fig. 171.—The Six-Cylinder Engine Crankshaft Arrangement.

The arrangement of the cranks in the six-cylinder engine is shown in Fig. 171. It will be observed that the two centre cranks are parallel, the second and fifth are also parallel and at 120° to the centre ones, and the two outer ones parallel and at 240° (reckoned in the same direction) to the middle pair. Every 120° (or one-third of a revolution) a pair of pistons will be at the top of their stroke, one on the compression, and the other the exhaust stroke. There are therefore three firing strokes per one revolution of the crankshaft. The usual firing order (Fig. 171) of the six-cylinder engine is: 1, 4, 2, 6, 3, 5. Another firing order, used on certain engines, is: 1, 5, 3, 6, 2, 4.

The balance of the six-cylinder engine is almost perfect, there

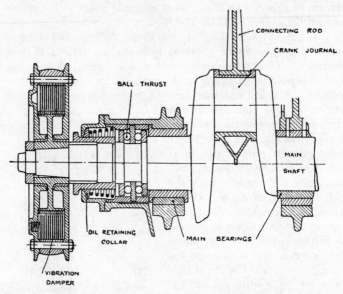

Fig. 172.—The Lanchester Vibration Damper, employed to damp out the vibrations due to the variable engine torque on the crankshaft. The damper contains a series of plates immersed in oil.

being no "secondary" forces unbalanced; there is only a very small unbalanced force, known as the *Sixth Harmonic*, which causes a slight vibration at six times engine-speed frequency.

The six-cylinder engine is now used for engines of 2 litres capacity and above, although more recently a six-cylinder engine of only 1·6 litre capacity, viz. the Triumph Vitesse, has been in production. On American cars the smallest engines are of the six-cylinder type, and of 2·4 to 3·0 litres. The Jaguar car engines, hitherto have all been of six-cylinders in capacities ranging from 2·48 to 3·78 litres with B.H.P.s from 120 at 5,750 r.p.m. to 265 at 5,500 r.p.m., respectively.

For small and medium powers, it is the practice to cast the cylinders in monobloc; this lessens the manufacturing costs, and enables a clean and compact design to be realized. For larger powers (e.g., 4 litres and above) it is usual to employ two sets of cylinder blocks of three cylinders each, otherwise the monobloc system would be too heavy and cumbersome.

Alternatively, the monobloc cylinder can be fitted with two cylinder heads, each head covering three cylinders; this method facilitates maintenance and overhaul of these engines.

Typical Six-Cylinder Engines.—With the exception of the six-throw crankshaft, twelve-cam camshaft and different inlet and exhaust manifold, the six-cylinder type ignition apparatus and, in general, a rather longer engine arrangement there is no marked difference between the general design of the six- and four-cylinder engines. In the less expensive engines the crankshaft has only four main bearings, so that there are two crankpins between each main bearing; otherwise there is a main bearing on either side of each crankpin, making seven main bearings, in all.

An example of good six-cylinder engine design is that of one of the B.M.C. Morris engines, shown in longitudinal sectional view, in Fig. 173. This engine is of the overhead push-rod and rocker arm operated valves type. It has a bore of 79·375 mm. (3·125 in.) and stroke of 88·9 mm. (3·5 in.), giving a cylinder capacity of 2,639 c.c. (161 cu. in.). The engine develops about 90 B.H.P. at 4,250 r.p.m. The inlet and exhaust valve diameters are 1·693 in. and 1·420 in., respectively and the inlet and exhaust valve seating angles are 30° and 45°. The valves are operated, initially, from a long camshaft, located well above the crankshaft, which runs in four replaceable white metal bearings in steel linings. The camshaft is driven by roller chain from the camshaft and has an automatic chain tension adjuster. The valve cams actuate relatively large hollow tappets which have conical cavities above for the ball-ends of the push-rods. T-slot aluminium alloy pistons with anodized surfaces each carry three compression and one oil control piston rings, all these being located above the gudgeon pins. The latter pins are offset in their pistons

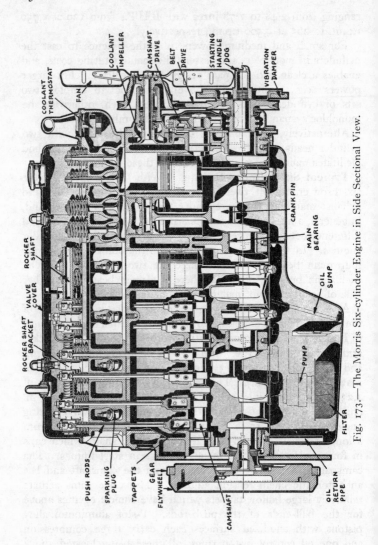

Fig. 173.—The Morris Six-cylinder Engine in Side Sectional View.

towards the thrust side and are clamped by pinch-bolts to the small ends of the connecting-rods. The crankshaft runs in four steel-backed white-metal half-shell bearings; the end thrust of the shaft is taken by split thrust plates on the second main bearing from the radiator end. The oil circulation pump and also the ignition unit distributor shaft are driven by helical gearing from the camshaft. The water-circulating pump, which is integral, on a single shaft with the cooling fan, is belt-driven from the crank-shaft by pulleys.

The valve timing of this engine is as follows:

The inlet valve opens at 5° before T.D.C. and closes at 45° after B.D.C.

The exhaust valve opens at 40° before B.D.C. and closes at 10° after T.D.C.

The ignition firing order from the front, or radiator end is 1–5–3–6–2–4.

An example of a somewhat larger six-cylinder engine is that of the B.M.C. model, shown in Fig. 174, in part-sectional view. This engine has a bore of 87·3 mm. (3·4375 in.) and stroke of 111·1 mm. (4·375 in.) giving a capacity of 3,993 c.c. (244 cu. in.). It develops a power of well over 100 H.P.

It is of the overhead push-rod and rocker-arm operated type, similar in general lines to the A.40 valve gear, with oil-cushioned tappet ends. The crankshaft has four replaceable "Thinwall" main bearings and is fitted with a torsional vibration damper of the rubber-metal pattern.

The camshaft is driven by a double roller chain, with double chain sprockets on the crankshaft and camshaft, respectively.

A single carburettor, fed with petrol from an A.C. mechanical fuel pump is used to provide the mixture. It is fitted with an air intake filter having oil-wetted gauze and air silencer. The inlet manifold has a "hot spot", exhaust heated part to facilitate cold starting. Alternatively to the single carburettor, triple S.U. synchronized carburettors can be fitted, where additional power output is required.

The engine has twin-line rubber block mountings at the front and a semi-circular rubber mounting cradle at the rear of the

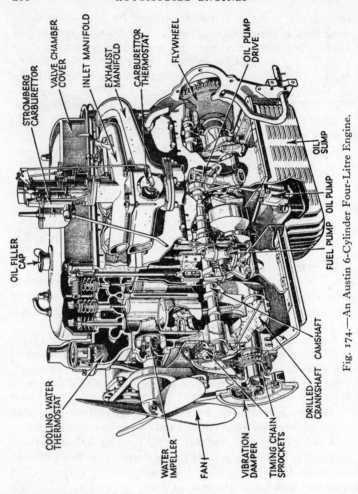

Fig. 174.—An Austin 6-Cylinder Four-Litre Engine.

attached gearbox unit, thus giving a three point power unit mounting.

The Armstrong Sapphire six engine illustrated in side sectional view in Fig. 175, is of the high camshaft overhead valve design, with the valve axes inclined to one another and to the cylinder

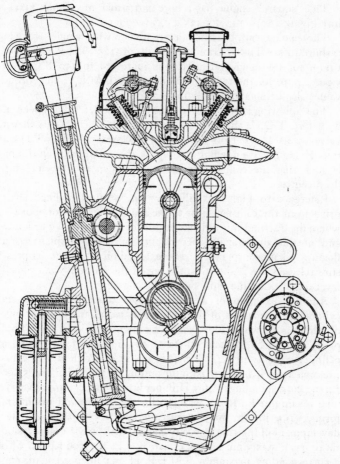

Fig. 175.—The Armstrong Sapphire 6-Cylinder 3·435 Litre Engine.

axis. A single camshaft is used to operate these valves by differently inclined push-rods and rocker arms. The sparking plug is located above and centrally in a deep cavity within the over-head valve cover unit.

The "square" engine has a bore and stroke of 90 mm. (3·543 in.), giving a capacity of 3,435·3 c.c. (209·6 cu. in.).

The engine can be fitted alternatively with single or twin carburettors. The maximum output of 125 B.H.P. at 4,750 r.p.m. for the single carburettor is increased to 150 B.H.P. at 5,000 r.p.m. for the twin carburettors, the B.M.E.P. and torque values also being appreciably higher.

The following is the valve timing: The inlet valve opens at 8° before T.D.C. and closes at 62° after B.D.C. The exhaust valve opens at 46° before B.D.C. and closes 18° after T.D.C., thus giving a valve opening overlap of 26°. The ignition timing is such that the contact breaker points open at 5° before top dead centre.

Reference to Fig. 175 will show the vertical inclined shaft drive from the camshaft gear to the ignition distributor above—where the distributor is fully accessible for inspection. The lower end of this drive shaft operates the oil pump—which has a floating oil intake member. The external oil filter with its pressure release valve is shown to the left of the oil pump and the oil level dip stick on the right side of the engine.

Another example of a well-designed and widely-used six cylinder engine is that of the *Jaguar XK model*, the cylinder unit of which is shown in Fig. 176. The engine depicted has a bore of 83 mm. (3·27 in.) and stroke of 106 mm (4·17 in.) giving a cylinder capacity of 3,442 c.c. (210 cu. in.). With an 8:1 compression ratio it had a maximum output of 160 B.H.P. at 5,200 r.p.m., corresponding to 46·5 H.P. per litre (0·762 cu. in.) capacity. The maximum B.M.E.P. was 140 lb. per sq. in. and maximum torque, 195 lb. ft., both occurring at 2,500 r.p.m. A later development of the same capacity engine, with an 8:1 compression ratio, has a maximum output of 210 B.H.P. at 5,500 r.p.m., with a corresponding maximum B.M.E.P. of 155 lb. per sq. in. and torque of 215 lb. ft. both at 3,000 r.p.m.

Referring to Fig. 176, the inlet valve is on the left and exhaust valve on the right. These valves are set at an included angle of 70° and, as stated earlier, are operated by two parallel camshafts. The inlet valve is of $1\frac{3}{8}$ in. diameter and the smaller exhaust valve, $1\frac{1}{4}$ in. diameter.

The valve timing is as follows:

The inlet valve opens at 15° before T.D.C. and closes at 57° after B.D.C.

The exhaust valve opens at 57° before B.D.C. and closes at 15° after T.D.C., thus giving a valve opening overlap of 30°.

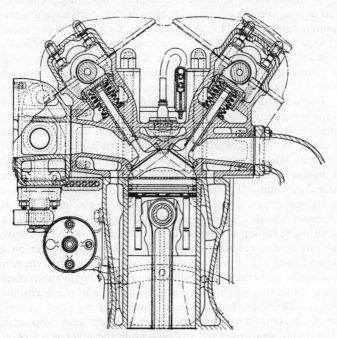

Fig. 176.—Cylinder and Valves of the Jaguar XK Six-cylinder Engine.

The Opposed Six Cylinder Engine.—This type of horizontal engine has a set of three cylinders on each side of the crankshaft. Each opposite pair has its own pair of crankpins, as shown at (*e*), in Fig. 148, with their cranks at 180°. In a recent automobile version of this type, namely the American Chevrolet Corvair engine, the crankpins are arranged in pairs, as described, but in a plane at 60° to that of the next pair. This arrangement provides good engine balance, and is claimed to obviate the necessity of counterweights.

The Corvair engine is air-cooled and mounted at the rear end of the car, with the flywheel at the front end, operating clockwise when viewed from the flywheel end. The engine has a bore of 3·375 in. and a shorter stroke of 2·60 in., giving a displacement of 140 cu. in. With a compression ratio of 8:1, it develops a maximum output of 80 B.H.P. at 4,400 r.p.m. for a mean piston speed of 1,900 ft. per min. The maximum engine torque is 125 lb. ft. at 2,400 r.p.m. and the maximum B.M.E.P., 136 lb. sq. in. at the same speed. A later supercharged Corvair engine had an output of 150 B.H.P. at 4,400 r.p.m.

The original engines* had cast-iron cylinders, each bank of three being secured in position by a cast aluminium alloy cylinder head, provided with cooling fins. The head and cylinder are secured to the crankcase by four long studs and nuts. The overhead-push rods and special pivoted rocker arms, are operated from a camshaft arranged below the crankshaft, driven by a 1:2 helical gear unit. The inlet valve seating is of nickel steel, and that of the exhaust, chrome steel. A special air-cooling system is necessary in this type of rear-mounted engine. It comprises a sheet metal casing, or shroud, over the whole assembly. Above the engine and centrally mounted is a centrifugal blower on a vertical shaft. The 11 in. diameter blower is driven from the rear end of the crankshaft, which has the drive pulley in the vertical plain, while the driven pulley of the blower is in the horizontal plane. Therefore, the belt-drive is taken of a pair of pulleys, arranged above the driving pulley, with their axes horizontal. One pulley also drives the dynamo while the other is a free or jockey pulley and has an adjustable arm mounting for altering the belt tension.

The air from the blower is directed by the sheet metal casing over the finned cylinder barrels and heads and passes rearwards and outwards below the car. There is an oil cooler which is cooled by the blower air. The rate of engine cooling is controlled by a thermostatic control, such that when the engine is cold, the blower air intake is closed, but opens gradually as the engine

* 1959.

cylinders warm up. The blower runs at about 1·6 times engine speed and, at 4,000 r.p.m. (engine speed) delivers 1,800 cu. ft. of air per minute. The engine has two Rochester single-barrel carburettors.

The Vee-Six Engine.—Although the six-cylinder "in line" engine is so widely employed, on account of its excellent engine balance and smooth torque, it has the drawback of greater overall length, and in consequence the use of a longer crankshaft which is more liable to torsional vibration effects than a shorter crankshaft. The inlet induction system, also, is longer and even mixture distribution to the cylinders is more difficult.

The more general substitution of the V-8 for the vertical eight-cylinder "in line" engine, in automobiles, for somewhat similar reasons has led designers to consider the V-6 arrangement of the cylinders instead of the vertical "in line" one. If this can be effected, satisfactorily, from the view points of good engine balance and torque, then a more compact and rigid engine will result, which should allow higher engine speeds and outputs without roughness or vibration troubles. The shorter engine length of the V-6 engine is a marked advantage in chassis and body construction.

There are, however, certain problems to be solved, if the V-6 engine is to compete successfully with its "in line" rival. One of these difficulties is in regard to the firing intervals of the V-6 engine. If a three-throw crankshaft, with cranks at 120°, as in the six-cylinder vertical engine is employed, then the cylinders must be set at an angle of 120°; this, in turn would necessitate a much wider car bonnet than usual. If, however, a 60° angle is used between the two cylinder banks, then a six-throw crankshaft is needed, for equal firing intervals; this indicates a longer crankshaft and engine than for the 120° V-six arrangement.

The V-type of engine has been used over a long period by the Italian Lancia Company (Turin). In addition to the compact vee-4 engine, described earlier, this Company later introduced the Aurelia car fitted with an interesting vee-six engine with cylinder blocks at 60°. The engine has a bore of 70 mm. (2·76 in.) and stroke of 76 mm. (2·99 in.) giving a cylinder capacity of 1,750 c.c. (106·9 cu. in.). Overhead valves are employed and these **are** inclined in the light alloy cylinder heads, being operated from a single camshaft by pushrods and rocker arms. The crankshaft, which has six throws, runs in four main bearings; the two banks of cylinders are staggered to provide for the mounting of the connecting rods on the separate crankpins. A triple roller chain is used for the camshaft drive and it is kept at correct tension automatically, by an hydraulic device.

The more recent Lancia Flaminia car engine, of the vee-six type has a bore of 80 mm. (3·15 in.) and stroke of 81·5 mm. (3·20 in.) giving a cylinder capacity of 2,458 c.c. (150·0 cu. in.). It has a compression ratio much higher than that used for the Aurelia engine, namely, 9·1:1 (in the Farina model), instead of 6·85:1. The maximum power output is 132 B.H.P. at 5,100 r.p.m. with a maximum torque of 137 lb. ft. and maximum B.M.E.P. of 138 lb. per sq. in., both at 3,500 r.p.m.

The Buick Vee-six Engine.—The successful introduction of the Buick aluminium vee-eight engine in 1961 led to a reconsideration of the possibility of replacing the vertical-six engines, by vee-six engines, of the aluminium cylinder block type. As a result of investigations and experimental work upon alternative possible arrangements, in regard to the crank pin and cylinder vee angles it was decided to concentrate upon a vee-six engine having cylinders at 90° and a crankshaft having three throws or crank pins, so that connecting-rods of opposite cylinders had their big-end bearings side-by-side on each crank pin—as is the common vee-eight engine arrangement.

The advantages of such an engine are its appreciably shorter length, and lower overall weight than for the comparable vertical-six engine. Thus the 90° vee-six engine was about 75 per cent. the length of the vertical six, but was necessarily much wider and of rather greater overall depth although in each case well within the engine bonnet space requirements.

The primary balance of the 90° vee-six engine was satisfactorily obtained by the use of counterbalance weights on the crankshaft, but there was an unbalanced secondary force giving rise to a rocking couple acting in a horizontal plane and at a frequency of twice engine speed. It was therefore decided to accept this secondary unbalance effect and to minimize it by using specially

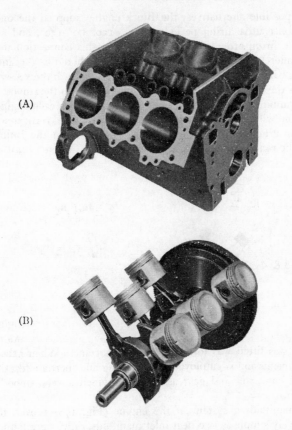

(A)

(B)

Fig. 177.—Some details of the Buick
Vee-six Engine.
(A)—The Cylinder Block. (B)—The Crankshaft
Connecting-rods and Pistons.

designed flexible engine mountings. It was found that the maximum mounting movement from peak to peak was of the order of only ·007 in.

The choice of a 90° angle also affects the engine's firing periods.

Of two possible alternatives, the Buick engine adopted the one giving consecutive firing periods of 90°–150°–90°–150°, and so on with a firing order of 1–6–5–4–3–2. In this connection the left cylinder bank from front to rear was numbered 1–3–5 and the right bank, 2–4–6   Another important factor with the vee-six engine is the effect of the *uneven firing intervals* upon the smoothness of running, since the variations of torque during each engine revolution will be appreciably greater than for the vertical-six engine—with its noted smooth torque qualities.   In the Buick engine the practical effect of this uneven spacing, when a heavier

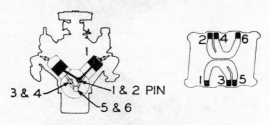

Fig. 178.—The Crankpins and Inlet Manifolds of the Buick Vee-six Engine.

flywheel was fitted was found to be unimportant.   When automatic transmission is employed the damping and inertia effect of the fluid converter and gear assembly provided a very smooth operation.

The manifolding system of the engine (Fig. 178) proved to be relatively simple as two neat inlet manifolds, only, were found necessary for the two-barrel carburettor employed.   It should be mentioned that with this engine the firing order alternated between the two banks of cylinders.

The Buick production engine has a bore of 3·625 in. and stroke of 3·20 in., giving a cylinder capacity of 198 cu. in.   The compression ratio is 8·8:1.   The engine develops a maximum output of 135 B.H.P. at 4,600 r.p.m., with a maximum torque of 205 lb. ft. at 2,400 r.p.m.   It has a dry weight of 414 lb. i.e. about 3 lb per B.H.P. or ·68 B.H.P. per cu. in. capacity.

The inlet and exhaust valve head diameters are 1·500 in. and 1·3125 in., respectively.

The valve timing is, as follows:

The inlet valve opens at 18° before T.D.C. and closes at 82° past B.D.C.

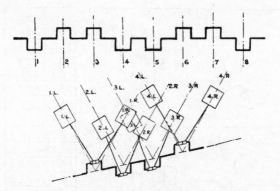

Fig. 179.—Illustrating Two Arrangements of Eight-Cylinder Engine Crankshafts.

Top—The "Straight Eight." Bottom—The Vee-eight type.

The exhaust valve opens at 62° before T.D.C. and closes at 38° after T.D.C. It will be seen that there is a relatively large inlet and exhaust valve overlap, namely, of 56°.

The Vertical Eight-Cylinder Engine.—This type possesses an advantage over the six-cylinder engine in the matter of more uniform torque and rather better acceleration, but its balance is not, generally speaking, so good, except in one particular "straight eight" crank arrangement.

The eight cylinders may be arranged either in one long line, or in two rows of four cylinders inclined to one another (i.e., similarly to four sets of vee-twin engines, one behind the other).

The former arrangement gives a rather long engine and therefore long and expensive crank- and camshafts. The latter arrangement is more compact, but the balance is not quite so good.

The Straight Eight Engine.—One popular arrangement of cylinders for the straight eight type is that shown in outline in Fig. 179 (top), consisting of two symmetrical pairs of four-cylinder engines. Evidently in this case the secondary up-and-down forces, which in the four-cylinder engine are unbalanced,

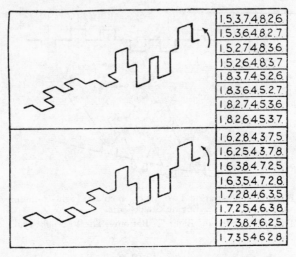

	1.5.3,7.4.8.2.6.
	1.5.3.6.4.8.2.7.
	1.5.2.7.4.8.3.6.
	1.5.2.6.4.8.3.7.
	1.8.3.7.4.5.2.6.
	1.8.3.6.4.5.2.7.
	1.8.2.7.4.5.3.6.
	1.8.2.6.4.5.3.7.
	1.6.2.8.4.3.7.5.
	1.6.2.5.4.3.7.8.
	1.6.3.8.4.7.2.5.
	1.6.3.5.4.7.2.8.
	1.7.2.8.4.6.3.5.
	1.7.2.5.4.6.3.8.
	1.7.3.8.4.6.2.5.
	1.7.3.5.4.6.2.8.

Fig. 180.—Two alternative Crank Arrangements for Straight Eight Engines (with Alternative Firing Orders on the right).

are in this type balanced, although there is a small rocking action due to these forces not acting in the same line.

There are several alternative arrangements of crankshafts for straight eight engines which give equal firing intervals, namely, four per revolution. It will be apparent that some of the cranks will have to be at right-angles to others; the particular arrangement adopted will depend also upon considerations of engine balance.

Fig. 180 illustrates two alternative crank arrangements that were used in motor-car engines. It will be observed that these may be regarded as two four-cylinder crankshafts at right-angles.

In the tables on the right-hand side of Fig. 180 are given the alternative firing orders for the cylinders that are possible.

The straight eight engine was previously used to a greater extent than the Vee-eight, in spite of its greater overall length and the difficulty in damping torsional vibrations. It gives better accessibility and more ample bearing surfaces for the connecting-rod and main crankshaft, and, generally speaking, a

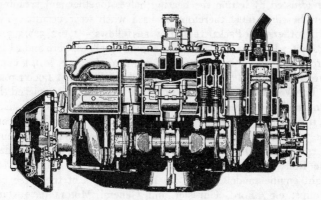

Fig. 181.—An earlier Packard Straight Eight Engine.

much cleaner design. It has, however, been entirely superseded by the 90° vee-eight engine.

An example of a high-powered straight eight engine is the Packard one shown in Fig. 181. This has a bore and stroke of $3\frac{1}{4}$ in and $4\frac{1}{4}$ in., respectively, corresponding to a cylinder capacity of 4,620 c.c. (282 cu. in.). The compression ratio was 6·43:1, but a special cylinder head was available to give a higher compression ratio for use with high octane fuel. The engine developed 120 B.H.P. at 3,800 r.p.m., with the lower compression ratio aluminium head. The valves were of the side pattern, the long camshaft having bearings between each pair of cams. The crankshaft had five main bearings so that there was a pair of cranks between each bearing. The arrangement of the cranks, as shown in Fig. 179, consisted of the two outside pairs as in a four-cylinder vertical engine and the four inside pairs, also arranged in a similar manner to the four-cylinder engine, but at right angles to the outer pairs. The four cranks of the left-hand cylinders are therefore looking-glass images or reflections of those of the right-hand ones; this arrangement gives very good engine balance and torque characteristics.

The camshaft, crankshaft and big-end bearings were all of the shimless steel-backed, white-metal lined pattern, which cannot be adjusted by letting the bearing halves together and scraping; new bearings must therefore be used when wear occurs. The firing order of the Packard engine was as follows:—1, 6, 2, 5, 8, 3, 7, 4.

A later Packard eight-cylinder engine of $3\frac{9}{16}$ in. bore and $4\frac{1}{2}$ in. stroke, with a cubical capacity of 5·9 litres (359 cu. in.), a compression ratio of 8·7 to 1 and developed 212 B.H.P. at 4,000 r.p.m.

The Vee-Eight Engine.—This type of engine has replaced the "in-line" eight-cylinder models, for the higher powered cars used in this country and in the U.S.A. In this connection, apart from the $1\frac{1}{2}$ litre high output engines used for road and track racing purposes, e.g., the B.R.M., Coventry Climax and Ferrari engines, the Rolls Royce and Daimler car manufacturers produce vee-eight engine models, while car manufacturers of the three big groups, e.g., Ford, Chrysler and General Motors have standardized this type for their larger output cars, as distinct from the "compact" car four and the standard six-cylinder engines.

The principal advantages of the vee-eight engine may be summarized, briefly, as follows:

(1) For eight-cylinder engines, other than radial ones, it provides the shortest engine.

(2) For the more popular 90° angle arrangement of the two cylinder banks, it enables a single camshaft to be located above the crankshaft, so as to provide a relatively simple valve gear arrangement, for either the side-valve or overhead valve type engine.

(3) By a suitable choice of crankshaft crank angles it is possible to obtain good engine balance. In this connection, if the two outer cranks are in the same plane and are at 90° to the two parallel inner pairs, very good balance results.

(4) The engine is not prone to the same torsional vibrations as the "in line" type.

(5) Only a four-throw (or crank) crankshaft is used, instead of an eight-throw one, as in the "in line" type. In this case two connecting-rods, from opposite cylinders operate on the same crank pin.

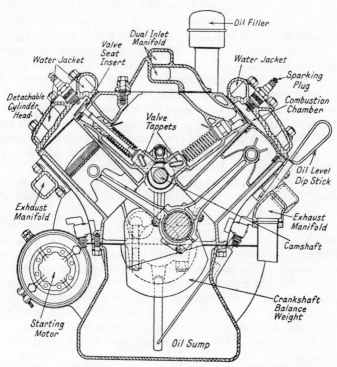

Fig. 182.—Sectional view of a Side-valve Vee-Eight Engine.

(6) Even firing intervals between the cylinders can be obtained with the two-plane crankshaft arrangement, previously mentioned.

In general the vee-eight engine can be made to operate smoothly and practically silently and will fit into the usual bonnet space associated with cars in the larger category. It can be made relatively light in weight by employing aluminium alloys, high-strength-to-weight alloy steels and other special materials, so that its ratio of weight to power output need not exceed 2·6 to 3·0 lb./h.p.

The Ford Side Valve Vee-Eight Engine.—The engine shown in Fig. 182 was a good example of a compact eight-cylinder unit which was a pioneer of vee-eight engines. This engine had a bore and stroke of 77·78 mm. and 92·25 mm. (3,621 c.c.), respectively. The engine developed 88·5 B.H.P. at 3,700 r.p.m.

The two cylinder blocks were cast integral with the upper half of the crankcase; detachable aluminium alloy cylinder heads were fitted. The engine employed side valves of the expanded or mushroom end pattern. With these valves it is necessary to use split valve guides for assembly and removal of the "fixed clearance" valves. A unique feature was the provision made for removing the complete valve, spring, collar and cotter unit, merely by the prior removal of a cotter held in place by the valve spring. The valves were of silicon-chromium nickel steel.

A single camshaft, situated in the vee of the cylinders above the crankshaft, was employed; it was driven from the crankshaft by helical gears.

A three-bearing crankshaft was used, for the engine was relatively short. The cranks were arranged as shown in Fig. 179 (lower illustration). Two pairs of connecting-rods operated on the same pin, the respective pairs being 1—5; 2—6: 3—7 and 4—8. The big ends of each pair were arranged side-by-side on their crankpin. The bearings were white-metal lined and the thrust was taken on the rear bearing. The crankshaft, which was of alloy cast iron, was accurately balanced so as to be both in static and dynamic balance. The gudgeon pins were fully floating in the pistons.

Lubrication was by gear-driven pump, feeding oil under pressure to the crankshaft main and big-end bearings and camshaft bearings. The cylinder walls and gudgeon pins were lubricated by splash. Cooling of the engine was by means of two belt-driven impeller pumps located at the front of each cylinder head. These pumps drew the heated water from the engine into the upper radiator tank, from which it flowed downwards through the radiator tubes into the lower radiator tank and back to the water jacket of the engine.

British Vee-Eight Engines.—Apart from the high output $1\frac{1}{2}$ litre vee-eight racing-car engines mentioned earlier in this volume and the $2\frac{1}{2}$-litre Coventry Climax mentioned later, two outstanding modern examples are the Daimler sports car and the Rolls Royce touring car and Bentley engines.

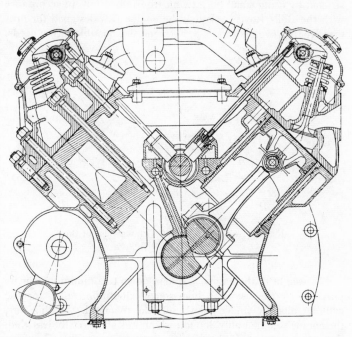

Fig. 183.—The Rolls Royce Vee-Eight Engine.

The Daimler engine has a bore of 76·2 mm. (3·0 in.) and stroke of 69·9 mm. (2·75 in.) giving a cylinder capacity of 2·549 c.c. (155·5 cu. in.). It has a compression ratio of 8·2:1 and develops 140 B.H.P. at 5,800 r.p.m., with a maximum B.M.E.P. of 150 lb. per sq. in. and torque of 155 lb. ft., both at 3,600 r.p.m.

The larger Rolls Royce (and Bentley) engine, designed to supersede the previous six-cylinder engine of 4·9 litres (299 cu. in.) has a bore of 104 mm. (4·1 in.) and stroke of 91·5 mm. (3·6 in.), giving a capacity of 6·23 litres (380 cu. in.). The compression ratio is 8·0:1 and, while no performance figures have been made public, it is known that the prototype engine had an output, at a fairly low "top" speed, of 40 B.H.P. per litre, the mean piston speed being 2,600 ft. per min. The output was purposely restricted by using small bore twin carburettors, but, in accordance with the Rolls Royce practice, the engine was designed for progressive improvement in its performance for possible future needs.

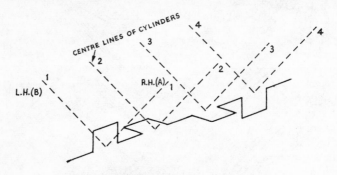

Fig. 184.—The Crankshaft Arrangement of the Rolls Royce Engine.

The weight of the engine was kept down to a minimum, for its purpose, by the use of a high-silicon aluminium alloy for the cylinder block, cylinder heads and inlet manifold. In this way, the engine weighed only 890 lb., which was about 10 lb. less than the previous six-cylinder engine with its gearbox. Cast-iron wet-type cylinder liners having one upper and two lower artificial rubber sealing rings are used in the blocks. Any possibility of corrosion by the coolant is prevented by anodizing the water jackets and treating the outer surfaces of the liners with plastic-bonded aluminium paint.

Fig. 183 which shows the engine in transverse section gives details of the valve-operating gear, cylinder head joints, the wedge-shaped combustion chamber, austenite steel valve seat inserts, cast-iron inlet valve guides (bronze is used for the exhaust valve guides), the connecting-rods and certain lubricating oil passages. The valve tappets used are of the hydraulic lifter automatic kind. The exhaust valves were faced with Stellite, as were the valve stem tips. This extremely hard coating greatly extends the useful life of these working faces. The inlet and exhaust valve head diameters are 1·750 in. and 1·50 in., respectively.

The valve timing used on the prototype engine was, as follows:

Inlet valve opens at 23° before T.D.C. and closes at 100° after B.D.C.

Exhaust valve opens at 55° before B.D.C. and closes at 26° after T.D.C. The timing tappet clearances were ·010 in.–·015 in.

The crankshaft, of heat-treated alloy steel was of the two-plane shaft type (Fig. 184), with the two outer cranks parallel and the two inner cranks also parallel but at 90° to the outer pair. With suitable crank pin balance weights, the engine balance is theoretically perfect (for the primary and secondary forces).

The Coventry-Climax High Output Engines.—Developed for car racing purposes the recent vee-eight engines have achieved a high reputation for high output and reliability. These include the 2·5 litre and more recently the 1·5 litre engines which, in 1962 achieved many notable international racing-car successes.

The F.P.E. 2·5 litre engine (Fig. 185) has a bore of 76 mm. (3·0 in.) and stroke of 68 mm (2·675 in.) giving a cylinder capacity of 2,477·3 c.c. (151 cu. in.). With a compression ratio of 11:1 and operating on special 115/145 P.N. aviation fuel the engine had a maximum output of 258 B.H.P. at 8,250 r.p.m., this being equivalent to 104 B.H.P. per litre or 1·71 B.H.P. per cu. in. The maximum piston speed was 3,535 ft. per min. It is of interest to note that the maximum output of this engine was equivalent to 4·56 B.H.P. per sq. in. of piston area.

Fig. 185 is a part cross-sectional view of the F.P.E. 2·5 litre engine, showing on the right the cylinder, piston assembly and valves. The valves are operated by two camshafts for each cylinder block, i.e., four camshafts in all; these were driven by gear trains from the crankshaft, as shown on the left, in Fig. 185.

The cylinders in each block are at 90° to one another and each inlet manifold has four carburettor chokes tube, for high volumetric efficiency. Some other special features of this engine, which weighs only 340 lb., i.e. 1·32 lb. per H.P. include a five-bearing crankshaft, with an opposite pair of connecting-rods on each crank pin journal, combined cylinder block and crankcase in R.R.50 (cast) aluminium alloy, fitted with wet-type cylinder liners, and cylinder heads in the same alloy. The valve seats in the head are of austenitic steel for the inlet valves and copper-

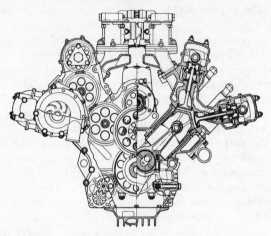

Fig. 185.—The Coventry-Climax High Performance
F.P.E., 2½ Litre Engine.

bronze for the exhaust valves. The valve guides are made of phosphor-bronze. Each cylinder block has its own cooling water pump. Another interesting feature is the provision of an eight-cylinder type magneto, located in the vee part of the cylinder blocks, for ignition purposes.

Recent American Vee-Eight Engines.—As mentioned earlier, the L-head side valve type of combustion chamber that had been in use over a long period, has been increasingly replaced by the over-head (parallel or inclined) valve engine; the F-head combustion chamber is also used. Further, in some cases the flat-topped piston and domed cylinder head has been used in certain engines. In general, the in-line overhead valve arrangement of combustion chamber and valves, similar in principle to the type shown at (B) in Fig. 40, is used. The inlet valve is made larger than the exhaust valve and the sparking plug placed centrally in order to give a short flame travel path.

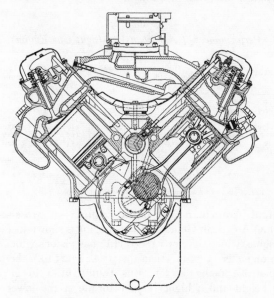

Fig. 186.—The Pontiac Vee-Eight Engine.

Turbulence is provided by the piston coming close to the flat part of the combustion chamber head.

It is notable that in the majority of American O.H.V. engines having flat-topped pistons, compression turbulence is provided by the "squish" method, i.e., by part of the piston on the opposite side to the sparking plug coming up close to the flat portion of the combustion head opposite to it—as in the original Ricardo method of the early 1920s.

The wedge-shaped combustion chamber shown in Fig. 186 is used on several overhead-valve engines with the sparking plug located centrally at the wider end of the wedge. Compression ratios of 8·5:1 to 10·5:1 are used and the combustion conditions obtained result in appreciably greater power output and lower fuel consumptions than in previous designs.

The general performances of some typical 1962 American vee-eight engines are given in Table 8.

TABLE 8

Performances of American Vee-Eight Car Engines

Make of Car	Bore in.	Stroke in.	Piston Displacement cu. in.	Compression ratio 'to 1'	B.H.P.	At r.p.m.	Maximum Torque	At r.p.m.
Buick (Le Sabre)	4·187	3·640	401	10·25	280	4,400	424	2,400
Cadillac	4·000	3·875	390	10·5	325	4,800	430	3,100
Chevrolet (300)	4·000	3·25	327	10·5	250	4,400	350	2,800
Ford (Galaxie)	4·130	3·78	406	10·9	405	5,800	448	3,500
Oldsmobile	3·500	2·80	215	8·75	155	4,800	210	3,200
Mercury (Ford)	3·500	2·87	221	8·7	145	4,400	216	2,200
Rambler	4·000	3·25	327	8·7	250	4,700	340	2,600
Studebaker (Lark)	3·562	3·25	259	8·5	180	4,500	260	2,800

Note.—Most of above cars are supplied with optional higher power engines.

Typical American Vee-Eight Engine.—A cross-sectional view of an American General Motors engine is shown in Fig. 186. The engine, which develops 180 B.H.P. has pistons which have a greater diameter (3·75 in.) than the stroke (3·25 in.), the ratio of bore-to-stroke being 1·15:1; this results in a lower overall engine height and a higher engine torque at the lower engine speeds. The angle between the two four-cylinder unit banks is 90°. A special feature of this engine is the use of ball-pivoted, sheet-metal rocker arms instead of the usual rocker arm shaft and plain rocker bearings. The central camshaft is arranged in the vee-space between the cylinder units, so that all of the sixteen valves are operated by this single camshaft. Hydraulic self-adjusting tappets, or lifters, are used between the lower ends of the tubular push-rods and the valve cams. The valves are fitted with double springs and the valve guides are slightly tapered.

The combustion chamber is of the wedge-section type and the sparking-plug is located near to the inlet valve to provide for the most efficient combustion conditions and also ensure adequate plug cooling by the relatively cool air-fuel inlet charge. The combustion chambers are machined all over, to ensure a large flame quench area, and provide accurate volume dimensions, so that the compressions of all of the cylinders are identical.

The pistons are of the slipper-skirt pattern with inbuilt steel struts to control the expansion. Their skirts are cam-ground and tin-plated. Each piston has three piston rings—as is common American practice. These consist of two upper tapered compression rings, the top one of which is chromium-plate to give greatly increased life, while the lowest ring is of the four-unit oil-control type, with chromium-plated oil-scraping edges.

The cylinder-piston arrangement is designed to give a desaxé layout with $\frac{1}{16}$ in. displacement. This ensures a gradual change in the thrust pressure and reduces piston slap, when cold.

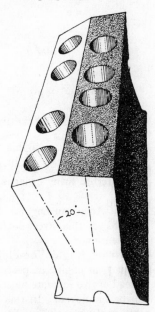

The crankshaft is of the five-bearing, counter-balanced kind machined from a steel forging and then heat-treated. A harmonic balancer is incorporated in the crankshaft pulley assembly. It consists of a 3-lb. 13 oz. steel weight, retained in the pulley by means of flexible springs. The $22\frac{1}{2}$ in. long camshaft is made from a cast-iron alloy, the cams being hardened. The camshaft-crankshaft drive is by means of a 60-link Morse chain and 42-tooth and 21-tooth sprocket wheels. In most other respects the engine detail designs follow standard American practice. The crank-

Fig. 187.—A Vee-Eight Monobloc Engine Cylinder Block.

case is provided with a special pressure-suction positive kind of ventilation whereby outside filtered cool air is circulated through the crankcase and under some of the pistons, leaving finally by a special baffling system into the open air again.

Firing Order of Vee-Eight Engines.—There are two principal alternative arrangements for the crank journals, namely: (1) A similar one to that used for the 4-cylinder vertical engine and (2) That shown in Fig. 184, in which the outer pair of cranks are in the same plane but are opposite, while the two inner cranks are in a plane at 90° to the outer cranks plane and are opposite one another.

If the cylinders on the left bank, from the front, or radiator end

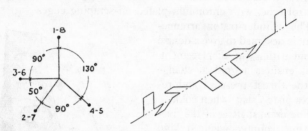

Fig. 188.—Crankshaft arrangement of Narrow Angle Vee-Eight Engine.

are numbered 1–3–5–7 and those on the right bank, 2–4–6–8, then the preferred firing order is, as follows:

$$1–8–4–3–6–5–7–2.$$

Most American engines use the crank arrangement shown in Fig. 184, as this gives much better engine balance than the first one mentioned.

Narrow Angle Vee-Eight Engine.—In order to obtain as compact an engine as possible, with a single monobloc cylinder instead of two separate blocks, the arrangement shown in Fig. 187 can be employed. In this case the cylinders are inclined at a small angle, namely, 20°, in order to keep the overall width of the cylinder block as small as possible. Further, the opposite cylinders are staggered in relation to one another so that a conventional, but much shorter design of crankshaft can be employed.

It is possible with this type of engine, using an eight-throw crankshaft suitably counter-balanced, to obtain proper balancing of the primary and secondary reciprocating inertia forces. If the engine is viewed from the forward end and the cylinders in the left bank are numbered 1, 3, 6 and 8, whilst those of the right bank are 2, 4, 5 and 7, then the arrangement of the cranks which gives correct engine balance will be as shown in Fig. 188. The firing order of the cylinder is then 1, 3, 2, 4, 8, 6, 7 and 5, and the intervals, in degrees, between the explosions will be 90, 70, 90, 110, 90, 70, 90 and 110, so that there is no very appreciable variation in the firing intervals.

The narrow angle engine in question, with its shorter crank-shaft, is less liable to torsional vibration effects. Another advantage is the increased stiffness of the whole engine resulting from the use of a single-cylinder block. On the other hand the valve operating mechanism for the inclined cylinders is usually more complicated than for the straight eight type.

The Opposed-eight Cylinder Engine.—This alternative type, which has the advantages of better balance and higher operational speeds, has been developed principally for racing car purposes, by Dr. Porsche, the designer of the Volkswagen opposed-four engine. In its $1\frac{1}{2}$ litre formula design, which was decided upon after experimental work on opposed-six and opposed-twelve cylinder engines, four double-barrelled Weber carburettors are used, with their intakes vertically above the engine, which is water-cooled and designed for rear-end mounting on the chassis. The cooling fan is mounted horizontally above the centre of the power unit, in a similar manner to the Chevrolet Corvair blower. It is, however, gear driven. Ignition is by two Bosch eight-cylinder units, to supply high voltage current to a pair of sparking plugs in each cylinder. The engine was designed for an output of 180 to 200 B.H.P. and maximum speeds of about 10,000 r.p.m. The cylinders each have a bore of 66 mm. (2·6 in.) and stroke of 56·4 mm. (2·28 in.) giving a cylinder capacity of 1,498 c.c. The compression ratio is 10:1.

Twelve-Cylinder Engines.—Evolved primarily for aeronautical use, this type has been used in some makes of car, notably in the Rolls Royce, Packard, Lincoln "Zephyr," and Daimler "Double Six." Although much superior in the matter of torque, while the balance is perfect, the extra complication and expense of manufacture hardly justifies its use in automobiles when such good results can be obtained from the smaller number of cylinder engines.

The common arrangement for the twelve-cylinder engine consists of two sets of six cylinders inclined Vee-fashion at an angle of 60° as a rule. The crankshaft has the same crank

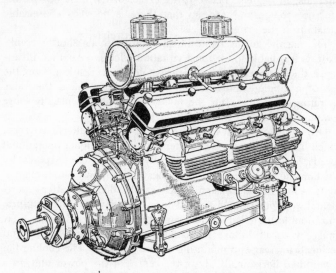

Fig. 189.—The Rolls Royce Twelve-Cylinder Engine.

disposition as in the six-cylinder engine, and six sets of forked and plain connecting-rod pairs are used. For the best results two magnetos, or coil-ignition units, two circulating pumps, and two carburettors are employed.

Two good examples of British 12-cylinder engines, since out of production, were the Daimler Double-Six and the Rolls Royce Phantom III types.

The former was a V-type engine of 81·5 mm. bore and 104 mm. stroke, giving a cylinder capacity of 6,500 c.c. This engine represented a reversion from sleeve to poppet-valve design. Special features of this engine include aluminium-alloy crankcase;

pent-roof combustion chamber; seven-bearing crankshaft, with enclosed automatically-lubricated torsion damper; overhead valves, operated by push-rods and rockers from two camshafts mounted in the crankcase; special cam contours permitting large tappet clearance with freedom from valve seat wear; four-point rubber mounting of engine; dual-type carburettor with air cleaner and silencer, and initial lubrication to the pistons and tappets.

The Rolls Royce 12-cylinder engine (Fig. 189) had two banks of six cylinders inclined to each other at 60°. The bore and stroke were $3\frac{1}{4}$ ins. and $4\frac{1}{2}$ ins. respectively, giving a cylinder capacity of 7,340 c.c.

Special features of this engine included a cylinder block cast integral with the upper half of the crankcase; cylinders fitted with cast-iron liners of the wet type; aluminium cylinder heads; overhead valves operated by a single camshaft mounted in the vee angle of the crankcase, through push-rods and rocker arms; automatic (hydraulic) tappet clearance adjustment; dual-ignition by battery and coil through two independent contact breakers, two distributors and two coils; four carburettors, placed in the vee angle, so arranged that two carburettors feed each row of cylinders; independent starting carburettor, quite separate from the others; large capacity air cleaner and silencer; extra oil to the walls of the cylinder for starting, etc.

Lubrication was by an engine-driven gear-type pump, the oil from the pressure side of which went through an oil filter before entering a cooling chamber consisting of a honey-comb matrix, the temperature of which was controlled by the water discharged from the circulating pump on its way to the cylinders. The oil then entered a relief valve which determined the three different pressures used, by means of spring-loaded release valves arranged in series. The crankshaft and connecting-rod bearings were fed at the full pressure of 50 lb. per sq. in.; the overhead rocker shafts at 10 lb. per sq. in., and the timing wheels at $1\frac{3}{4}$ lb. per sq. in.

Engine cooling was by centrifugal water pump and fan-cooled radiator; the latter had thermostatically-controlled shutters to maintain the water at constant temperature.

Sixteen-Cylinder Engines.—Three commercial model cars have used two sets of "straight eight" cylinders inclined at an angle or vee, to give a still more even torque than that of the twelve-cylinder engine, and a practically top-gear performance. This type of engine was made by firms specializing in the "straight eight." Thus, the General Motors, Ltd., once made an eight-cylinder Buick and sixteen-cylinder Cadillac.

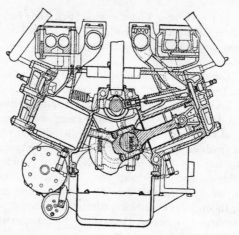

Fig. 190.—The Cadillac Sixteen-Cylinder Engine.

A front sectional view of the Cadillac sixteen-cylinder engine is shown in Fig. 190. The engine had a bore and stroke of 88·9 mm. (3·5 in.) each, and a cylinder capacity of 7,060 c.c. (431 cu. in.). The engine developed 185 B.H.P. at 3,600 r.p.m. The cylinders are arranged in two banks of eight cylinders each, inclined at a relatively wide angle, so that it approached the horizontally-opposed engine arrangement and gave a wide design of engine. Both cylinder banks and the greater part of the crankcase are in a single casting. The connecting rods are shorter than in most other designs using a smaller number of cylinders; this is because of the short stroke of the engine. The crankshaft has nine main bearings and two connecting rods have bearings on each crankpin, namely, side by side. A feature of the pistons and connecting-rods is the method of clamping the gudgeon pin to the small end of the rod.

An advantage of the wide angle arrangement of the cylinders is the relatively large clear space at the top of the engine which is used for mounting most of the accessories, whilst providing ready access to the valve springs. The valve tappets are of the hydraulic compensating type for maintaining the correct clearance automatically.

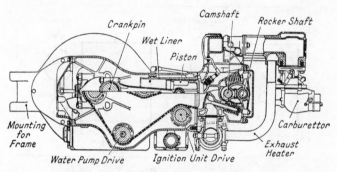

Fig. 191.—The Pancake Engine.

The Flat or Pancake Engine.—The amount of space occupied by the ordinary six-cylinder engine of the commercial vehicle reduces that available for passengers or goods, so that in several instances this difficulty has been overcome by placing the engine below the floorboard level. For this purpose it becomes necessary to use a "flat" type of engine such as the four-, six- or eight-cylinder opposed ones consisting of pairs of cylinders on opposite sides of the crankshaft as shown in Fig. 148 (d). Alternatively, as is the case with the Leyland A.E.C., Gardner and G.M. engines, an ordinary six-cylinder or other vertical model engine can be arranged so as to work in the horizontal position similar to that shown in Fig. 191. By suitable design of the chassis members the parts of the engine, e.g., the sparking plugs, valves and carburettor requiring maintenance attention can be rendered accessible without difficulty.

Several makes of compression-ignition engine for large motor vehicles and for rail cars are now made in opposed-cylinder flat engines.

Horizontal Six Engine.—The petrol engine shown in Fig. 191 is a six-cylinder in-line horizontal one of 5 in. bore and 6 in. stroke. It develops 180 B.H.P. at 2,200 r.p.m. The engine is mounted underneath the floor of the motor vehicle.

The upper part of the illustration shows part of the cylinder

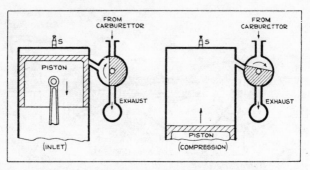

Fig. 192.—Principle of Rotary Valve Engine.

and piston, and it will be noted that a wet type cylinder liner is fitted. Overhead inclined valves operated by means of overhead camshaft and curved rocker arms are used. The camshaft is driven by means of a relatively long chain which passes over four sprocket wheels; two of the latter drive the water-circulating pump and ignition cam and distributor rotor.

The carburettor is shown on the extreme right-hand side where it is mounted in an accessible position; it has an exhaust gas-heated jacket and is controlled by means of a governor.

Some Other Engine Types.—Ever since the evolution of the poppet-valve engine, attempts have been made to replace the poppet valves and their operating gear by some more simple and quieter means for attaining the same object. In some cases *piston valves*, resembling those of steam-engine practice, were used to govern the inlet and exhaust operations; in others *rotating valves* of cylindrical form, with ports cut in them, as in the Darracq rotary-valve engine, were employed. The principle of the Darracq engine is illustrated in Fig. 192, the inlet and compression operations being shown. The rotary valve consisted of a section of a cylinder rotating in a cylindrical casing having ports communicating with the cylinder, carburettor and exhaust pipe, respectively, at one-half engine speed. As shown on the left, the piston is just about to descend on its suction stroke, the rotary valve being about to open the carburettor port. At the end of the inlet stroke, the valve shuts off both the carburettor and exhaust ports, thus sealing the cylinder. The piston is arranged to cover the port leading to the rotary valve, during the firing period so that the valve is not subjected to the maximum explosion pressure. When the piston is about three-quarters of the way down its firing (or expansion) stroke the rotary valve opens the exhaust port; the latter remains open during the next upward stroke of the piston. In the Itala engine, which was once used on cars, a vertical rotating valve, with specially designed ports, somewhat on the lines of the usual piston valve arrangement, was employed.

The *cuff-valve* engine arrangement, which had a short vogue, consisted of a pair of concentric slotted sleeves situated above the cylinder, the outer sleeve serving as the exhaust, and the inner sleeve as the inlet valve. These sleeves were provided with conical seatings and were operated up and down similarly to an ordinary poppet valve; in effect the arrangement was equivalent to a pair of concentric, but hollow, poppet valves of large area.

Although none of the engine types mentioned above have survived, certain later types such as the Edwards, Aspin and Cross have given promising results. The disadvantages of the early pattern engines previously mentioned were that, in the majority of cases, it was found that the effect of the hot exhaust gases in the common valve distributor reduced seriously the volumetric efficiency, and also caused lubrication and distortion troubles. Rotary and rocking valves which were exposed to the exploding and expanding gases were found to wear badly. In some cases the extra complication of the valve gears did not justify their use, and in others the explosion event tended to blow the valves off their seats.

Sleeve Valve Engines.—There is one type of engine, namely, the sleeve valve engine, which has certain advantages over the poppet valve one. Briefly speaking, this type comprises a single liner, or two cylinder liners which can slide up and down (and also rotate in one case) relatively to the piston. Ports cut in the upper portions are arranged at the correct intervals. In effect, then, we have the piston working in a cylinder liner which itself can slide in another liner or in the outer cylinder casting to control the inlet and exhaust operations.

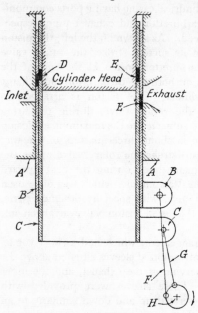

Fig. 193.—The Double Sleeve Valve Engine Principle.

The advantages claimed for the sleeve-valve engine are: (1) Quietness in operation, there being no noisy cams, tappets and valves. (2) Simplicity of construction and a considerably smaller number of working parts; this results in longer life and better reliability. (3) Higher volumetric and thermal efficiency.

Knight Double Sleeve Engine.—The principle of this once popular Daimler engine is illustrated in Fig. 193. It consisted of a pair of concentric sliding sleeves between the cylinder barrel and the piston. Each sleeve was given a small up-and-down motion by means of a crank, driven at one-half engine speed, and a short connecting-rod. The upper ends of these sleeves were arranged to slide between the cylinder barrel and cylinder head, and were provided with ports to register at the appropriate

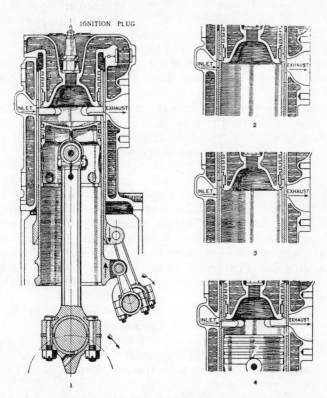

Fig. 194.—Illustrating the operation of the Double Sleeve-Valve Engine.

1. Beginning of suction stroke.   2. End of compression stroke.
3. Firing stroke.   4. End of exhaust stroke.

moments in the cycle of operations, with the fixed inlet and exhaust valve ports in the cylinder block, as indicated diagramatically in Fig. 193. In this diagram A represents the cylinder and B and C the two thin sleeves for valve opening and closing purposes. The sleeves have inlet ports D, and exhaust ports E, such that by a pre-determined arrangement or timing of the movements of the sleeves, the inlet ports and exhaust ports of each sleeve respectively coincide at specified times with the fixed inlet and exhaust ports. The sleeves are operated by connecting-rods F and G, which are actuated by two cranks arranged at different phase-angles from the secondary shaft H running at half engine speed.

The sleeves therefore have an up-and-down motion, giving zero velocity at the two ends and a maximum velocity at about mid-stroke. With this arrangement the valve ports could be given quick opening and closing operations.

The advantages of this type of engine over the poppet valve one were that it ran *much quieter*—since there were no valve cams, tappets, valves or other impact members—and, on account of its pocketless combustion chamber, gave a *higher thermal efficiency*; moreover, by suitably designing and apportioning the sleeve and cylinder ports *higher volumetric efficiencies* could be obtained than with the poppet valve type. Another advantage was the ability of this type of engine to stand up to long periods of operation without maintenance attention, other than replenishing the oil for lubrication.

The principal drawback of the double-sleeve engine has been its somewhat high rate of oil consumption on account of the relatively large area of sleeve surface to be lubricated. Another difficulty has been that of sleeve inertia effects which have limited the maximum engine speeds and outputs to values well below those of poppet valve engines.

The Knight double sleeve-valve engine was adopted by the Daimler Company in 1908, after exhaustive tests, one of which consisted of an official R.A.C. test, during which a four-cylinder engine of 124 mm. bore and 130 mm. stroke ran continuously for 5 days 14 hours, giving an average B.H.P. of 54·4, and a petrol consumption of 0·679 per B.H.P. hour. Following this it was fitted to a car and ran for over 2,150 miles on the Brooklands track at an average speed of 42·4 m.p.h. (occupying 45½ hours). It was then given a final bench test of 5¼ hours. A subsequent examination of the pistons, sleeves and other working parts revealed no perceptible wear, nor did the ports show any signs of burning or wear.

The cylinder head of the Knight engine was detachable; it had an extension, as shown in Fig. 194 (1) not unlike a stationary piston, which projected into the cylinder, and was provided with piston rings to ensure gas-tightness. The ignition plugs were situated centrally in the cylinder heads, and the combustion chamber was spherical in form. It has been shown by Ricardo that sleeve-valve engines, with their almost ideal combustion chamber shapes, are less prone to misfiring and detonation for a given fuel than poppet-valve engines. The two sleeves are grooved at regular intervals, with special serrations also near the ports to ensure adequate lubrication.

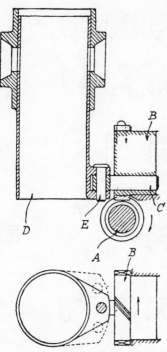

Fig. 195.—Principle of Single Sleeve Valve Engine.

The sleeve-valve principle was also embodied in the twelve-cylinder Daimler "Double Six" engine,* a vee-type engine having two sets of six cylinders at an angle to each other.

The Single Sleeve Engine.—In this type of engine, which was built under the Burt M'Collum patents, and has been used by Messrs. Argylls, Barr & Stroud, and Wallace (Glasgow), and for aircraft, motor-cycle and car engines, a single sleeve only is employed, and the various operations are accomplished by rotating the sleeves slightly, in addition to giving it an up-and-down motion; the ports are also given a special shape.

* The later Daimler engines now use poppet valves.

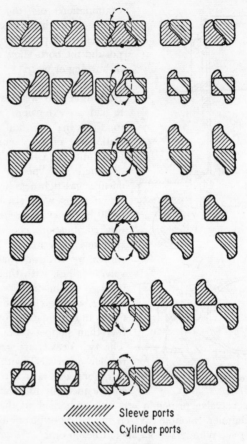

/////////// Sleeve ports

\\\\\\\\\\\ Cylinder ports

Fig. 196.—Illustrating the Operation of the Single Sleeve
Valve Engine.

The principle of the single sleeve valve engine is illustrated,
schematically, in Fig. 195. This diagram shows only the single
sleeve or working liner and the part of the cylinder containing the
valve ports. The piston is arranged to reciprocate in this sleeve,
and the latter is constrained to move both up and down and

sideways, so that any point on it describes an ellipse. One method of obtaining this motion is that shown, whereby the gear-shaft A, driven at one-half engine speed by chain or gear-wheels from the crankshaft, engages with the sleeve operating crank gear B, through spiral gearing, for the axes of A and B are at right angles. The gear B carries a tee-headed sleeve crankpin C, which is attached to the sleeve valve D by a pin E and held between suitable lugs on the sleeve. The crankpin C is free to rotate, and can also slide endwise in a bearing contained in the gear B; it can also swing sidewise about the pin E anchored in the sleeve lugs. This mechanism causes the sleeve to have the desired "elliptical" motion for the valve port events. The sleeve near its upper end is provided with ports of special shape, usually varying from four to six in number.

The general arrangement of these valve ports is such that one half of them form the inlet ports on one side, and the other half the exhaust ports on the other side of the sleeve. There are, however, several alternative arrangements for the ports.

Fig. 196 shows the port arrangements of a sleeve valve unit having two inlet ports on the right, a common port in the centre and two exhaust ports on the left. The cylinder unit has six ports arranged symmetrically, as shown. The movement of the sleeve ports is indicated by the central arrowed ellipse, the black spot on which shows the centre of the bottom edge of the middle common port.

The six diagrams cover the complete cycle of operations for a four-cycle engine and they are, as follows, namely: (1) The inlet ports beginning to open. (2) Inlet ports fully open—as shown by the white spaces on the right. (3) Inlet ports closed and compression occurring. (4) Top of compression, firing stroke about to commence; in each case the cylinder ports are sealed by the plain surfaces of the sleeve. (5) Approaching end of expansion stroke; exhaust ports about to open. (6) Exhaust ports fully open, as indicated by the white spaces on the left.

The valve timing is fixed by the shapes and positions of the ports and the valve areas can be chosen to give a high value for the volumetric efficiency. The inlet ports can be designed so as to give a degree of rotation to the inlet charge, i.e. turbulence.

The results of a lengthy investigation of a high-power single sleeve valve engine revealed the fact that there was practically no wear between the piston and sleeve, and the sleeve and cylinder proper, and only an inappreciable amount on the eccentric driving joint. The type of engine not only yields high thermal efficiencies and power outputs, but has also been shown to wear better than the poppet-valve type. In regard to performance it may be of interest to mention that an Argyll racing-car engine using single sleeve valves broke a number of world's records at Brooklands in 1913. Rated at 17·3 H.P. it had four cylinders each of 83·5 mm. bore and 130 mm. stroke; the stroke of the sleeves was 1·5 in. The compression ratio used was 5·6:1. The engine gave 75 B.H.P. at 3,200 r.p.m. and a B.M.E.P. of 115 lbs. per sq. in, at 2,400 r.p.m. The inlet valve ports opened on top dead centre, and closed 30 degrees after bottom dead centre. The exhaust ports opened at 60 degrees before bottom dead centre, and closed 15 degrees after top dead centre.

With the exception of a single sleeve valve engine once made by Messrs. Vauxhall Ltd., there have been no commercial engines made for motor cars, in this country.

Ricardo Sleeve Engine Design.—Fig. 197 illustrates a single sleeve valve engine (Ricardo) of the two-stroke fuel injection type, capable of giving a good performance. The single sleeve S is operated by the rocking lever L, and it had not only an up-and-down motion but also a small amount of rotary movement, as shown by the dotted path at the left-hand end of the lever L.

The sleeve has two sets of ports cut through it. The upper set when moved downwards so as to come opposite to the ports E are used for the exhaust. Whilst these ports are open, scavenging air is admitted through the ports P, and corresponding ones in the sleeve (these are shown just below the piston), so that a charge of air under pressure sweeps along the cylinder and scavenges the remaining exhaust gases through the ports E. The cylinder head H forms the turbulent combustion chamber and also acts as a guide for the upper part of the sleeve; it is provided with the two piston rings shown for this purpose. The four-cycle sleeve valve engine works in a somewhat similar manner, except that the inlet ports are at the top on the one side, with the exhaust ports also at the top and on the opposite side. A later design dispenses with the ports in the upper part of the sleeve and uses the top edge of the sleeve itself to cover and uncover the exhaust ports.

Bristol Single Sleeve Valve Engines.—The more recent Bristol aircraft engines used single sleeve valves. As a result of a relatively long period of development it has been possible to obtain appreciably better performances than from poppet-valve engines of similar dimensions. Thus, in the case of the "Perseus" sleeve-valve engine of $5\frac{3}{4}$ in. bore and $6\frac{1}{2}$ in. stroke a B.M.E.P. of 190 lbs. per sq. in. was obtained at 2,800 r.p.m. on 87-octane fuel, whereas for the "Mercury" four-poppet valve engine of similar cylinder dimensions the B.M.E.P. of 165 lbs.

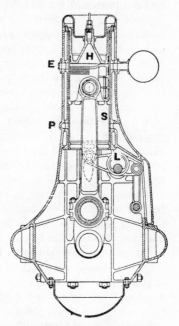

Fig. 197.—The Ricardo Sleeve Valve Engine.

per sq. in. was obtained at 2,650 r.p.m. When using 100-octane fuel the "Perseus" engine gave a B.M.E.P. of 220 lbs. per sq. in. at 2,800 r.p.m. and the "Mercury" engine 185 lbs. per sq. in. at 2,750 r.p.m. The latter results give the B.H.P. per sq. in. of piston area as 5·05 and 4·18, respectively.

This type of engine lends itself well to supercharging and in this connection mention may be made of the results of some tests made by the Bristol Aeroplane Company which, it is stated, could not have been obtained from an engine having poppet valves under similar conditions. With a supercharge of 14 lbs. per sq. in. and a cylinder of 168·8 cu. in. capacity, the B.M.E.P.'s obtained were 305 lbs. per sq. in. at 2,400 r.p.m. and 272 lbs.

per sq. in. at 3,000 r.p.m.   The corresponding gross powers per unit piston area are 6·0 B.H.P. per sq. in. and 6·7 B.H.P. per sq. in.

Some interesting results have been obtained by Ricardo from a small sleeve valve engine as long ago as 1930, when an experimental engine ran heavily supercharged for several hundred hours on ordinary aviation spirit at a B.M.E.P. of over 400 lbs. per sq. in. and a specific output at least three times as great as that of existing poppet-valve aircraft engines, without any signs of distress, or overheating of the engine.   With a special high octane fuel the engine would maintain a B.M.E.P. of 550 lbs. per sq. in.   With these *high degrees of·supercharging* the exhaust is discharged from the cylinder at pressures of 300 lbs. per sq. in. and above so that the question arises as to whether it would not be feasible to *utilize some of this pressure energy* by arranging for the exhaust to operate in a low-pressure cylinder In this connection, the compounding of the engine is possible since the same sleeve valve can be arranged to operate as both inlet and exhaust valve to the high-pressure and as the inlet valve to the low-pressure cylinder.   In this way, greater fuel economy could be obtained and a high power-to-weight ratio.   The Ricardo engine, with a bore of $2\frac{7}{8}$ in. and stroke of $3\frac{1}{2}$ in. ran at speeds of 5,000 to 6,000 r.p.m. and gave well over 100 B.H.P. per litre; the compression ratio used was 6·8 : 1.

Advantages of Single Sleeve Valve Engine.—The principal advantages of the single sleeve-valve engine may be summarized, briefly, as follows, namely, (1) Absence of hot exhaust valve with its detrimental effect upon the limitation of the maximum power of the engine.   (2) Considerable reduction in the number of working parts owing to absence of the valves, springs, cotters, timing gear, etc.   (3) Use of higher compression ratios due to better combustion chamber shape and absence of ·hot exhaust valve. (4) Greater power output per sq. in. of piston area.   (5) Lower fuel consumption per B.H.P. hour.   (6) More silent operation than for poppet-valve engine owing to absence of timing gears, tappets and valves.   (7) Reduced maintenance attention due to (2).   (8) Reduced height of cylinder owing to absence of overhead valve mechanism.   (9) Greater reliability due to absence of valves and valve mechanism.

Automobile Types. In the automobile field, developments in sleeve-valve engines have been made with the object of reducing the weight of the sleeve and its operating mechanism, to improve the high-speed performance, or permit of higher operating speeds. While it cannot be disputed that the well-designed sleeve-valve engine is more efficient under similar operating conditions than the equivalent poppet-valve engine, its design is apt to make it appreciably more costly for mass production.

Of the more promising later sleeve-valve engines, one example has here been selected for brief mention, namely the Swiss Schmid engine.

The Schmid Engine. The principal features of this automobile engine, shown in Fig. 198 are the thin split sleeves and light aluminium or magnesium alloy cylinder block. The sleeve, which has a rectangular, as distinct from oval path of the Bristol sleeve, has a narrow slit along its whole length. The sleeve is about 1 mm. thick and the gap of the slit when in the cylinder (cold) is 0·03 mm. When hot the leakage path is of negligible proportions.

There are no junk rings in the cylinder head and the piston rings do not pass the inlet and exhaust ports. The rectilinear motion is given to the sleeve, by two eccentric shafts, one of which runs at engine speed and the other at one-half engine speed. These shafts operate cross links engaging with two lugs on the base of the sleeve. The extremities of each pair of links are connected by bridge pieces. The sleeves are fully floating and free to expand or contract within the cylinder.

Lubrication of the cylinder walls and sleeves is provided for by using a slightly smaller diameter at the base of the sleeve and by ten sloping oil grooves; the piston is lubricated through two small holes near the bottom of the sleeve. In the experimental four-cylinder engine, of 60 mm. bore and 90 mm. stroke (1,017 c.c.), the sleeve was machined from chrome-carbon steel, which was heat treated and ground to precision limits, it weighed only 0·6 lb.

In the engine the sleeve had a stroke of only 11 mm. (0·43 in.). The area of the ports controlled by the sleeve was 18 per cent of the piston area, there being four ports for the induction and four for the exhaust. The port openings are covered during the compression stroke by the almost stationary sleeve; moving

downwards from the compression position the sleeve then uncovers the exhaust ports.

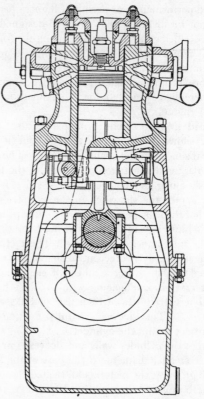

Fig. 198.—The Schmid Sleeve Valve
Engine.

In regard to performance, the engine—with a compression
ratio of 7·5 to 1 developed 39 B.H.P. at 4,000 r.p.m.

The Cross Rotary Valve Engine.—This example of the
rotary valve class met with a fair measure of success as a result
of having overcome the practical difficulties connected with the

maintenance of gas-tight joints and lubrication of the valve member.

Important advantages are claimed for this rotary valve. Thus, it operates at a lower temperature than the poppet valve (the exhaust one of which often runs at a red heat), and presents a

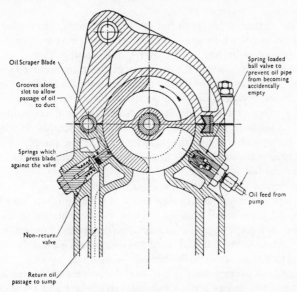

Oil Scraper Blade

Grooves along slot to allow passage of oil to duct

Springs which press blade against the valve

Non-return valve

Return oil passage to sump

Spring loaded ball valve to prevent oil pipe from becoming accidentally empty

Oil feed from pump

Fig. 199.—The Cross Rotary Valve Engine.

cool, polished surface to the flame. This enables a higher compression ratio to be used. Further, it operates quietly and does not require periodic adjustment as with the tappet clearances of poppet valves. Running cooler, this valve also gives higher volumetric efficiency.

The rotary valve is arranged horizontally over the top of the cylinder (Fig. 199) and, in the single cylinder type has a diagonal rib to separate the inlet and exhaust sides. In the side of the tubular valve, ports are cut. The rotation of this tube alternatively brings the inlet and exhaust ports in communication with the

cylinder; during the compression and firing strokes the plain
portion of the tube seals the cylinder; the tube runs at one-half
engine speed for four-cycle operation.

By making the tubular valve of larger diameter it can be arranged
to run at one-quarter engine speed. The valve may be liquid-
cooled for car engines, but for motor-cycle ones a simple contact-
cooled one is satisfactory.

The difficulty, experienced in early engines, in connection
with making the tube gas-tight has been overcome by chamfering
the phosphor-bronze sleeve ports and setting them inwards
slightly, so that when the valve is in position these edges are
sprung and thus seal the joint. Lubrication of the sleeve is carried
out by supplying a liberal quantity of oil to the surface of the
rotary valve at about the lateral centre line and removing the
surplus oil from the other side by means of a scraper device and
non-return valve so that the oil is returned to the sump. The oil
supply is controlled by the throttle opening, being greater at the
larger throttle openings.

In regard to the performance of this engine which has been
developed over a period of about 20 years, high outputs have been
obtained from small engines. Thus, in the instance of an air-
cooled 247 c.c. engine (unsupercharged), using a low grade fuel
of 65 octane value, a B.M.E.P. of 165 lbs. per sq. in. was developed
at 5,350 r.p.m. and 176 lbs per sq. in. at 6,000 r.p.m.

The high performance is due to the relatively cool combustion
chamber which has no hot exhaust valve heads and no pockets.
It is usual to employ compression ratios of 10:1 to 11:1 with
normal commercial grades of petrol, since no detonation occurs
with these compressions; the corresponding poppet valve engine
would only be able to employ a compression ratio of about 6:1
with such fuels.

A special feature of the Cross engine is its relatively low fuel
consumption due to the high thermal efficiency which is associated
with the compression ratios previously mentioned. Thus, the
247 c.c. engine referred to gave a fuel consumption of only 0·35
lbs. per B.H.P. per hour at 4,000 r.p.m. It is necessary to employ
efficient cooling means for this type of engine so that special
attention is needed in the design of the cooling fins or water jackets

in order to maintain the valve unit and its housing sufficiently cool under all circumstances.

A later development of the Cross engine is the controlled valve loading system which overcomes the objection sometimes put forward against rotary valve engines, namely, that the valve is

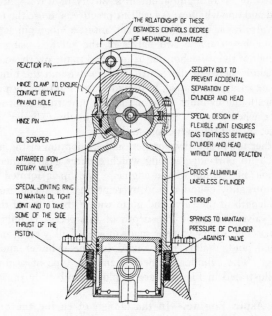

THE RELATIONSHIP OF THESE DISTANCES CONTROLS DEGREE OF MECHANICAL ADVANTAGE

REACTION PIN

HINGE CLAMP TO ENSURE CONTACT BETWEEN PIN AND HOLE

HINGE PIN

OIL SCRAPER

NITRARDED IRON ROTARY VALVE

SPECIAL JOINTING RING TO MAINTAIN OIL TIGHT JOINT AND TO TAKE SOME OF THE SIDE THRUST OF THE PISTON

SECURITY BOLT TO PREVENT ACCIDENTAL SEPARATION OF CYLINDER AND HEAD

SPECIAL DESIGN OF FLEXIBLE JOINT ENSURES GAS TIGHTNESS BETWEEN CYLINDER AND HEAD WITHOUT OUTWARD REACTION

'CROSS' ALUMINIUM LINERLESS CYLINDER

STIRRUP

SPRINGS TO MAINTAIN PRESSURE OF CYLINDER AGAINST VALVE

Fig. 200.—Improved Cross Engine Arrangement.

exposed to the full explosion pressure and must therefore tend to bind in its housing or to experience excessive wear on its surface and bearings; further, differential expansion effects between the valve and its housing are eliminated.

The loading of the valve is controlled by the designer so that only sufficient pressure is exerted upon the valve to maintain its contact with the sealing port-edge lip.

The controlled valve loading is brought about by utilizing the gas pressure in the cylinder to provide the necessary force to exert pressure on the two halves of the valve housing and controlling the amount of this force by the principle of leverage or mechanical advantage. (Fig. 200).

The system of mechanical advantage or leverage to reduce this pressure can be brought about in a variety of ways. A simple method and one which works well in practice is to use the top half of the valve housing as the lever and insert a hinge pin at one of the joints between the two halves of the valve housing. The upper half of the valve housing is attached to the overhead stirrup at a position between the centre of the valve and the hinge pin.

It is necessary, of course, to provide some light spring pressure underneath the cylinder to maintain the upward pressure of the cylinder when there is no gas pressure in the cylinder, as, for instance, during the induction stroke.

Another important development is the adoption of the method of raised lip port-edge sealing which ensures gas-tightness over the whole life period of the engine. The lip is formed in the aluminium alloy cylinder head by removing the metal around it for the length of the valve and up to within about 10° of the split in the valve housing. The depth of this recess is only about ·004 in. The rotary valve is made of special alloy steel, nitrogen hardened, and Y-alloy is favoured for the cylinder and valve housing. The cylinder is linerless and the type of pistons and rings illustrated in Fig. 200 are employed with fully satisfactory results.

The Aspin Engine.—In this design of engine the ordinary poppet valves are replaced by a rotating conical member having ports for the inlet and exhaust operations. The engine in question has a relatively high output due to the fact that it can employ much higher compression ratios than those used in poppet-valve engines without detonation effects occurring. The absence of hot exhaust valve heads and the beneficial shape of the "pocketless" combustion chamber are the chief contributory factors in this connection.

Referring to Fig. 201* which shows the cylinder head, rotating valve and port arrangement in detail, the rotary valve unit, shown

* *The Automobile Engineer.*

at F is a Nitralloy steel shell filled with a light alloy and coned externally to 60°. Attached to this is a cylindrical member B running on roller bearings at A and C. Above is the half-speed gear wheel E engaging at the left with the smaller gear wheel H which is driven by the vertical shaft N. Referring, again, to the left-hand sectional view the cavity in the rotating

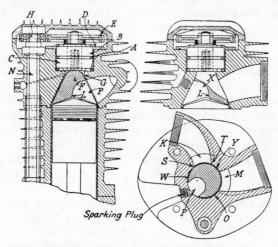

Fig. 201.—Illustrating Principle of the Aspin Engine.

valve F constitutes the compression space, the piston crown at top dead centre coming close to the lower conical surface of the rotary valve; the contour of the port at L is stated to be an important item in the valve design.

The valve F rotates anti-clockwise and, as shown, is approximately in the charge ignition position and remote from any hot area. After ignition the port P rotates past the plain surface at O, during the firing stroke and thence reaches the angular position opposite M, the exhaust port, when the exhaust operation occurs. After this the inlet port K is uncovered and overlaps M to a degree which is determined by the length of the section Y and by the location of the point T, which represents the start of the induction stroke. The point S determines the valve closing lag, after which the port P passes over the blank surface W to give the compression operation and thence to the ignition position near the end of the compression stroke shown in the left-hand illustration.

Owing to the shape of the combustion chamber formed in the rotating valve and to the absence of the hot-spots associated with poppet-valve engines, it is possible, in the Aspin engine, to employ higher compression ratios without detonation effects. It is believed that after ignition there is an initial flame-front movement only, the whole of the remaining charge being then simultaneously projected radially from three sides and at varying but very high velocities upon the flame nucleus as the piston reaches its top dead centre, thus prohibiting any specifically directional movement since all avenues of frontal movement are then closed and complete combustion occurs very quickly, but progressively.

Performance Figures. An experimental single-cylinder air-cooled engine, of 2·64 in. (67 mm.) bore and 2·78 in. (70·5 mm.) stroke, having a capacity of 15·2 cu. in. (249 c.c.) developed 5 B.H.P. at 1,800 r.p.m., 25 B.H.P. at 7,000 r.p.m. and 32 B.H.P. at 11,000 r.p.m. It was found possible to employ a compression ratio of 14·1:1 with an ordinary fuel of 0·74 S.G. and heating value 18,810 B.T.U.'s per lb. It is stated that low octane fuels, such as paraffin oil, can be used at this high compression ratio without detonation and that the engine, which was not supercharged, could be operated at a speed of 14, 850 r.p.m. without failure.

Mention may be made of two later types of engine, namely, a small car model and a commercial vehicle one, have been built and tested on the bench and road. The smaller model is a four-cylinder, water-cooled engine of 937 c.c. (57 cu. in.), which under test developed a maximum output of 37 B.H.P. at 4,800 r.p.m., with maximum torque occurring at 2,800 r.p.m. It has a compression ratio of 12:1 and ran on ordinary petrol, without any signs of detonation. When fitted into an 8 H.P. car chassis, the fuel consumption at 25 to 30 M.P.H. was 62 to 64 M.P.G.; at 35 to 40 M.P.H., 50 M.P.G. and at 45 to 50 M.P.H., 42 to 45 M.P.G. The ordinary poppet-valve engine normally used in the same car had a fuel consumption at 40 M.P.H. of about 35 M.P.G.

The larger commercial vehicle four-cylinder engine, of $4\frac{1}{4}$ in. bore and 5 in. stroke, (4,650 c.c. or 285 cu. in.) had a compressino ratio of 9·2:1 and it developed 110 to 114 B.H.P. at 2,500 r.p.m. with a minimum fuel consumption of 0·42 lbs. per B.H.P. hour—a

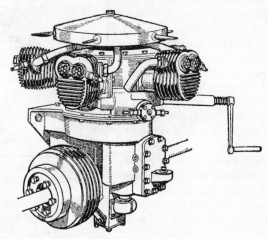

Fig. 202.—A $1\frac{1}{2}$-litre Five-Cylinder Radial Engine which was used in a Motor Car.

value comparable to that of a Diesel engine. The maximum torque of 265 lbs. ft. was developed at 1,750 r.p.m.

The Radial Engine.—In this type the cylinders are arranged in star fashion about the crankshaft's axis, so as to lie in the same plane. The connecting-rods all operate upon a single throw crank.

The two-, three- and five-cylinder air-cooled radial engines have been used, in the past, for automobile purposes, and also a continental racing car was fitted with a radial engine. Engines with 7, 9, and 14 cylinders have been used for aircraft.

The advantages of this type lie in its compactness, low weight per horse-power and accessibility. Against these merits must be offset the disadvantage of larger frontal area, more complicated exhaust pipe system and the difficulty of ensuring that all the cylinders receive equal mixture charges and lubrication; following successful aircraft radial engine practice, however, this difficulty can be surmounted.

The simplicity of the single-throw crankshaft and the single cam ring for operating the valves are other attractive features of this type of engine.

The usual arrangement for radial engines is to employ an odd number of cylinders, such as 3, 5, 7 and 9 in order to obtain equal firing intervals as expressed by equal crank-angle degrees.

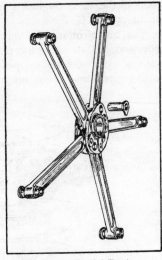

Fig. 203.—Radial Engine Connecting-rod Arrangement.

A nine-cylinder air-cooled radial C.I. engine was employed in military tanks made in the U.S.A. This engine known as the Guiberson, is similar to the aircraft engine made by the same manufacturers.

In regard to the *Connecting-rod* arrangement for radial engines this consists of one main rod, known as the *Master Rod* which is usually provided with a white-metal lined bearing running on the single crankpin. The big-end of this rod is flanged to a U-section shape and bearing pins

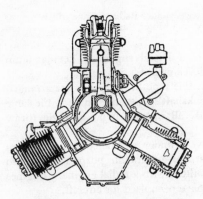

Fig. 204.—A Three-Cylinder Single Sleeve Valve Radial Engine.

pass through the flanges; these pins form the bearing members for the big-end bushes of the other connecting-rods, which are known as *Link Rods*. In the illustration of a typical five-cylinder connecting-rod system the upper left-hand rod is the master rod and the other four the link rods.

The crankshafts of radial engines are built up, so that one crank member is detachable and the master rod can slide over the crankpin; it is not necessary to employ a split big-end bearing.

The Wankel Engine.[*]—This purely rotary type of engine, which has been under development, from ideas originated with Felix Wankel, since 1924, and has since been taken up by N.S.U., of Germany (1951) and later by the American Curtiss Wright Company (1958) is based upon the rotation of what may be termed a rotor having a triangular shape, but with curved sides, within a casing of a special geometrical shape, known as a two-node epitrochoid, derived from a family of curves based upon the path of a point on the radius of a circle rolling on the outside of a fixed circular shape. It is well known that there are many forms of rotor (geometrical) shapes which can revolve within geometrical casings, such that one or more parts of the rotor will always remain in contact with the casing, and this is a special example.

The earlier D.K.M. engine had a rotor and chamber (casing) that rotated in the same direction but at different speeds; in the ratio of 2:3, but in all later models the chamber unit is fixed and the rotor revolves about its own axis, as indicated by the smaller dots in Diagrams *D, E. F* and *G* (Fig. 205), so that this axis is eccentric to the centre of the casing, which is shown by the larger black dots. The end of the rotor is provided with an internal gear or annulus, concentric with its centre and this annulus rotates around a smaller fixed pinion gear, mounted on the side cover of the casing. The rotor revolves about the geometrical centre of the casing and it also rotates about its own centre, at $\frac{2}{3}$ of the angular velocity of the main shaft. In order to give the correct ratio that is necessary for the correct operation, namely 1:3 for the rotor and main shaft, the epicyclic gear mentioned was introduced. The main shaft carries an eccentric or equivalent "crank", on which the rotor rotates on roller bearings.

[*] For more recent information see Chapter XI.

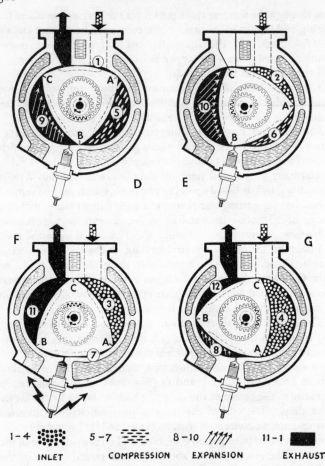

INLET · COMPRESSION · EXPANSION · EXHAUST

1-4   5-7   8-10   11-1

Fig. 205.—Illustrating the Operation of the
Wankel Rotary-type Engine.

Cycle of Operations.—Referring to Fig. 205 (*D*) this shows the
mixture inlet and exhaust ports (on the right and left, respectively),
the water-cooled chamber, sparking plug location, rotor and
epicyclic gearing. The rotor rotates clockwise and during each

complete rotation three complete Otto cycles occur, viz., Induction, Compression, Expansion (following ignition) and Exhaust. If we confine our attention, firstly to the induction process, by following the movement of the side CA of the rotor, then at (1) in Diagram (D) the exhaust and inlet ports are sealed. At (2) in (E) the inlet port has opened—since there is a recess in the "sliding" side of the rotor as indicated by the dotted line between CA. At (3) in (F) the space between CA and the casing is partly filled with mixture. This corresponds to the piston being part way down on its induction stroke in the orthodox piston-cylinder engine. At (4) in (G), the mixture completely fills the space corresponding to that of the ordinary cylinder and combustion volume.

Next, consider the compression sequence, by referring to the rotor side AB, which is shown at the beginning of the compression at (5) in (D). The next stage is shown at (6) in (E) and the final, or full compression stage—corresponding to just before top dead centre of the piston in the orthodox engine—at (7) in (F), when the ignition spark occurs. The expansion of the burnt gases proceeds in the stages (8) in (G), (9) in (D) and (10) in (E), at which point the tip C of the rotor commences to open the exhaust port which is indicated in black on the left in (D). The exhaust process then occurs during (11) in (F), (12) in (G) and ending at (1) in (D). It will thus be seen that during one complete revolution of the rotor, three complete cycles have taken place.

The Rotor Seals.—In common with the other rotary engines employing rotors that revolve within special casings, including the eccentric rotor having a series of vanes which contact the circular casing of the fixed member, problems have arisen in the case of the Wankel engine regarding the efficient sealing, over long operational periods of the three rotor tips and also the two side faces of the parallel-sided rotor. In this connection a considerable amount of experimental work has been carried out on various sealing methods and materials, the result of which has been to achieve a fair measure of success. The side seals consisting of segments of piston rings with light beryllium springs behind, assisted by gas pressure have proved effective. The rotor tip seals, consist of sliding vanes moved outwards against the chamber wall by centrifugal force but it has been shown that variation of the gas and friction forces on the vanes at different parts of the working cycles tend to cause an oscillatory or "flutter" effect in the vanes. The use of light expansion springs and the plating of the curved chamber surface with chromium have helped to overcome this difficulty. Endurance tests* on Wankel engines indicate that, at the time, operational periods of 500 to 800 hours, without failure were possible with the apex and side seals described earlier. To compare with piston ring endurance, the total operational period should be at least 1,000 hours.

Power Output of Wankel Engine.—It can be shown that the Wankel engine, while acting on the four-cycle principle has a power output of a two-cycle engine having three cylinders, but with its output shaft being driven with a 3:1 step-up ratio. The power output is thus given as follows:

$$\text{B.H.P.} = \frac{3\ V\ \text{B.M.E.P.}}{6,475} \times \frac{N}{3} = \frac{V\ N\ \text{B.M.E.P.}}{6,475}$$

Where V = capacity of one rotor face swept chamber, in litres, N is the revolutions per minute and the B.M.E.P. is in pounds per sq. in.

It will be seen that this is also the same formula as that of a single cylinder two-cycle piston engine of the same capacity but, as is mentioned later the B.M.E.P. of the Wankel engine and also its maximum r.p.m. are much higher than those of the piston-type two-cycle engine.

Some Experimental Results.—Several experimental engines have been built on the Wankel principle, for test and subsequent development purposes; also to overcome some of the practical problems in connection with the rotor end and side seals and certain thermal and mechanical problems. These engines have been built in relatively small sizes, namely, the KK models of 125 c.c. and 250 c.c. capacities.

The results of some tests on the 250 c.c. engine showed that B.M.E.P. values of 80 to 175 lb. per sq. in. were obtained with a maximum output of about 35 B.H.P. at about 9,000 r.p.m. and with a B.M.E.P. of about 105 lb. per sq. in. The full-load

* Oct. 1960.

torque was about 30 lb. ft. at 5,500 r.p.m. The fuel consumptions for the speed range of about 5,000 to 8,500 r.p.m. were from about 0·55 to 0·65 lb. per B.H.P. per hour.

A duration test of 100 hours on a 250 c.c. KKM engine of improved design was made with the engine developing 31 B.H.P. at 5,500 r.p.m. and with a B.M.E.P. of 143 lb. per sq. in. The fuel consumption ranged from 0·70 lb. per B.H.P. per hour at the start of the test falling to about 0·60 lb. at the end of 100 hours. Examination of the engine components after the test indicated a small amount of wear at the tops of the rotor vanes, and on the chamber surface and erosive wear at the exhaust port. With the use of chromium-plated aluminium chambers and other improvements it was anticipated that the wear could be reduced appreciably in future engines.

In the test engines the rotors, which are subjected to high average temperatures, due to the occurrence of the equivalent of twelve piston engine strokes, per revolution, were oil-cooled and a certain amount of oil was admitted to the chamber to assist the rotor seals in their operation, so that the oil consumption was appreciably greater than in the equivalent petrol engine. It was also found that exhaust gas temperatures were much higher at full output, namely, about 1,600° F. (872° C.) than in piston engines producing the same mean cylinder pressures.

While owing to space considerations it has been possible, to give only a brief outline of the Wankel engine, the following summary may be of interest to readers.

Advantages

(1) The Wankel engine, for a given power output is relatively small in size and weight. Thus, a 250 c.c. test engine had an outside diameter of about 9·5 in. and overall width of 7 in. The weight of the KKM 250 c.c. engine with a cast iron casing was 73 lb. and with an aluminium alloy casing, 48 lb. This is equivalent to about 2½ lb. per B.H.P. Later results show that weights of 1 lb. or less are obtainable.

(2) The power output per litre was about 120 B.H.P. at 9,000 r.p.m.

(3) The engine contains no reciprocating parts and any small unbalanced rotary items are readily balanced by counterweights,

so that this type of engine runs practically free from vibration up to the highest operational speeds. Practical road tests appear to confirm this.

(4) It has fewer working parts than the equivalent piston engine with its pistons, connecting-rods, valve gear components, etc. As with the two-cycle engine ports in the casing take the place of valves and valve gearing.

(5) With modern machining methods and materials, it is claimed that it can be designed for mass production more cheaply than the equivalent petrol engine.

(6) High volumetric efficiency.

Certain Disadvantages.

(1) Lower torques at lower road speeds than the piston engine.

(2) Lower engine braking effect in automobiles.

(3) Rotor sealing difficulties and chamber surface wear.

(4) Higher oil consumption per B.H.P. per hour.

(5) The much higher engine speed range introduces transmission problems, so that new transmission designs are required.

(6) Higher specific fuel consumptions over the lower speed range.

(7) Ignition troubles which, in experimental engines, necessitate more frequent plug changes. Later developments, however, by Curtiss Wright in America have shown that a transistor power, condensor discharge ignition method will overcome this drawback.

(8) Possibility of chamber unit distortion due to close proximity of inlet and exhaust ports.

The attractions of this small, light and high output engine are such that a considerable amount of research and development work is in hand, with the object of overcoming the difficulties encountered in earlier engines and in the subsequent production of larger output engines for automobile, aircraft and stationary power applications. The principle appears applicable to Diesel engines, since the rotor design can be adapted to give compression ratios up to at least 20:1.

Notes on Rotary Expansion Engines.—Apart from the present Wankel engine there have been several previous attempts at designing or building unusual types of engines using either a rotor of special geometric shape, rolling within a ported chamber of curvilinear geometric section. So far back as 1901 patents were

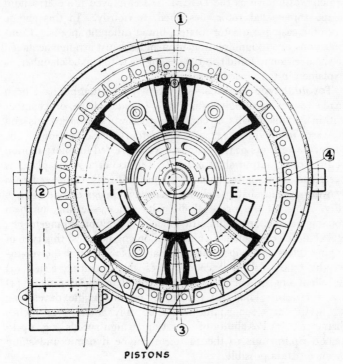

PISTONS

Fig. 206.—Showing the Principle of the Bradshaw Omega Toroidal Engine.

(1) Expansion stroke.   (2) Pistons at bottom of inlet stroke.
(3) End of exhaust stroke.   (4) End of firing stroke.   *I* denotes Inlet ports and *E*, the Exhaust ports.

taken out for a rotary expansion engine, but this and subsequent types met with no practical success. It is only since about 1938 that any real progress was made, following the patents of De la Vaud who in 1938 investigated the possibilities of trochoidal curved units for rotary positive displacement air compressors and pumps.

From about 1947 onwards work was recommenced by Wankel on his rotary expansion engine, the first practical working model of which was known as the D.K.M. and employed the epitrochoid shape three-sided rotor, described previously. In this engine both the casing and rotor rotated, but at different speeds. Complications due to ignition induction and exhaust arrangements led to the decision to build this type of engine with a fixed chamber, as explained in the preceding section.

Toroidal-type Engines.—In recent years attempts have been made to develop engines having a toroidal shape of chamber within in which pairs of curved pistons worked backwards and forwards to provide the strokes of the four-cycle operation. The power movements of the pistons were converted to rotary motion at the power output shaft either by the use of a swash plate, as described later, or by pairs of connecting-rods connected to a two-throw crankshaft, as in the Bradshaw Omega engine of 1956 (Fig. 206). If the practical difficulties of gas-sealing, uniform cooling, satisfactory volumetric efficiency, ignition arrangements, etc. could be overcome, on a production engine basis, this type of engine offers the important advantages of: (1) High power-to-weight ratio. (2) Absence of valves and operating gear. (3) Excellent engine balance over the complete speed range. (4) Compactness. Thus an 1,100 c.c. (67 cu. in.) engine, developing 56 B.H.P. at 5,000 r.p.m. would be only about 9–10 inches diameter. (5) Possibility of operating at high speeds, e.g. up to 12,000 r.p.m., due to the complete engine dynamic and static balance that is possible.

In the Selwood Orbital engine (Fig. 207), which operates on the two-cycle principle the reciprocating movements of the two opposed pistons, is transmitted *via* gudgeon pins to the swash plate the inner part of which runs on two rows of ball bearings. The experimental engine had six curved (toroidal) cylinders, each having a pair of opposed pistons bolted to a trunnion block with the gudgeon pin, to connect the piston with a spider arm on the swash plate assembly. As in the case of the Bradshaw Omega engine the chamber in which the pistons reciprocate, rotates. It is arranged for diametrically opposite "cylinders" to fire at the same time, so as to cancel out end thrust on the main

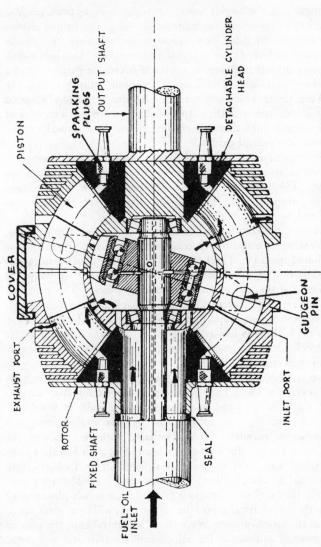

Fig. 207.—The Selwood Orbital Engine, with Swash-plate Method for Conversion of Piston Motions to Crankshaft Rotation.

bearings. As the engine works on the two-cycle principle and there can be no crankcase compression—as in the simple piston-type engine—air must be supplied to the cylinders during the induction period by an external compressor. The lubrication difficulty inherent in such an engine is overcome by the use of a petrol-oil and air. The ignition method employed for the sparking plugs which are arranged on the end cover a round circle, concentric with the output shaft, is to provide a number of fixed contacts such that as each plug passes these a spark crosses across to the central electrode.

The experimental 700 c.c. (43 cu. in.) engine was designed to operate at 10,000 r.p.m. and with its air compressor, weighed 60 lb. The bore and stroke of the pistons were 1·75 in. and 1·5 in., respectively. The outside diameter of the unit was 10·5 in.

The engine has the advantages of lightness, compactness, excellent balance and the absence of valves and valve gear. It is capable of operation at high speeds with a consequent big improve-ment in output. But, as with other engine-ported two-cycle engines the thermal efficiency is relatively low. The principal drawback is that inherent with the swash-plate principle, namely, that since a certain proportion of piston thrust is directed on the main shaft, there is less available for actual torsion on the shaft; the mechanical efficiency is thus lower than for the orthodox piston-type engine.

The Swash-Plate Engine.—Amongst the numerous experi-mental engines which have been suggested, or built, from time to time, the axial or swash-plate engine (Fig. 208), is worthy of mention. Briefly, the arrangement of this engine comprises a number of parallel cylinders, situated with their centres, or axes, on a circle. Each piston has a rod provided with a ball-end which can work in a socket situated in an inclined plate known as the *Swash-Plate*; the latter can rotate about its axis. Usually the block of cylinders is fixed and the swash-plate rotates about the symmetrical axis of the system. It will be evident, then, that as the swash-plate revolves about its central axis, the pistons will move in and out, as the ball-ends of the rods are, as it were, guided around the swash-plate periphery. Each piston completes two strokes per revolution of the swash-plate so that the ordinary four-stroke cycle can be followed. The length of the piston stroke is given by the distance between the perpendiculars from the ball-ends on to the shaft axis at the extreme positions. In this manner the piston thrusts due to the firing and expansion loads are converted directly into turning effort, without the use of a crankshaft.

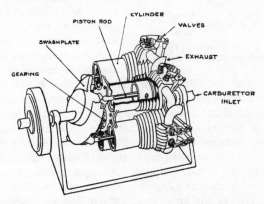

Fig. 208.—Illustrating the Revolving Swash-Plate Engine
Principle.

If, instead of fixing the cylinder unit, the swash-plate is fixed rigidly to a suitable mounting, then the *cylinders could rotate about the swash-plate* axis. This arrangement would have the advantages of (*a*) Better cooling for the cylinders. (*b*) Enhanced flywheel effect, due to the weight and rotation of the main engine unit. The carburation, ignition and exhaust connections would, however, introduce practical problems.

It has always been doubtful whether the claimed advantages of this engine, can be offset against the increased frictional losses due to the thrusts of the rods on the swash-plate and the other practical difficulties mentioned.

The Bristol Axial Engine.—A particularly neat and compact engine of the multi-cylinder type was built and put into

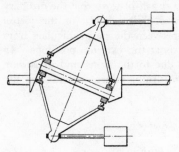

Fig. 209.—Showing Principle
of Bristol Axial Engine.

experimental service by the Bristol Tramways and Carriage Company. This engine has nine cylinders equally spaced and arranged axially about the crankshaft axis. It belongs to the engine class known as the *Wobble Plate* type, and differs from the swash plate model by the fact that the member which transmits power from the pistons to the crankshaft does not revolve. The wobble plate, or star member, is mounted upon a Z-shaped crankshaft (Fig. 209). The slanting crankpin is set at an angle of $22\frac{1}{2}$ degrees to the line of the main shaft with the centre of the crankpin in line with the centre of the main shaft. The star member has ball-bearings, and is mounted upon the crankpin so that the shaft can be revolved within the star member, thus imparting a "wobble" to it. A torque member, or stabilizer, is attached to the star member in such a way as to allow universal motion but, at the same time, prevent the star member from revolving upon the shaft.

The connecting-rods and pistons of the nine cylinders are connected to the star member by spherical bushes, which encircle the ball-ended arms arranged around the periphery of the star member.

The engine is provided with a large rotary valve system. The rotary valve has four pairs of inlet and exhaust ports which register alternately with a single port in each cylinder end. The rotary valve rotates at one-eighth engine speed, and a perfect seal was claimed to be obtained between the rubbing faces of the valve and the fixed valve plate on the cylinder ends.

The engine described was built in the 7-litre class, and was actually 5 cwt. lighter than the corresponding 7-litre vertical cylinder engine; in addition it occupied considerably less space.

The engine had a higher mechanical efficiency than the ordinary type of axial engine owing, it is believed, to the smaller number

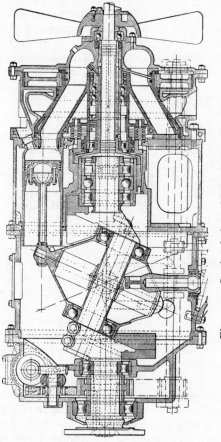

Fig. 210.—Sectional view of Bristol Axial Engine.

of principal bearings and absence of connecting-rod angularity
and its side thrust upon the piston.

It is stated by the makers that this engine required only one-
third of the power of an ordinary engine to motor it around.

The axial engine (7-litre) gave a maximum B.H.P. of 150
at 3,000 r.p.m. as against 117 B.H.P. at 2,600 r.p.m. for the
7-litre comparable vertical engine.   The maximum engine torque
was 320 lb. ft. as compared with 294 lb. ft. of the latter engine.

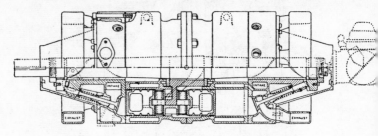

Fig. 211.—The Herrman Barrel Engine.

The petrol consumption over the whole speed was about 15 to
17 per cent less than that of the vertical engine.

Another interesting feature of the Bristol engine was that, by
the provision of two counter-balance weights to the crankshaft, it can be given a theoretically perfect balance.

The Hermann Barrel Engine.—In this design two cylinder
blocks are employed, each having six parallel cylinder bores.
They are joined together by flanges and bolts.   The pistons are
cast in pairs connected by a solid web carrying two rollers.   These
are in contact with the cam plate which forms two sine waves
(Fig. 212).   One revolution of the engine shaft corresponds to
four strokes of each piston or to a complete cycle of four-cylinder
unit.   All valves are located in the cylinder heads, being slightly
inclined to the cylinder axes so that the stems converge towards
and are in contact with bevelled cam faces on a plate mounted on
the mainshaft.   Each plate has two cam lobes, the inner for the
exhaust and the outer for the inlet valves.   An annular induction
manifold is formed round the main bearing having radial passages

to the valve pockets, the combustion chambers being shaped to obviate any detonation tendency. The cylinders fire in direct sequence all round the block.

The later engine had twelve cylinders, the bore and effective stroke being 77·6 mm ($3\frac{1}{16}$ in.) and 95·3 mm. ($3\frac{3}{4}$ in.) respectively, giving a capacity of 5·407 litres (330 cu. in.). The compression ratio was 7·5:1 and the fuel was 100 octane rating. The performance results were obtained under sustained operation conditions but without accessories. The maximum output was 144 B.H.P. at 1,600 r.p.m. and the maximum B.M.E.P., 120 lbs. per sq. in. at 900 r.p.m. The minimum specific fuel consumption was 0·5 lbs. per B.H.P. hour.

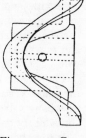

Fig. 212. — Cam Plate used on the Herrmann Barrel Engine.

The frictional losses are only 75 per cent of those of a similar size poppet-valve engine. The engine is less liable to detonation also, and it has been shown possible to operate on fuel of 62 octane rating with a compression ratio of 6·5:1.

When supercharged, a similar engine developed a maximum output of 245 B.H.P. at 1,500 r.p.m., using fuel of 100 octane rating. During an endurance test of 25 hours this engine maintained an output between 208 and 213 B.H.P. at 1,500 r.p.m.

The dry weight of the engine was 248·75 lbs. and it measured 30 in. in length by 12·75 in. diameter.

Engines Using Gas as Fuel.—Apart from the ordinary petrol and C.I. engines, which employ liquid fuels, there are automobile engines in use operating upon combustible gases, such as town lighting (or coal) gas and producer gas.

The use of the vehicles to which these systems are applied is limited chiefly to commercial goods carrying ores, where fuel costs are of primary importance.

Coal gas is also used for running-in new petrol engines.

Coal gas has been used on a small scale over a considerable period, more particularly for both goods vehicles and private cars during the war periods, when petrol was both dear and limited in supply.

More recently a high-pressure gas system has been used, in which coal gas from town supply sources is first compressed to 5,000 lbs. per sq. in., and stored in strong receivers. It has been used to supply commercial vehicles which are fitted for the purpose with light, high strength "Vibrac" alloy steel cylinders, with gas at 3,000 lbs. per sq. in. A typical vehicle installation would consist of six steel "bottles" connected by strong piping together and to a special pressure-reducing valve for supplying the engine with gas at a low pressure.

The engine used is the ordinary petrol type, but in place of the usual carburettor there is a gas inlet with a variable air supply or mixing valve. There is also a suction-operated device which shuts off the gas supply when the engine stops.

It was shown that when using coal gas of 475 to 500 B.T.U.'s per cubic foot, calorific value, the overall running costs of an engine (including cost of compressing and storing gas) are equivalent to petrol about 8d. to 9d. per gallon.

Another type of vehicle has its own gas generator in the form of a kind of "boiler" mounted on the chassis. The fuel used is either anthracite, coke or wood. Air is drawn through the ignited or glowing fuel in a smaller amount than necessary for complete combustion, and water vapour is also drawn into the heated fuel. The resultant product drawn out of the "producer" is a gas containing hydrogen, methane, carbon-monoxide, nitrogen and usually a small amount of carbon-dioxide. This gas has a high calorific value, and can be used in a similar manner to coal gas.

Petrol Injection Spark Ignition Engines.—In this type of engine no carburettor is employed. Instead, pure air is drawn into the cylinder during the suction stroke, the fuel being injected either directly into the cylinder head during the compression stroke or on to the inlet valve from the manifold side and ignited towards the end of the latter by means of a sparking plug. This type of engine does not operate on the compression-ignition principle and has a much lower compression ratio, namely, about 8·0:1 to 9·5:1.

Development in this class of medium-compression engines in this country was confined largely to aircraft engines, where

maximum power per unit weight and minimum fuel consumption
are of primary importance.   There is little doubt that the well
designed fuel-injection system gives higher volumetric efficiency,
better mixture distribution to individual cylinders and lower
fuel consumption.

For automobile purposes, however, the injection system has

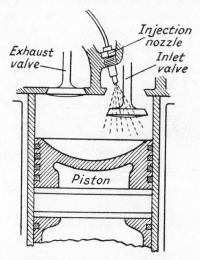

Fig. 213.—The Thornycroft Petrol-Injection Method.

certain drawbacks to offset the advantages previously mentioned.
Thus, instead of a relatively simple carburettor of a well developed
design, there is substituted a complicated and much more costly
fuel injection pump made to high precision limits, together with
individual fuel injection nozzles and fuel pipe lines to the cylinders;
these are, of course, additional to the ignition system, including
the H.T. leads and sparking plugs.   Moreover, the small size
of petrol-injection pump required for the average car engine
would require very careful design and construction to meter the
extremely small quantities of petrol required—more particularly
at slow running and cruising speeds.

Other disadvantages include the extra weight of the injection equipment and the additional items that would require maintenance and servicing, since the fuel injection nozzles are prone to carbon up if of the direct cylinder injection pattern.

Altogether, the advantages at present are well in favour of the much more simple, light and relatively inexpensive carburettor for production cars, at least. The additional power output of the petrol-injection system would be allowed for in the car-

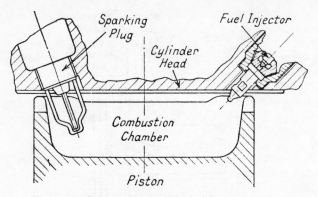

Fig. 214.—Cylinder Head of Typical Petrol Injection, Spark-Ignition Engine (Allis-Chalmers).

burettor engine by a slight increase in cylinder dimensions, or multiple carburettors.

Successful petrol-injection systems are, however, now used for *racing car and some car engines* as better fuel distribution and higher outputs per cylinder are thus obtained.

Diesel Engine Conversion.—It is also of interest to note that some promising performance results have been obtained from a converted Thornycroft high speed Diesel engine of the NR6 type. This Dielsel engine (Fig. 213) has six cylinders of $4\frac{1}{8}$ in. bore and 6 in. stroke, giving 481 cu. in. (7,880 c.c.) cylinder capacity. The compression ratio is 16:1 giving maximum compression and combustion pressures of 560 lb. and 960 lb. per sq. in. respectively. The output is 100 B.H.P. at 1,800 r.p.m. with a maximum B.M.E.P. of 105 lb. per sq. in., and torque, 337 lb. ft.

In order to operate on the petrol-injection method, the compression ratio was reduced to 6·92:1. The system used was to inject a 30° spray cone of petrol into the inlet valve port (Fig. 213) during the first 80° or so of the induction stroke. A special C.A.V. fuel injection pump was employed. The modified engine developed 150 B.H.P. at 1,900 r.p.m. and gave a maximum B.M.E.P. of 140 lb. per sq. in. and torque of 445 lb. ft. (both at 800 r.p.m.).

In regard to fuel consumptions the Diesel engine had an equivalent of 8·25 m.p.g. on the 22-tons Thornycroft vehicle, whereas the petrol-injection converted engine, under similar conditions, gave 6·2 m.p.g. As compared with a carburettor engine giving 100 ton miles per gallon the petrol-injection one had a figure of 130 and the Diesel vehicle 180-200.

The advantage of this petrol-injection engine is that it enables vehicles to be used in countries where Diesel fuel is not obtainable and to operate at much higher speeds.

It may be mentioned that a commercial Waukesha engine is made that can readily be converted to run as a petrol engine, with a carburettor; to work as a Diesel engine, or to operate on a hydrocarbon gas, such as methane or butane.

Allis Chalmers Engine.—Fig. 214 shows the cylinder head of the Allis Chalmers engine, which uses a cavity-type of piston. The compressed air is given a certain degree of turbulence before injection, so that it moves rapidly past the sparking plug points, at the moment of ignition.

The Hesselmann engine also works on the same principle; it has been widely used on American commercial vehicle and stationary engines.

The advantages claimed for the fuel-injection spark ignition engine may be summarized as follows:—(1) Better distribution of fuel and air (mixture) to the cylinders, since the air is drawn into each cylinder direct and not through a carburettor. (2) Higher volumetric efficiency on account of the absence of carburettor and inlet manifold; these cause a certain amount of obstruction to the mixture flow and result in the cylinders not receiving their full charges. (3) Elimination of backfires. (4) Freezing, or icing up troubles in very cold weather are avoided;

in the case of aircraft engines this is a notable difficulty. (5) Better atomization of the fuel. (6) The use of alternative fuels of lower flash-point than petrol is practicable. (7) Absence of the modern complicated carburettor with its various automatic mixture compensation devices, float chamber, etc.

Fuller information on the subject of petrol-injection engines is given in Volume 2 *Carburettors and Fuel Injection Systems*, of this Series.

CHAPTER IV

SUPERCHARGED ENGINES

Supercharged Engines.—The power output of an ordinary four-cycle petrol engine depends upon the quantity of mixture drawn into the cylinder, other conditions remaining the same. Now the quantity of mixture induced will be governed by its temperature and the difference between the pressure (about $14 \cdot 7$ lbs. per sq. in. absolute at sea level) outside and that within the cylinder. In the ordinary engine the pressure is atmospheric outside, so that no matter how well designed the engine is from the carburettor and valve ports point of view, there is a definite limit set to the power output for a given cylinder capacity. In order to obtain more power, it is necessary to supply air to the engine at higher pressure than atmospheric. The power developed increases roughly as the density of the air. A more exact relation is as follows:—

$$\text{H.P.} = k \, (\rho)^{1.1}$$

where H.P. is the horse power, ρ the air density, and k a constant.

The power output of any internal combustion engine is *proportional to the weight of air per min.* that is used by the engine. It is also proportional to the thermal efficiency under the given air inlet pressure conditions.

Usually, the thermal efficiency of engines that are supercharged by means of engine-driven air pumps is rather lower than for the unsupercharged engine; this is partly due to the lower compression ratios employed, to the power used in the compressor and to certain thermodynamic factors.

A drawback of the ordinary suction-induced charge engine is that at the higher engine speeds less and less charge is drawn into the cylinders, owing to the increased frictional and other air resistance effects.

The supercharged engine employs some form of air compressor or pump to force air at a higher pressure than atmospheric into the cylinders during the induction strokes, so that at the end of each induction stroke the cylinder has a greater weight of air-and-petrol charge. There is, hence, a greater amount of potential heat energy in the charge and, within certain limits, a greater amount of power is developed at the same engine speeds.

In practice the supercharging devices are arranged to give varying degrees of increased pressure to the cylinder charge (as measured just at the commencement of the compression stroke) from low supercharge pressures of 3 to 5 lbs. per sq. in. above atmospheric up to higher supercharges of 20 to 25 lbs. per sq. in. above atmospheric; the latter pressures are sometimes used for high output racing engines.

Boost Pressures.—It is now usual in aircraft engine supercharging practice to specify the degree of supercharging in terms of "boost" pressures, namely, the air pressures in the inlet manifold above atmospheric pressure. Thus a boost pressure of 5 lbs. per sq. in. indicates that the pressure in the inlet manifold or intake pipe is 5 lbs. per sq. in. above that of the outside atmosphere. At ground level the pressure in the inlet manifold would therefore be $14.7 + 5 = 19.7$ lbs. per sq. in.

Supercharging of petrol engines is now employed only for (1) Boosting the power of aircraft engines to compensate for the falling off in power output, as the altitude increases; and, thereby, to obtain the greatest output for unit weight of the engine, and (2) For some racing car engines, where maximum output for a given cylinder capacity is required. (3) For certain Diesel engines.

Supercharging is now seldom used for ordinary car engines, except indirectly for two-cycle engines of the air-scavenged kind. Any increase in output is obtained more economically from the manufacturing cost viewpoint, by an increase in cylinder size.

Supercharging versus High Compression.—Of the possible methods for increasing the power output from an engine of given capacity, and fixed maximum speed there are two important ones that can be employed in practice, namely, (1) By the use of higher compression ratios. (2) By supercharging. In each case the result is to increase the brake mean effective pressure, and since the speed has been assumed constant *the power output will increase directly as the mean pressure*.

The disadvantage of the high compression method is that with the increase in mean pressure obtained the maximum pressure of combustion also increases. With supercharging however, as the pressure of the supercharged mixture in the cylinder at the commencement of the compression stroke increases the maximum cylinder pressures do not increase at the same rate and therefore a much higher value of B.M.E.P. can be obtained for the same or lower values of the maximum cylinder pressure.

Some Supercharging Results.—Another important factor is that of cylinder temperatures which are generally lower for a given B.M.E.P. for supercharged engines than for high compression ones; further, it is not so necessary to employ fuels of such high octane values with normal types of supercharged engines.

The results of some tests made on supercharged engines have shown that in a given instance an engine having no supercharge gave a B.M.E.P. of 110 lbs. per sq. in., a maximum cylinder pressure of 495 lbs. per sq. in. and maximum combustion temperature of 4400° Fah. (abs.) (2427° C.).

With a supercharge of 25 per cent above atmospheric, the B.M.E.P. and maximum pressures were 132 and 600 lbs. per sq. in. respectively, and maximum temperature of 4525° Fah. (2718° C.). For a 50 per cent supercharge the corresponding pressures were 162 and 735 lbs. per sq. in., respectively, with a maximum cylinder of 4590° Fah. (2352° C.).

When the supercharge was 100 per cent the mean and maximum pressures were 195 and 1000 lb. per sq. in., respectively with a maximum temperature of 4675° Fah. (2580° C.). It is seen that the temperatures increase at a relatively slow rate.

An interesting comparison between the supercharged and high compression pressure types of engines giving about the same power output from a cylinder of given bore and stroke at a given speed and mixture strength has been made by A. H. R. Fedden. For the purposes of comparison the performance of the unsupercharged engine was also considered and the following results obtained:

TABLE 9

Comparison between Supercharged and High Compression Engines

	Unsupercharged Engine A	High Compression Engine B	Supercharged Engine C
Compression Ratio	5:1	11:1	5:1 (Supercharged to 50 per cent)
Indicated Mean Effective Pressure (lbs. per sq. in.)	126	154	157
Maximum Cylinder Pressure (lbs. per sq. in.)	455	1000	595

These results show clearly the advantage derived from supercharging for the given power output as represented by the higher mean pressure which is obtained for a maximum pressure of only 595 lb. per sq. in. as compared with 1000 lbs. per sq. in. for the high compression engine. The latter would have to be built much heavier in order to withstand the higher pressures.

In regard to the degree of supercharging employed in the engine in question it should be explained that the original pressure in the cylinder for the unsupercharged engine was 14·7 lbs. per sq. in. and for the 50 per cent supercharge, 22·0 lbs. per sq. in. The effect of supercharging, however, results in a higher compression pressure, the respective values for the three engines *A*, *B* and *C* given in the preceding table, being approximately, 140, 265 and 210 lbs. per sq. in., respectively.

Power for Driving the Supercharger.—In most supercharged systems the air compressor is driven from the engine crankshaft by a gear train or other means of transmission, so that at each engine speed a certain amount of the extra power developed as a result of the supercharging is absorbed in driving the compressor and this power must therefore be deducted from the total power developed in order to give the useful or shaft power of the engine. On the other hand the pistons of the engine during the "suction" stroke do not have to draw air into the cylinder but are actually under the air pressure effect of the supercharger so that they perform useful work.

Gear-driven superchargers of the centrifugal and blower types usually absorb from 5 to 10 per cent of the total power output to drive them so that this power must be subtracted from the total power in order to obtain the net or useful output.

When the supercharger is of the *exhaust driven centrifugal compressor type* the engine does not have to supply additional power to drive the compressor and the total power developed therefore requires no deduction on this account. The system in question is therefore *the more efficient one*. In this connection it has been shown that for aircraft engine purposes exhaust-driven superchargers—although introducing certain practical difficulties in connection with the high temperatures involved in the exhaust system—enable an aircraft fitted with a given engine to attain higher altitudes than for the gear-driven supercharged engine of similar dimensions.

Fuel Consumptions of Supercharged Engines.—Although it is necessary to use a greater quantity of fuel per unit time with supercharged engines, yet owing to the greater power outputs obtained the amount of fuel used per B.H.P. per hour is usually only a little greater than for the same engine unsupercharged; this applies to moderate degrees of supercharging. When, however, engines are supercharged to give outputs exceeding about 30 to 40 per cent, the mixture is made richer, to reduce detonation and pre-ignition tendencies, so that rather higher fuel consumptions per B.H.P. hour result. With the exhaust gas turbine supercharger, owing to the utilization of some of the energy of the exhaust gases that, otherwise, would be wasted, higher thermal efficiencies with appreciably lower fuel consumptions have been obtained than for the unsupercharged engine of similar dimensions and nominal compression ratio.

Automobile Engine Results.—Fig. 215 illustrates the effects of supercharging an American eight-cylinder automobile engine. The engine in question was fitted with a centrifugal air compressor, gear and friction driven at six times the crankshaft speed, i.e., at speeds up to 24,000 r.p.m.

The full lines in Fig. 215 show the B.M.E.P., B.H.P. and engine torque values of the unsupercharged engine, whilst the dotted line diagrams indicate the improvements obtained by supercharging. It will be seen that the horse power is increased over the whole speed range, and the maximum B.H.P. obtained by supercharging is about 148 as against 112 for the unsupercharged engine.

Another interesting point shown by these curves is that the average crankshaft torque is increased over the whole range, the maximum value of the torque being about 230 lbs. ft. as against the previous value of 210 lbs. ft.; moreover, the torque curve is flatter over the speed range so that the top gear performance of the engine is much better.

It is of interest to note that the Auburn passenger cars fitted with superchargers were certified to do at least 100 miles per hour.

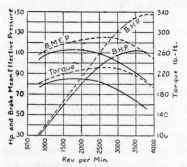

Fig. 215.—Effects of Supercharging on Performance.

In some cases the supercharger is provided with a clutch so that it need not be used at low to medium engine speeds, but only to increase the power output above the medium speed values for acceleration, improved hill-climbing and top-gear performance; the clutch is then usually interconnected with the throttle so as to come into operation after the throttle has opened a certain amount. One model of the Mercedes car employed such an arrangement.

Later *Studebaker-Packard Golden Hawk* Vee-eight engines were fitted with centrifugal superchargers. The supercharger was fitted with a variable-ratio vee-belt drive from the crankshaft pulley to the supercharger pulley which was provided with one fixed and one sliding pulley which gave a low speed of 5·7 times crankshaft speed and a high ratio of 10·1. The engine was of 289 cu. in. and used a rather lower compression ratio, namely

7·5:1 than the standard engine (non-supercharged) of 8·0:1. The supercharge pressure at the outlet was 5 lb. sq. in. above atmospheric.

The Corvair Supercharged Engine.—A supercharged version of the Chevrolet Corvair six-cylinder opposed air-cooled engine has enabled the maximum output of the improved original engine of 100 B.H.P. at 4,400 r.p.m. to be increased to 150 B.H.P. at 4,400 r.p.m. The full boost pressure used was 10 lb. sq. in. above atmospheric. No supercharging occurred at speeds below

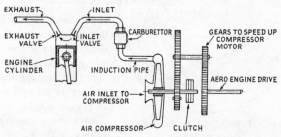

Fig. 216.—The Gear Driven Centrifugal Air Compressor type of Supercharger.

about 2,400 r.p.m. The system used was that of *turbocharging*, in which an exhaust gas turbine operated a centrifugal-type air compressor, running at speeds up to about 80,000 r.p.m.

Position of Supercharger.—The supercharger can be fitted in two alternative positions, namely: (1) between the carburettor and the engine; (2) outside, with the carburettor between the supercharger and the engine (Fig. 216). The former arrangement (Fig. 220) is that more generally preferred, for, in the latter case, owing to the increased air pressure on the petrol jet in the carburettor, it becomes necessary to fit a compensating device to increase the pressure in the float chamber. This is done by connecting the top of the float chamber (which is otherwise sealed) to the carburettor (compressed) air-inlet pipe.

The advantage of the arrangement shown in Fig. 220 is that a normal design of carburettor can be used, since the mixture is not compressed until after it leaves the carburettor; moreover, there is a cooling effect due to the evaporation of the fuel.

Inlet Safety Valve.—It is usually considered advisable to fit a *safety-valve in the inlet pipe*, so that in the event of a blow-back or back-fire no damage is done to the inlet pipe or super-charger, for the blow-off valve opens direct to the atmosphere. It is spring-loaded and adjusted to open at a pressure just above the maximum supercharge value. In certain instances, however, the designer omits this blow-off valve, relying upon the fact that one of the inlet valves is usually open and can release the excessive pressure into its cylinder.

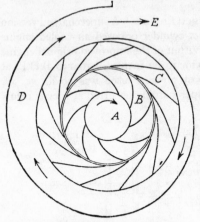

Fig. 217.—Principle of the Centrifugal Supercharger.

Charge Intercooling.—Another factor that is generally taken into consideration is the *increased temperature of the charge* due to its heat of compression. This surplus heat tends to reduce the quantity of charge forced into the cylinder, and it is necessary to get rid of it by cooling the compressed charge in an *intercooler* fitted between the supercharger and the engine. Aluminium inlet manifolds provided with air-cooling fins or inter-cooling units are used for this purpose.

Types of Supercharger.—There are three principal classes of air-compressor employed for supercharging engines, as follows: (1) The *Centrifugal Air Pump*; (2) the *Eccentric Drum* and *Vane Pump*; and (3) the *Rolling Drum* or *Root's Blower*. The ordinary reciprocating air pump is not used on high-speed petrol engines on account of its size and weight.

The Centrifugal Air Compressor.—The principle of the centrifugal air compressor is illustrated in Fig. 217, which shows a rotor *A* carrying a vaned portion *B*; this rotates within a casing containing a set of fixed vanes *C*, the latter being mounted within an outer casing *D* of gradually increasing cross section, i.e., it is of volute or spiral shape, the largest section being at the exit portion E. The air inlet to the compressor is at the centre A, the air flowing axially into this region. The rotor—which requires high speeds of rotation, namely, from about 18,000 to 25,000 r.p.m. for aircraft engines and up to 80,000 r.p.m. for automobile engines of 2 to 4 litres capacity—draws air from A and in virtue of the centrifugal effect on the air forces it outwards so that at the tips of the rotor B the air is delivered under pressure tangentially to the outer circular periphery of B and then enters the set of vanes C, which are arranged tangentially and give an expanding section towards their outer periphery. These are termed "diffusers" and their purpose is to convert the velocity head of the air into pressure head by slowing down the velocity and thus building up the pressure. The air is then delivered into the volute chamber D in increasing volume, which the gradually expanding casing of the compressor accommodates.

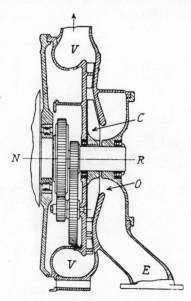

Fig. 218.—Centrifugal Compressor Gear Drive.

This type of compressor is usually driven by a train of gears such as that shown in Fig. 218. The engine crankshaft N carries a larger gear which meshes with the smaller one, shown immediately below, and a larger gear on the same shaft as the latter meshes in turn with a smaller one at the end of the rotor shaft R, so that a double step-up in gear ratio is thus obtained. This arrangement enables the rotor shaft axis to be in line with that of the crankshaft. The air or mixture from the carburettor below is drawn in at E and thence enters the central opening O of the rotor casing. It is discharged into the volute chamber V and finally leaves the latter at the upper end as indicated by the arrow; the compressor rotor is shown at C.

A single stage compressor of this type will give from 5 to 8 lbs. per sq. in. pressure, but if higher pressures are required two or more stages must be used, each rotor unit discharging into the central inlet of the next one, thus building up the pressure.

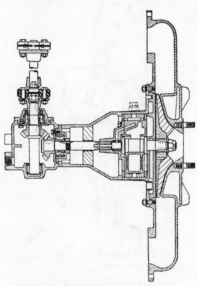

Fig. 219.—Showing the Auburn Centrifugal Supercharger and Bevel Gear Drive.

Exhaust Turbo Compressor.—Another method mentioned earlier is to utilize the energy of the exhaust gases to drive a high-speed gas turbine, known as the Rateau type (Fig. 220). The rotor shaft of this turbine also forms that of the centrifugal pump so that the latter is driven at the same speed as the exhaust gas turbine, viz., at 25,000 to 80,000 r.p.m. according to the rotor diameter. The turbine in question consists of a narrow wheel tapering inwards from the centre to the periphery—in order to give the necessary dimensions for strength purposes—and is provided with curved blades around its periphery. The high velocity exhaust gases meet and pass through this blade system, thus giving up some of their velocity energy to the turbine wheel.

There is an increased exhaust back pressure when a Rateau turbine is used, but the practical difficulties associated with this effect have been surmounted; in particular the cooling of the exhaust valves has been improved considerably.

The advantage of the exhaust-driven compressor is that, in the progressive method of increasing the power output of a given capacity engine, the step-by-step power outputs obtained by such methods as compression-ratio and maximum engine speed increases give only limited percentage increases, whereas by supercharging much greater percentage outputs can be obtained

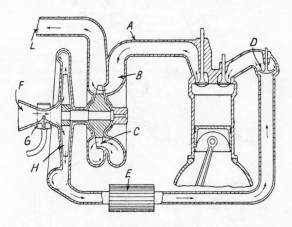

Fig. 220.—Exhaust Driven Supercharger with Outside Carburettor.

A—Exhaust gas outlet.   *B*—Turbine casing.   *C*—Turbine rotor.
D—Automatic blow-back prevention valve.   *E*—Mixture intercooler.
    F—Air entry to carburettor *G*.   *H*—Centrifugal pump rotor.

Note.—The arrows show the paths of the compressed mixture and exhaust gases.

in a single step, e.g., up to at least 50 per cent. with acceptable turbo-compressor units. It is usual, however, to reduce the compression ratio when supercharging a standard engine.

Eccentric Drum Compressor.—The second type of supercharger previously mentioned operates on the well-known eccentric drum and moving-vane principle. It has been used for car engines in this country, and is represented by several

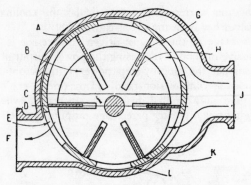

Fig. 221.—The Cozette Supercharger. *A*—Outer casing. *B*—Slotted rotor. *C*—Centre line of outer casing. *D*—Centre line of rotor. *E*—Perforated rotating drum. *F*—Delivery port. *G*—Sliding blade. *H*—Blade slot in rotor. *K* and *L*—Air tight portion of drum and rotor.

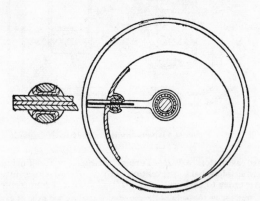

Fig. 222.—The Powerplus Supercharger, showing one Oscillating Blade and Slot.

alternative designs of which two, namely, the Cozette and Power-plus models are illustrated in Figs. 221 and 222.

The ordinary moving vane eccentric drum type of pump is not suitable for high-speed car engines, on account of the friction and wear of the blade ends. The modern eccentric drum pumps embody improvements for obviating wear of the blade ends and minimising the wear in the drum slots; ball-bearings for the rotor, effective end seals and an adequate lubrication of the moving parts are other improvements.

It is usual to mix oil with the petrol when this type of supercharger is fitted, unless, of course, an independent lubrication device is employed for the rotor blades. The bearings are almost invariably lubricated from the engine system.

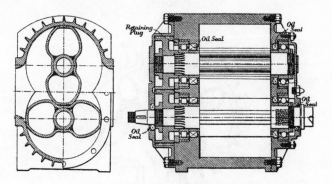

Fig. 223.—The Root's Supercharger.
(*A*) Cross Section.   (*B*) Side Section.

The Cozette model has a barrel mounted eccentrically with the drum, and a revolving slotted casing. Six radial blades are held against the inside of the latter by centrifugal force and rotate with this casing, moving in and out of their guides in the revolving drum as they rotate. Fig. 221 illustrates, in sectional view, this type of supercharger. It is possible to obtain pressures up to about 20 to 25 lbs. per sq. in. with rotary compressors of this type.

The Powerplus supercharger (Fig. 222) works on the same principle, but utilizes blades of the oscillating slot type to minimize slot wear; the blades are always radial to the lay-shaft; further, there is a fixed clearance between the blade and the casing.

The Root's Blower.—The third class of air-compressor, known as the Root's blower type, is a very old type that has been modernised in design for higher speed operation. In its simplest form it consists of a pair of rotors, each of a double-loop or "figure-8" section, the shapes of these rotors and positions of their axes of rotation being so arranged that, no matter what the positions of the rotors, they are always in line contact. They are driven positively by gearing so as to rotate in opposite directions and are provided with a casing which practically touches the rotors. The rotors rotate at about $1\frac{1}{2}$ times engine speed.

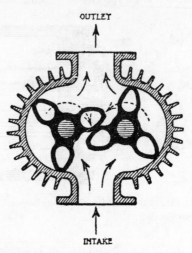

Fig. 224.—The Root's Type Supercharger, with Three Lobe Rotor.

Thus, as the rotors revolve—the top one clockwise and the bottom one anti-clockwise—they draw in the air (or charge) from the left-hand opening (Fig. 223) and discharge it on the right-hand side.

The Marshall-Drew design of Root's supercharger, shown in Fig. 223, has rotors ground to involute tooth form, the tips being grooved to improve the air sealing. The rotors run in ball and roller bearings and are driven by spur gears. The outer casing is of aluminium or magnesium alloy, and provided with cooling fins.

The supercharger illustrated measures 9 × 8 × 5 inches, will deliver 60 cu. ft. of free air per minute, and give a charge pressure of 5 lbs. per sq. in. above atmospheric, at 3,000 r.p.m. This supercharger is suitable for an engine of 52 cu. in. (852 c.c.), and will increase its power by about 30 to 40 per cent.

The clearances between the rotor and casing are kept very small, being of the order of ·003 to ·004 inch. The gears are lubricated from the engine, but no lubrication is required for the rotors.

Another design of Root's compressor uses three-lobed rotors as shown in Fig. 224. This pattern gives a more uniform pressure and better sealing of the charge in its passage through the compressor.

Notes on Root's Compressor.—The Root's compressor is still widely used, since it is of relatively simple construction; has a high volumetric efficiency; can deal with large volumes of air; can readily be balanced, to avoid vibrations and since it has no wearing parts, other than the simple gears and the end bearings it has a relatively long life. This type of compressor must, however, be carefully designed and machined, since the working clearances must be accurately maintained, to avoid pressure air leakages. It is, however, somewhat liable to pulsating flow, due to the periodic discharges of air on the delivery side.

CHAPTER V

TWO-CYCLE ENGINES

Two-Cycle Engines.—This type of engine, the principle of which is described on pp. 42 to 46, has always been an attractive one to inventors, since it has a higher output for a given cylinder capacity, gives a more uniform torque, is lighter, and in general owing to the much smaller number of components, in its simpler designs, is less complicated to manufacture and therefore cheaper to produce. The majority of two-stroke engines which have been produced to date do not appear to have overcome all of the disadvantages associated with this type, namely: (1) Lower m.e.p. than in the case of the four-stroke, thus giving appreciably less than twice the power output for a given cylinder capacity; usually the increase is only from 40 to 60 per cent. (2) Appreciably higher fuel consumption, namely, from 15 to 20 per cent., due usually to an escape of charge into the exhaust when both inlet and exhaust ports are open together, and to the lower compressions which are used. (3) Tendency to overheat if run on open throttle for long periods, due to almost twice as much heat being generated within the cylinder as in the case of the four-stroke. (4) Want of flexibility, the result being a much lower speed range as a two-stroke; at low speeds this engine tends to run as a four-stroke and thus loses part of its possible power. (5) Lubrication difficulties, leading to sparking plug troubles, piston ring sticking and blocked exhaust ports.

Although a large number of different designs of two-cycle engines has been patented, most of these engines can be classified into the following categories: (*A*) Crankcase compression. (*B*) Double piston. (*C*) Differential piston; and (*D*) Separate compressor type engines.

(*A*) **Crankcase Compression Engines.**—This class includes the simplest and, incidentally, the cheapest types of engines as used on small motor-cycles, lawn mowers, and in some cases light cars. In this design the under-side of the piston is employed to draw the mixture into the crankcase—as the piston ascends— and then to compress it on the down-stroke of the piston.

Suitable ports are arranged in the cylinder walls, which are uncovered by the top edge of the piston when the latter is near the bottom of its stroke, for the exhaust escape and also for the transfer

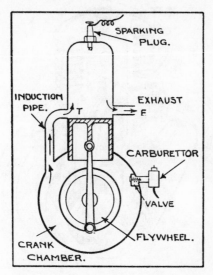

Fig. 225.—The simple Two-port Engine.

of the mixture from the crankcase to the upper side of the piston. In all cases the piston on its next ascent closes, first the transfer and then the exhaust ports, thus sealing the cylinder for the mixture compression stroke.

The various designs of two-cycle engine in this class differ in the arrangement, number and dispositions of their ports. They are known, respectively, as two-port, three-port, four-port and six-port engines according to the equivalent number of ports in the cylinder wall and in some cases, the piston.

The principle of the two-port engine is illustrated in Fig. 225. In this example the carburettor is connected with the crank chamber by means of a port having a non-return valve. The upward stroke of the piston draws the charge through this port into the crankcase, where it is compressed on the next down stroke and, as the piston uncovers the transfer port T, the charge flows into the cylinder, there to be compressed by the piston on its

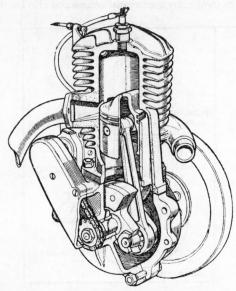

Fig. 226.—Part sectional view of Three-port Engine.*

next ascending stroke. The exhaust port E is uncovered by the piston towards the end of its next descending (or firing and expansion) stroke.

This type although extremely simple and cheap to build is inefficient as the cylinder cannot be given its full charge, owing to the limited period of the charging stroke and to the escape of part of the charge through the exhaust port.

The three-port engine, the operation of which is illustrated in Fig. 11, represents an improvement on the two-port type, mechanically, but its volumetric efficiency also is low.

* Courtesy *Motor Cycle*.

The six-port engine (Fig. 227), of which the Levis is an example, employs a port in the piston itself to introduce the crankcase charge into the transfer port; the chief object of this is to allow the ingoing charge to cool the piston crown—as this part becomes hotter in two-cycle than in four-cycle engines. This cooling effect, however, is obtained partly at the expense of heating the charge and thus reducing the volumetric efficiency.

Fig. 227.—The Levis Six-port Two-cycle Engine.

Practically all engines of the class described have pistons provided with deflector crowns, the object of the latter being to deflect the charge upwards towards the cylinder top so as to provide a longer leakage path, as it were, to the exhaust; this arrangement also assists in scavenging the burnt gases through the exhaust port. Generally speaking the crankcase compression engine is of low efficiency, and gives the lowest cylinder mean pressure values of two-cycle types.

Lubrication of Two-Cycle Engines.—The most widely used method of lubricating the smaller sizes of two-cycle engines, e.g., those of motor cycles, lawn mowers and certain of the smaller cars is by mixing lubricating oil with the petrol. This mixture is drawn into the crankcase during the outward stroke of the piston, where it is required to lubricate the connecting-rod and main bearings and also the lower part of the cylinder walls. During the firing stroke the oil in the charge is burnt and generally tends to build up deposits in the exhaust port, pipe and silencer.

The petrol-oil proportions on which crankcase-compression engines can operate range from about 10:1 to 40:1; in some cases engines have run satisfactorily over long periods on a 60:1 mixture. The recommended range, however, for automotive engines is between 15:1 and 25:1, or roughly, from about 1 pint of oil to

2 gallons of petrol in the former case, and 1 pint to 3 gallons, in the latter. This method is known as the *Petroil* one.

The proportion of oil in the petrol depends upon the viscosity of the oil, so that if oils of the higher S.A.E. viscosity 40 are used instead of S.A.E. 30, the proportion of oil can be reduced to about one-half, and from S.A.E. 40 to S.A.E. 50, by about the same proportion. The drawbacks associated with the Petroil system include: (1) Increased oil consumption per mile. (2) Inefficient lubrication of the connecting-rod big- and small-end bearings which, with bush bearings leads to more rapid wear. (3) Fouling of the sparking plug by carbon deposit from the burnt oil on the plug insulator and by bridging across the electrodes gap—a common fault, known as "whiskering". (4) Build up of carbon deposits in exhaust port and rest of the system.

The fouling of the exhaust system can be reduced to an appreciable extent by using compounded oil, containing metallic detergent additives, but it has been found that while effective in preventing car-bonizing of the exhaust system, the ash residues from the burnt oil, tend to promote plug fouling to an increased extent. However, by using an oil additive that does not produce ash deposits, not only were carbon deposits controlled but sparking plug fouling over appreciable running periods was avoided. Sparking plug troubles, in two-cycle engines can be minimized or avoided by the use of special ignition systems, as described in Volume No. 6 of this Motor Manual series, and the employment,

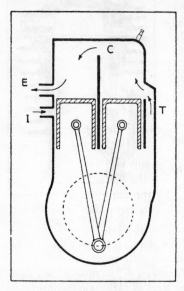

Fig. 228.—Double Piston Engine.

C—Common combustion head.
E—Exhaust port.   I—Mixture inlet port.   T—Transfer port.

with existing coil-ignition systems, of special designs of sparking plug, e.g. the Lodge "Golden" plug, which has an inbuilt condenser and small air gaps.

The D.K.W. Lubrication System.—The Petroil lubrication system has now been replaced in the German Auto-Union and D.K.W. car engines by a method which employs an engine-driven

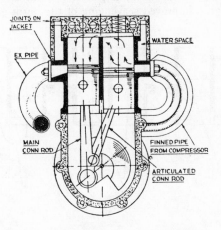

Fig. 229.—The D.K.W. Two-Cycle Engine.

oil pump to deliver a measured amount of oil from a separate reservoir and to inject it into the petrol in the correct proportion, close to the carburettor main jet. When the engine is running at idling and slow speeds, very little oil is injected but, as the throttle is opened, the quantity of oil is increased. In this way not only is the blue smoke associated with slow running eliminated but the overall oil consumption is reduced to between one-half to one-third that of the usual Petroil amount. The oil reservoir has a capacity for 1,500 to 2,000 milse of running, and when the oil level is getting low a red light on the instrument board warns the driver.

(*B*) **Double-Piston Engines.**—In this class there are two cylinders, with a common combustion chamber, and two pistons which move up and down practically together. The transfer or charge entry port is arranged near the bottom end of one cylinder (Fig. 228), whilst the exhaust port is near the bottom of the other cylinder. This arrangement it will be seen places the transfer and exhaust ports as far away as possible, so that the incoming charge sweeps right through the two cylinders, always in the same direction, i.e., there is a uni-directional flow of the charge. Improved scavenging of the burnt gases is thus obtained.

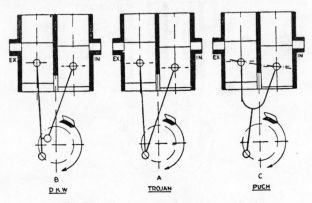

Fig. 230.—Typical examples of Double Piston Engines.

Some typical engines of the double-piston double-cylinder pattern are the D.K.W., Trojan, Puch and Zoller. Fig. 229 illustrates the general layout of the D.K.W. engine and shows the special arrangement of the connecting-rods employed to enable one piston to be in advance of the other for more effective scavenging and charging; it will be noted that the right-hand connecting-rod is hinged to the left-hand one for this purpose. The D.K.W., Trojan and Puch engines are shown, schematically, in Fig. 230 in order to indicate the connecting-rod arrangements employed for the purpose of giving the exhaust port piston a lead over the charging port one; this ensures the earlier closing of the exhaust port before the inlet or charging port.

The port timing diagrams of an ordinary and a double piston

two-cycle engine given in Fig. 231 indicate the differences in the
timing arrangements of the two types. In the later engine the
exhaust port opens at $77\frac{1}{2}°$ before B.D.C. and is open for a crank
angle of 25° before the transfer port opens. The latter remains
open for a total angle of 130° and the exhaust port for the same
period or angle. After the exhaust port has closed the transfer
port remains open for another 25°. This arrangement gives a

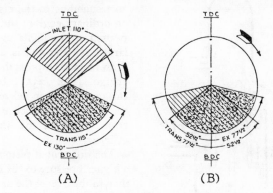

Fig. 231.—Port Timing Diagrams. (A) Ordinary Engine.
(B) Double Piston Engine.

longer cylinder charging period and a lower charge loss than that
of the ordinary two-cycle engine.

Instead of having the two cylinders side-by-side, a long straight
cylinder having two pistons working in opposite directions can
be used, as in the Junkers (Fig. 234B) and Werry engines. In
this case the combustion chamber is formed by the space left
between the pistons when the latter are nearest together. The
inlet and exhaust ports are arranged at opposite ends, so as to
obtain a uni-directional effort.

Whichever arrangement is employed, the principle is the same.
The double-opposed piston scheme necessitates two crankshafts,
which must be coupled together by connecting-rods or gearing.

The uni-directional charging and scavenging method can be
used with crankcase compression or with a separate pump for
compressing the charge.

Owing to the better scavenging and to the fact that less charge is lost through the exhaust, this type of two-cycle engine is more efficient than that previously described.

(*C*) **The Differential Piston Engine.**—In this class of two-cycle engine the piston is made with two parts of different diameter, the smaller portion corresponding to the piston of an ordinary engine; the larger portion works in its own enlarged cylinder and is used to draw in and compress the fresh charge (Fig. 232). Usually this charge is then delivered to the crankcase, whence it is transferred to the smaller cylinder.

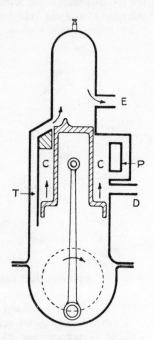

Fig. 232.—Differential Piston Engine.
C—Compression chamber.
D—Mixture inlet.
E—Exhaust port.
P—Transfer port for compressed mixture to crankcase.
T—Transfer port to cylinder.

The differential piston gives a higher pressure to the charge than in the case of the ordinary crankcase compression engine. Thus, the Dunelt engine gave about 50 per cent. increase in charge pressure.

The differential piston engine is larger and heavier than the normal type; but it definitely gives a bigger output and shows a higher overall efficiency.

(*D*) **Separate Compression Type Engine.**—The charge, in this type of engine, is forced into the cylinder by means of a separate pump, so that crankcase compression is obviated.

There are many variations of this method, although the principle is the same. For example, a reciprocating or rotary air pump can be used to force the charge either through the usual ports at the bottom of the piston's stroke, or through a mechanically-operated or simple non-return valve in the cylinder head.

Similarly the engine can be of the double-cylinder type previously described, and the air-pump can deliver the fresh mixture through the inlet port at the base of one cylinder, whilst the exhaust gases are ejected through the port at the base of the other cylinder. This method is used in the Junkers engine.

The principal advantage of the separate pump method is that the engine can be supercharged, so that a greater amount of power can be obtained from a given size engine. In order to obtain the best results, some efficient scavenging of the burnt gases is necessary; this also assists in cooling the cylinders. This can be done by the supercharger although sometimes a separate scavenging air-pump is used, the mixture being introduced afterwards.

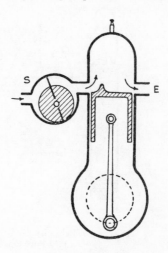

Fig. 233.—Separate Compressor
Engine (Two-port)
E—Exhaust. *S*—Compressor.

Air Scavenging Considerations.—As mentioned earlier an important requirement for the satisfactory performance of the two-cycle engine concerns the elimination of the burnt gases remaining inside the cylinder when the piston is near the lower end of the firing stroke and the exhaust port or valve is opened. The fresh mixture or a separate air charge is employed to get rid of these gases, or scavenge the cylinder, but the scavenging operation must be carried out during the very limited period available for this purpose. Thus, in the case of an engine operating at 1,800 r.p.m., with a scavenging period of say 100° of crank angle—and this is a liberal allowance—the whole process of scavenging must be completed within a period of 0·009 sec. Unless the residuals are effectively scavenged the fresh air charge will be diluted by exhaust gases and thus reduced in amount and the engine will not develop its full power.

It is not possible to scavenge efficiently engines of the type shown in Fig. 233, the only satisfactory method being to arrange for the air scavenge (or charging) and the exhaust ports to be at opppsite ends of the cylinder, so that uniflow scavenging is obtained.

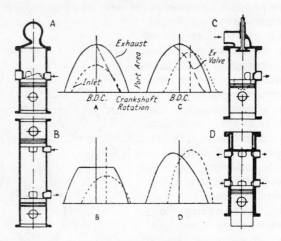

Fig. 234.—Types of Air Scavenged Two-Cycle Engines
and Port Opening Diagrams.

Referring to Fig. 234*, the arrangement shown at *A* is equivalent to that previously given in Fig. 233 for exhaust and scavenge ports near the bottom of the cylinder. The corresponding air inlet and exhaust port opening diagrams are shown to the right of the cylinder diagram. Although cheaper to manufacture than other designs, this type of engine suffers from the disadvantages of low volumetric efficiency and higher fuel consumption previously explained.

The most promising alternative designs are those illustrated in diagrams *B*, *C* and *D*, Fig. 234; in each of these uniflow

* "High Speed Two Stroke Oil Engines". P. E. Biggar.

scavenging is provided for. The method used in diagram *B* employs two opposed pistons in the same cylinder. One piston controls the exhaust ports at its end of the cylinder, whilst the other piston governs the air inlet ports at its end. This arrangement, which was used on the Junkers' motor vehicle and aircraft two-cycle C.I. engine, permits efficient uniflow scavenging and avoids excessive loss of scavenging air since the exhaust port controlling piston is given a lead over the inlet port and closes just before the latter—as shown by the port opening diagram; the residual exhaust gases are thus reduced to a minimum.

The arrangement shown in *C* (Fig. 234) has the air inlet ports at the lower part of the cylinder and a poppet type exhaust valve in the cylinder head; as in the case of the opposed piston engine, the exhaust valve opens well before the inlet ports—which are, of course, piston-controlled—and it closes in advance of the inlet ports. This system provides for much better piston and exhaust port cooling than would be possible if the upper valve were the air inlet and the lower ports the exhaust ones.

The engine shown, diagrammatically, at *D* (Fig. 234) represents a single-sleeve valve engine, with the exhaust ports near the top of the cylinder controlled positively by the sleeve valve; the inlet ports near the lower end of the cylinder are also controlled positively by the sleeve valve. The exhaust has a lead over the inlet, as shown by the port opening diagram to the left at *D*, in Fig. 234, and it is possible with this controlled exhaust and inlet port to obtain the most efficient scavenging action.

Quantity of Air Required.—It has been shown experimentally, that the power developed by a two-cycle engine increases with the amount of air supplied for scavenging at any given speed. There is, however, a limit to this quantity of air since the size of the compressor and the power required to drive it both increase with the amount of air delivered and with the pressure of the air supply.

For a satisfactory performance the quantity of air delivered by the compressor for scavenging is from $1\frac{1}{2}$ to 2 times the cylinder volume for an air pressure supply of 4 to 6 lb. per sq. in. above atmospheric pressure; the pressure in question seldom exceeds 7 lb. per sq. in.

The power required to supply air to the cylinder under these conditions, allowing for the unavoidable wastage of 25 to 30 per cent. through the exhaust ports, is of the order of 10 to 15 per cent. of the total or gross output of the engine. In general, for a direct-driven air compressor, about 1/60th of the gross output will be absorbed for each lb. per sq. in. of scavenging air pressure. The weight of the air compressor, using modern light alloys and alloy steels, would be about 0·1 to 0·25 lb. per B.H.P.

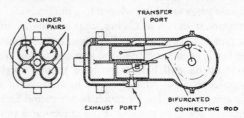

Fig. 235.—The Trojan Two-Cycle Engine.
(Four Cylinder.)

The Trojan Engine.—An interesting two-cycle engine was, at one time fitted to a light car. It is here included on account of its unconventional, yet quite practical design. The engine employs two pairs of water-cooled cylinders, each pair consisting of two parallel cylinders communicating with one another at the combustion ends. The two pairs of cylinders are quite separate, and are side by side. Each pair of cylinders has its own pistons, but there is a common connecting-rod of Y-shape terminating in a single big bed. The two cylinders act as a single one and each pair fires alternately.

The Trojan engine shown in Fig. 235 had a bore and stroke of 63·5 mm. and 117·5 mm. respectively, and was designed to give its maximum power over an appreciable range of speeds, so that its torque curve was a fairly flat one. Crankcase compression was employed, each pair of cylinders having a common crankcase. The mixture is drawn into the crankcase while the pistons are moving outwards (from the crankshaft), and is pumped into the upper cylinder of the pair (the cylinders lying horizontally one above the other) as the pistons move inwards. It enters the second cylinder through a port in the combustion head. The lower piston, which controls the exhaust port, moves in advance of the upper piston, which governs the transfer port, so that exhaust occurs before the transfer port is opened. As the ports are situated in opposite cylinders of a pair, there is, as stated previously, much reduced charge loss through the exhaust. The exhaust port closes before the transfer port, and owing to their differential motion at the top of the stroke, when ignition occurs, the gases rush through the port connecting the two cylinders, thereby causing a turbulent action which is beneficial to the combustion process. It is interesting to note that this Trojan engine contained only seven moving parts.

A later development of the Trojan two-cycle vehicle engine, had two pairs of working cylinders, with a common combustion chamber of modern design. The bore of each cylinder was 65·5 mm. and its stroke, 88 mm., giving a total cylinder capacity of 1,186 c.c.

As in the earlier version, one piston of each pair controlled the exhaust port opening and the other the inlet port opening. By suitable timing, using different relative piston positions, the exhaust port was opened before the mixture admission port, thus providing a good scavenging action but with minimum charge loss through the exhaust port. Instead of using crankcase compression an additional charge compressing cylinder was provided for each pair of working cylinders. Mixture was drawn into each charging cylinder from the carburettor and, at the appropriate timing period, supplied to the working cylinders.

The engine had a maximum output of about 24 B.H.P. at 2,250 r.p.m. but its special feature was the particularly uniform torque, which varied only by about 24 lbs. ft. over a speed range from 700 to 2,500 r.p.m.; this was of special advantage for commercial vehicle purposes.

The Foden Air-Scavenged Engine.—The Foden two-cycle engine employs an efficient air-scavenging and cylinder charging method which results in a high output per given cylinder capacity. Although this method is used on the Foden C.I. Engines the principle is also applicable to petrol engines. It is of interest to note that outputs up to 32 B.H.P. per litre have been obtained with normal commercial Foden engines, using the method described.

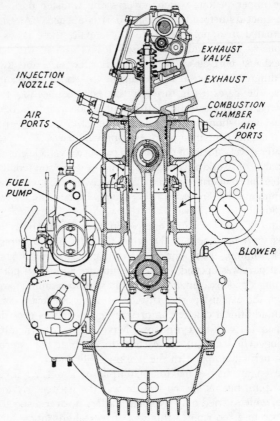

Fig. 236.—The Foden Two-Cycle Air Scavenged Engine.

The Foden operates on the Kadenacy principle in which a low-pressure rotary type compressor delivers air to ports arranged at the bottoms of the cylinders, whilst the exhaust gases are ejected through poppet valves in the cylinder heads.

The momentum effect of these gases is utilized to create a partial vacuum in the exhaust manifold, thus assisting the scavenging and cylinder charging by air admitted under pressure from the

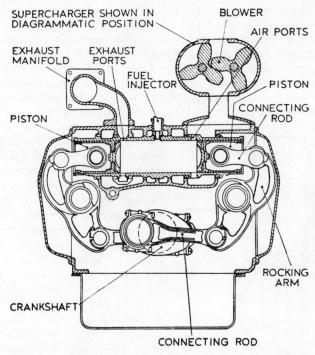

Fig. 237.—Illustrating the Principle of the Rootes Opposed-Piston Pressure-Charged Two-Cycle Engine, as used on Commercial Vehicles.

piston uncovered ports at the bottom of the cylinder. It has been proved that for the same output the Kadency type operates at appreciably lower cylinder pressures and, owing to the efficient scavenging of the exhaust gases is cooled satisfactorily, moreover, the fuel consumption is lower and the power output, for a given engine size, is greater than for the ordinary two-cycle engine.

The G.M.C. engine made by General Motors Company, of America, operates on a somewhat similar principle, namely, with exhaust valves in the cylinder head and air ports in the cylinder which are uncovered by the descending piston.

The Foden six-cylinder vertical engine, of 85 mm. bore and 120 mm. stroke, developed 126 B.H.P., which is equivalent to about 32 B.H.P. per litre. This engine has two mechanically-operated exhaust valves per cylinder.

A more recent two-cycle engine used on Rootes vehicles works on the opposed piston principle shown in Fig. 237. It is made in four and six (double) cylinder models. The pistons move inwards and outwards symmetrically. When moving together compression occurs and when closest, fuel is sprayed into the heated air charge, and ignites, driving the pistons outwards, where by the connecting-rods and rocking arm mechanisms they rotate the crankshaft below. The gases escape through exhaust ports in the cylinder. The engines are both of the 85 mm. (3·34 in.) bore and 120 mm. (4·72 in.) stroke dimensions and develop, respectively, 84 B.H.P. and 126 B.H.P., at 2,000 r.p.m., respectively. It will be seen that the air is supplied to the cylinders, at the appropriate moment through ports from the overhead compressor. This air also assists in scavenging the exhaust gases through their cylinder ports and provides some degree of supercharging.

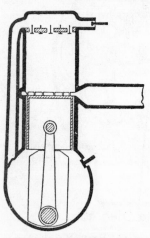

Fig. 238.—The Kylen Air-Scavenged Two-cycle Engine.

The Kylen Engine.—An interesting engine, known as the Kylen, of Swedish origin, is provided with a number of poppet valves in the cylinder head opening inwards automatically (Fig. 238). The exhaust gases are ejected through ports uncovered by the piston near the end of its stroke and the momentum of the escaping gases, owing to the special design of the exhaust system, results in a negative pressure within the cylinder. As the piston in descending has compressed the air within the crankcase, and there is a direct connection between the latter and the space above the cylinder head, the effect is to cause the automatic air inlet-valves in the head to open and allow scavenging air to flow into the cylinder and thus clear out most of the remaining exhaust gases. The satisfactory operation of this system depends upon careful design of the exhaust system to give the correct pulsations resulting in a negative pressure when the exhaust ports are uncovered; also upon the correct weights and spring pressures of the air inlet valves. Experimental engines operating on this system show better output and fuel economy than the normal types of three-port engines, with good flexibility. The best results appear to be obtained with petrol-injection into the cylinder and the use of compression ratios of 7:1 to 8:1.

Sulzer Opposed-Piston Engine.—An interesting and practical type of two-cycle engine, due to Sulzer Bros. of Switzerland and later adopted by other countries is shown in Figs. 151 and 239, for the Diesel engine version.

It will be seen that both the Sulzer and Foden two-cycle engines use a similar crankshaft operating mechanism.

The pistons move almost symmetrically .in their common horizontal cylinder and each of their reciprocating movements is converted into rotary motion by means of a pair of connecting rods A and B, a rocking lever C with its fixed bearing at D, and a crankshaft E, having its two cranks at 180°. The left piston F uncovers inlet ports G through the cylinder wall, to allow compressed air to flow into and, at first, through the complete cylinder unit, in order to sweep out, or scavenge, the exhaust gases from the previous firing stroke, through the exhaust ports H, as indicated by the arrows.

An important feature of this engine is that, due to the fact that the two pistons are intentionally placed a little out of phase, by means of the engine mechanism, the exhaust ports open first and after both inlet and exhaust ports are open together the exhaust ports close, leaving the inlet ports still open. The result is that after the air has scavenged the exhaust gases, air continues to pass into the cylinder to raise the initial air charge pressure above atmospheric, thus giving a degree of supercharging.

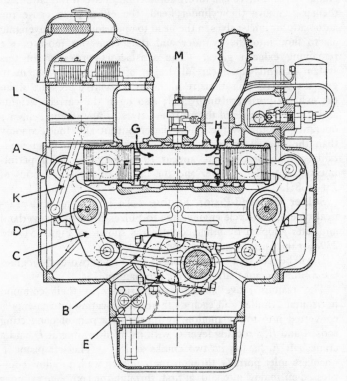

Fig. 239. The Sulzer Opposed-Piston Two-Cycle Engine.
A and *B*—connecting-rods. *C*—rocking lever. *D*—rocking lever bearing. *E*—crankshaft. *F*—piston. *G*—air inlet ports. *H*—exhaust ports. *J*—piston. *K*—air pump operating rod. *L*—air pump piston. *M*—fuel injector.

This type gives two firing strokes per revolution and is therefore equivalent to a four-cylinder four cycle engine (with twice the number of cylinders and piston assemblies). The engine can be well balanced and it runs very smoothly. It has, of course, more and heavier reciprocating members than a normal vertical engine, so that the maximum crankshaft speed is limited, in practice to about 2,500 r.p.m.

The Commer double, three-cylinder C.I. engine, operating on this principle, develops 90 B.H.P. at 2,400 r.p.m. for a cylinder capacity of 3·26 litres, which is equivalent to an output of about 27·6 B.H.P. per litre. In this engine the scavenging and charging air is supplied by a three-lobe Roots compressor, which runs at 1·8 times engine speed and supplies about 1·5 times the cylinder volume of air at 6 lb. per sq. in. pressure.

Reverse Flow Two-Cycle Engines.—Instead of using a deflector on the piston, in order to guide the incoming charge, the latter can be directed in an oblique direction upwards and towards the opposite side of the cylinder wall from whence it is reversed to the exhaust ports by specially cast ports. Two inlet ports are usually employed, so that the two charge streams meet above the level of the exhaust port and, travelling upwards, tend to force the exhaust gases downwards and through the exhaust ports. A later model Villiers engine of 125 c.c. employs this method. It has a flat-topped piston, two exhaust and four inlet ports. The *reverse flow scavenging* effect described is also arranged for in the D.K.W. engine (of German origin) shown in Fig. 240.

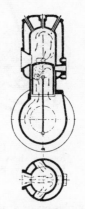

Fig. 240.
Reversed flow
D.K.W. Engine.

The beneficial effects of this method of scavenging the exhaust is indicated by the relatively high mean pressures and low fuel consumptions obtained.

The D.K.W. Three-Cylinder Engine.—This German two-cycle engine has been used over an appreciable period on D.K.W. cars and vans, having been introduced in 1954 and subsequently developed in later models.

The three-cylinder arrangement may be regarded as equivalent to one-half of a six-cylinder vertical engine with cranks at 120°, in order to give equal firing intervals. It can be shown that whereas the six-cylinder engine is in perfect balance for its primary reciprocating forces, and also its secondary forces, the three-cylinder engine the primary and secondary forces are balanced but their lines of action are separated so that rocking

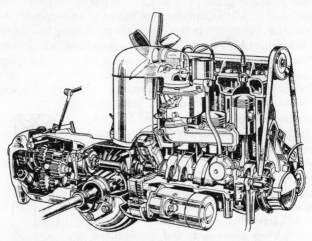

Fig. 241.—The D.K.W. Three-Cylinder Two-Cycle Car Engine showing also, the Gearbox and Differential for the Front Drive Shafts. The Starting Motor is shown below the Crankshaft.

couples occur at the engine and twice engine speed or frequency. These couples do not occur in the six-cylinder engine.

The D.K.W. engine (Fig. 241) has a cylinder capacity of 896 c.c. and with a compression ratio of 6·5:1 develops 35 B.H.P. at 4,000 r.p.m., giving a maximum torque of 51 lb. ft. at 2,000 r.p.m., so that top gear performance is good over the usual average speed range. The engine is of the *reverse scavenged* kind, being water-cooled and with a six-bladed cooling fan is shaft driven by a belt and pulleys from the crankshaft at a higher speed than that of the crankshaft. The cylinder head is of a light alloy and the water jackets are designed so as to surround each cylinder over its complete length.

The crankshaft is mounted on four roller bearings, while the big ends use needle bearings. With the petrol and oil mixture used there is little risk of bearing failures that have occurred in other engines using plain bearings. The reverse scavenging system used necessitates the use of a pressurized engine sump, so that special crankshaft sealing is necessary. The pressure fluctuations in the sump are taken advantage of to work a fuel lift pump to supply fuel to the carburettor.

The engine, as will be seen in Fig. 241 is of the front wheel drive type with the clutch and transmission at the front end and the cardan drive shafts behind the transmission.

The ignition system employs a distributor which is mounted at the back of the engine and driven at engine speed. Each cylinder is supplied with its own ignition coil.

The recent Auto-Union-D.K.W. two-cycle car engines include a three-cylinder model having a bore of 74 mm. (2·91 in.) and stroke of 76 mm. (2·99 in.) giving a cylinder capacity of 981 c.c. With a compression ratio of 7·2:1 the engine develops 44 B.H.P. at 4,500 r.p.m. With a compression ratio of 8·0:1 the output is 62 B.H.P. at 4,500 r.p.m. with a maximum torque of 68·6 lb. ft. and maximum B.M.E.P. of 86 lb. sq. in., both occurring at 3,500 r.p.m. This performance for an engine of slightly less than one litre, namely, the equivalent of 63 B.H.P. with the relatively high B.M.E.P. value, is noteworthy for an automobile two-cycle engine.

Mention should here be made of the German-designed three-cylinder two-cycle engine, known as the SAAB, fitted to Swedish cars of that name. It resembles fairly closely the D.K.W. engine, previously described. The recent model SAAB 96 has a bore of 70 mm. and stroke of 73 mm. (841 c.c.) and develops 42 B.H.P. at 5,000 r.p.m. It has a maximum torque of 61 lb. ft. and B.M.E.P. of 83 lb. per sq. in., both at 4,250 r.p.m. It employs a compression ratio of 7·3:1 and is of the front wheel drive design.

Future Two-Cycle Engine Developments.—In 1962 it was announced that improved and specially-tuned versions of the D.K.W. three-cylinder engines were being developed for fast touring and also racing cars. These included an 810 c.c. engine giving 55 B.H.P. and a road speed of 90–95 m.p.h. and a 1,000 c.c.

engine developing 65–68 B.H.P., giving a speed of about 105 m.p.h. The maximum torque of the 810 c.c. would be about 80 lb. ft., with possible development to 90 lb. ft. at 5,000 r.p.m. and power output equivalent to 100 H.P. per litre at 6,000 to 6,300 r.p.m. The petrol-oil mixture, using the new separate oil pump feed to the carburettor, previously described, was 30:1 for fast touring cars and 25:1 for racing cars.

Another three-cylinder two-cycle engine of 420 c.c. capacity, for small cars, was the Excelsior model.

Direct Injection Engines.—In order to overcome the drawback of loss of fuel in the portion of the charge that escapes through the exhaust ports in many designs of carburettor two-cycle engine the engine can be designed to operate with a charge of air only, during the transfer, charging or scavenging processes. After the exhaust valve or port is closed the petrol can be injected into the charge during the compression stroke, so that in this way no fuel is lost.

This principle was applied as long ago as 1910-12 by the late Prof. W. Watson to a converted Day two-cycle petrol engine and has since been used in experimental and production engines.

The direct injection engine in question uses compression ratios of the same order as in ordinary petrol engines and employs a sparking plug to ignite the charge so that it should not be confused with the compression-ignition engine which utilizes the heat of compression of the air charge to ignite the fuel.

CHAPTER VI

COMPRESSION IGNITION ENGINES

KNOWN also as the high-speed oil, or Diesel engine, this type operates on an entirely different principle to the ordinary petrol engine; the latter works on the Otto or *Constant Volume* cycle, whereas the former follows approximately what is known as the *Constant Pressure Cycle* of operations although many modern engines operate on the combined Otto and constant pressure cycles.

The compression-ignition, or C.I. engine requires no carburettor or ignition apparatus, and it uses a heavier and cheaper type of fuel, known as Diesel oil, instead of petrol.

The earlier C.I. engine was hitherto debarred from automobile work on account of its excessive weight for a given horse-power output; moreover, it generally worked at relatively low speeds, namely, below 300 r.p.m.

As a result of intensive experimental work it was shown that the C.I. engine could not only be run at speeds comparable with those of the petrol engine, namely, 2,000 to 4,000 r.p.m., but by using the same design methods and materials as for petrol engines it could be made only a little heavier than the petrol engine; actually the lower compression ratio petrol engine must always be the lighter type.

As a result of these developments, the C.I. engine has proved a formidable rival to the petrol engine in heavier motor vehicle, stationary and rail-car applications. Hitherto, the C.I. engine has been used chiefly in automobiles of the passenger-carrying and goods types, for it shows a considerable saving in the matter of fuel costs as compared with the similar size petrol engine vehicle; moreover, it has other advantages in connection with lower maintenance costs, freedom from fire risks, better engine torques, quicker starting and pulling from the cold.

Later Car-Type Engines.—More recently, in addition to the Citroen car that was fitted with a Comet-head engine as an alternative to the standard petrol engine, other makes of car have offered C.I. engine alternatives, notably the, B.M.C. (1½ litre), Standard, Rover, Mercedes Benz, Borgward (1·8 litre), and Fiat (1·9 litre). For taxicab purposes, also, the Standard, B.M.C. and smaller model Perkins engines have established their claims for marked reduction in fuel costs per mile, reliability and reduced maintenance attention.

The alternative C.I. car engine is heavier than its petrol rival and, at the same time, costs appreciably more, so that the car prices are higher, to the extent of £100 to £200. It can be shown that with existing fuel costs the C.I. engine car would have to cover total road distances of 30,000 to 40,000 miles before the increased initial cost of the C.I. car is wiped off by the reduction in fuel cost per mile of the C.I. car. In recent small automobile engines, as is mentioned later, a much smaller road mileage is necessary for balancing costs, e.g. 10,000 miles for the B.M.C. (1½-litre) engine.

It is usual to fit cars and taxicabs with C.I. engines of 1·8 to 2·0 litre capacity, developing maximum outputs of 38 to 44 B.H.P. at speeds up to 3,500 r.p.m. in some cases.

Usually, if the petrol engine car has a consumption of 25 m.p.g., the C.I. engine in the same car will give 35 to 40 m.p.g.

The C.I. engine car is heavier, but has very good torque characteristics, which enable it to run better in top gear at the lower engine speeds. These engines will start readily from cold and develop their full power outputs much quicker than the petrol engine.

The maximum road speeds, at present, are usually from 60 to 67 m.p.h. as compared with 70 to 80 m.ph. for the equivalent petrol engine car; this is because of the smaller output engines and heavier weight. By increasing the engine capacity by 15 to 20 per cent., identical power outputs are possible.

The C.I. engine does not run so quietly as its petrol engine equivalent, and is inclined to thump at low or idling speeds. The pre-combustion and Comet-types of combustion chamber is preferable to the direct injection one for the smaller sizes of engine, for quieter operation and higher engine speeds.

On account of its higher thermal efficiency, the heat losses of the C.I. engine are smaller than for the petrol engine, so that *smaller car radiators* can be used.

High Efficiency.—The C.I. engine operates at much higher compression ratios than the petrol engine. Thus, it employs compression ratios of 14:1 to 23:1, whereas the ordinary motor car engine uses ratios of 7·51 to 10·5:1.

It is well known that the efficiency of an engine becomes higher as its compression ratio is raised so that the high compression C.I. engine must be much more efficient than the lower compression petrol engine.

A direct result of this higher heat efficiency is the fuel consumption for a given horse-power output; the C.I. engine generally uses from 30 to 45 per cent. less fuel per H.P. on this account. Further, it can use a cheaper and heavier grade of fuel oil having a higher flash point (and, therefore, reduced fire risk).

How the Four-Cycle C.I. Engine Works.—The C.I. engine resembles the petrol engine in its general design as it has the same kind of trunk piston, connecting-rod, inlet (but for air only) and exhaust valves, etc. It differs, however, in the manner of introducing the fuel into the cylinder head and in the method of igniting the fuel.

The principal difference in the two types lines lies in the design of the combustion head and in the fuel supply arrangements to this head. Before, however, describing the four-cycle engine let us consider, briefly, the principle of the C.I. engine.

Instead of introducing a mixture of fuel and air into the cylinder, the C.I. engine draws in pure air only. This air is then compressed by the ascending piston to a considerably higher pressure than that used in the petrol engine; thus the air is compressed to $\frac{1}{14}$ to $\frac{1}{20}$ of its original volume, giving a compression pressure of about 450 to 550 lbs. per sq. in.

As a result of its high compression the temperature of the air is raised considerably—usually to about 500° to 550° C. Now, the temperature required to "self-ignite" the Diesel oil fuel is from 350° to 450° C. according to the grade. It will be evident, therefore, that the effect of injecting the Diesel oil in a fine spray into the combustion space containing this highly compressed and heated air charge will be to cause the fuel to burn very rapidly.

There is thus no necessity for any electric spark to ignite the fuel spray and compressed air. Immediately after the given amount of fuel has been injected the temperature of the combustion products is raised considerably, namely, to about 2,000° C. to 2,700° C., but the pressure does not rise as much above the compression pressure as in the case of the petrol engine; usually, it reaches a maximum value of 750 to 950 lbs. per sq. in. Thereafter the gaseous products expand, forcing the piston downwards thus providing the energy for the power stroke.

It should here be mentioned that the fuel is injected by means of a special fuel pump, through a fine nozzle into the combustion chamber, so that it issues in the form of a highly atomized spray to mix with the air already in the combustion chamber.

Reverting now to the four-cycle C.I. engine (Fig. 242), Diagram A shows the piston at the beginning of the air inlet stroke, the combustion chamber being filled with the remnant exhaust gases from the previous stroke. As the piston descends, the air-intake valve is opened mechanically, allowing air to flow into the cylinder. The air intake is usually provided with an air-cleaner and silencer.

When the piston reaches the bottom of its stroke the inlet valve is closed and since the exhaust valve is also kept closed, on its succeeding upward stroke (Diagram B) it compresses the air charge, finally giving the latter—as previously mentioned—a pressure of about 450 to 550 lb. per sq. in. and a temperature of about 500° to 550° C. Just before the piston reaches the top of its compression stroke (Diagram C), a mechanically-operated plunger pump is timed to force the fuel oil for combustion, under a high pressure of usually, 2,000 to 3,000 lbs. per sq. in., through the fuel injector, where it emerges into the heated air charge as a conical highly "atomized" spray. It ignites almost at once and continues to burn all the time the fuel is forced through the injector; this usually continues for a period equivalent to the rotation of the crankshaft through an angle of 15° to 30°.

On the next descending stroke the highly heated exhaust gases expand from their initial combustion pressure of about 750 to 950 lb. per sq. in. down to atmospheric pressure (14·7 lb. per sq. in.) which occurs when the exhaust valve opens, towards the end of expansion stroke. The inlet valve is closed during the expan-

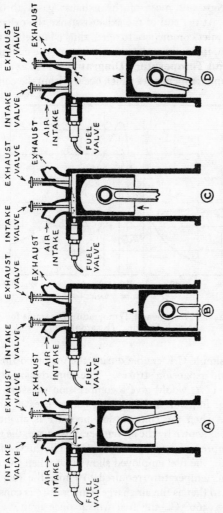

Fig. 242.—Principle of Four-Cycle C.I. Engine.
A—Suction. B—Compression. C—Fuel Injection. D—Exhaust.

sion and also the exhaust strokes (Diagram *D*), when the piston in ascending sweeps out most of the exhaust gases through the exhaust port. At the end of the exhaust stroke the exhaust valve closes and the inlet commences to open, thus completing the four-cycle operation, in two complete revolutions of the engine.

Pressure and Temperature Diagrams.—Fig. 243 illustrates the pressures and temperatures which occur within the combustion

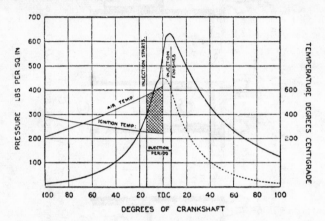

Fig. 243.—Pressure and Temperature Diagrams for C.I. Engine. (Ricardo).

chamber of a simple C.I. engine during the major portions of the compression and expansion strokes. The compression ratio used, namely, about 15:1, would give a compression pressure of about 450 lb. per sq. in., but fuel injection commences at about 14° before T.D.C. when the compression pressure is about 300 lb. per sq. in. and, as shown by the air temperature curve, the temperature of the compressed air is about 560° C. The curve of ignition temperatures for the fuel employed shows that when the injection commences the temperature required to ignite the fuel spray is about 260° C., so that as the air charge temperature is considerably higher, namely, 560° C., the fuel will at once ignite. The injection continues until about 7° past T.D.C. crank position. During this period, owing to the combustion of the fuel the pressure rises until it attains the value of about 630 lb. per sq. in. at the finish of the injection period. Thereafter, the hot gases of combustion expand progressively, doing useful work on the piston, as shown by the upper right hand line. The dotted line below is the expansion line for the compressed air charge when no injection takes place.

The Two-Cycle C.I. Engine.—This type operates in a similar manner to the petrol engine types, the only differences being as follows:—

(1) The C.I. engine employs pure air for charging the cylinder, instead of a mixture of petrol and air. The air charge is compressed and introduced by exactly the same means as in the various two-cycle petrol engines.

(2) The C.I. engine uses a much high compression ratio, namely, from 14:1 to 18:1.

The two-cycle C.I. engine is relatively more efficient than the petrol type, since there is no loss of fuel through the exhaust ports. On the other hand it is not yet so efficient, in automobile sizes, as the four-cycle C.I. engine.

Types of C.I. Engine.—There are numerous examples of high-speed C.I. engines used for automobiles and heavier vehicles. Whilst these all operate on the compression ignition principle, they differ widely in the designs of their combustion chambers and in the fuel injection systems employed.

It is only possible in an elementary book of the present nature to give a very brief outline of the principle types, but those seeking fuller information are referred to the author's book* on the subject.

The various designs of C.I. engines may be grouped into three broad classes as follows: (1) *The Direct Injection*; (2) *The Auxiliary Pre-Combustion Chamber System*; and (3) *The Turbulent Head* types. There are, however, certain other designs that utilize the advantages of two or more of these types.

(1) **The Direct Injection Engine.**—This is so called from the fact that the fuel is sprayed directly into the cylinder itself, there being no separate combustion head. (Fig. 241).

The piston *R* is shown nearly at the top of its compression stroke,

High Speed Diesel Engines. A. W. Judge, (Chapman & Hall Ltd.).

both the inlet valve *I* and exhaust valve *E* being closed. Commencing at the fuel supply end, fuel is drawn through *a fuel filter F* on the suction stroke of the fuel pump *P*, in order to rid it of all solid particles; the presence of the latter would not only cause excessive wear of the moving parts of the fuel pump but

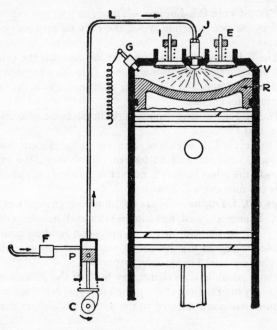

Fig. 244.—The Direct Injection Engine.

would also tend to choke the fine passages in the fuel injection nozzle *J*.

The fuel pump accurately measures out the predetermined quantity of oil and at the correct moment delivers it, past a non-return delivery valve into the fine bore fuel pipe *L*. The pump, which is mechanically driven from the engine, at one-half engine speed, in the four-cycle engine, is designed to give a high pressure to the fuel in order to force it through the injection valve in a wide angle spray form against the compression pressure existing in the cylinder at the time of injection. The usual fuel pressures given by the fuel pump are about 1,500 to 3,000 lbs. per sq. in., although pressures up to 8,000 to 10,000 lbs. per sq. in. have been used on some engines.

The usual form of injection nozzle *J* is that of a spring-loaded plunger having a conical or specially-shaped end, normally held on a seating, located near the wall of the combustion chamber, by its spring-pressure. During the period of injection, however, the high fuel pressure given by the pump overcomes the spring pressure and lifts the valve off its seating, thus allowing the oil to enter the combustion chamber. The nozzle is designed to give the desired shape of spray, while the injection pressure is made high enough to project the spray right across to the farthest parts of the combustion chamber, in order to ensure complete burning of the fuel.

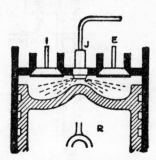

Fig. 245.—Direct Injection Engine with Special Piston Head.
E—Exhaust. *I*—Inlet.
J—Injector. *R*—Piston.

There are other shapes of piston crown and cylinder head, in direct-injection engines, but all operate on the same basic principle.

The plug shown at *G* (Fig. 244) contains a small coil of resistance wire which can be heated to redness by means of a current taken from a battery. This *Heater Plug* is only used for cold starting purposes and is switched off once the engine has started. Modern direct-injection engines will start from the cold without the aid of heater plugs, which are now confined to auxiliary combustion chamber type engines.

The direct-injection engine uses a multiple-hole injection nozzle and requires a higher fuel pressure than most other types of C.I. engine.

The Auxiliary Pre-Combustion Chamber System.—This method, which has now been replaced largely by the system (3), described later, was based upon the injection of fuel into an auxiliary chamber, which had a narrow neck communicating with the cylinder head. The ignition of the compressed air charge in the auxiliary chamber resulted in the projection of a mixture of

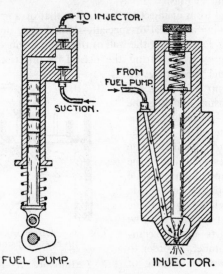

FUEL PUMP.

INJECTOR.

Fig. 246.—Illustrating Principle of Fuel Pump and
Injection Valve.

burnt gases and unburnt fuel through the neck into the cylinder head where the fuel was burnt finally.

The pre-combustion chamber engine as mentioned has been supplanted by the turbulent-head and direct-injection types, for it has the disadvantages of requiring a high-compression ratio; is more difficult to start, (heater plugs being necessary), and is not so efficient on account of the heat loss from the burning fuel to the throat of the combustion chamber. Moreover, it is more prone to "Diesel knock" when idling and also when under maximum power conditions.

(3) **The Turbulent-Head Engine.**—Whilst it is true that there is a certain amount of movement or agitation of the compressed

air in the combustion chamber just prior to the injection of the fuel, in practically all types of engine, in many cases this turbulence is more or less accidental and of relatively low degree. The type of combustion chamber under consideration is that in which the designer has deliberately arranged to impart a moderate-to-high degree of turbulence to the compressed charge for the purpose of securing a satisfactory admixture of the air and fuel particles. The injector, in this case, is usually of the single hole type, having a relatively large hole. It is placed in the combustion chamber so that the turbulent air stream sweeps rapidly past the outlet hole and each particle of fuel is given its proper air supply for combustion soon after it emerges from the nozzle. There are various methods of producing this turbulence, one common process being that in which the piston forces the compressed charge into an auxiliary chamber. The piston, which may be plain or with a central projection, is arranged almost to touch the cylinder top at the end of its stroke, thus forcing the remaining air charge through comparatively narrow spaces into the auxiliary chamber; the air is thus given a high degree of turbulence.

The Ricardo types of turbulent head, namely, the different designs of Comet heads are more widely used in the smaller, or automobile engines, as they are reliable, easy to start with heater plugs and can be run at higher engine speeds than the larger direct injection engines.

Some Representative Types.—The diagrams given in Fig. 247 illustrate the principles of some of the leading British turbulence-type C.I. engines.

The earlier Ricardo rotational-swirl system shown at A, consists in giving the air in the turbulence chamber a rotational swirl by means of specially inclined air-inlet ports. The piston almost touches the cylinder top and at the end of its compression stroke forces the swirling air charge into the smaller cylindrical turbulence chamber, whence the fuel is projected from the nozzle E in a direction across the air stream. This system was used in the larger single sleeve valve engines.

The Comet turbulent head is shown at B. This employs a small connecting throat arranged tangentially to the spherical

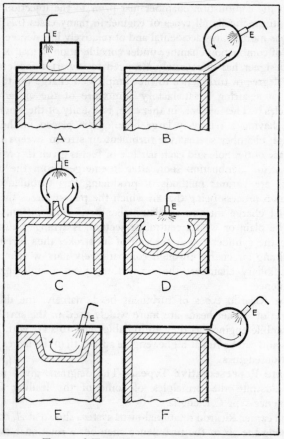

Fig. 247.—Types of Turbulent Combustion Chamber Engines.
A—Ricardo Rotary Swirl. *B*—Ricardo "Comet." *C*—Clerestory. *D*—Saurer Dual Turbulence. *E*—Leyland Cavity Piston. *F*—Spherical Anti-chamber.

chamber shown. The injector *E* projects the fuel across the swirling air charge. This type of head is used in a large number of British and Continental makes of engine. The clerestory head shown at *C* gives a similar turbulence effect; the combustion chamber is cylindrical, i.e., with flat ends in which the inlet and exhaust valves are seated. The projection on the piston is arranged off-centre in order to give a tangential entry to the air.

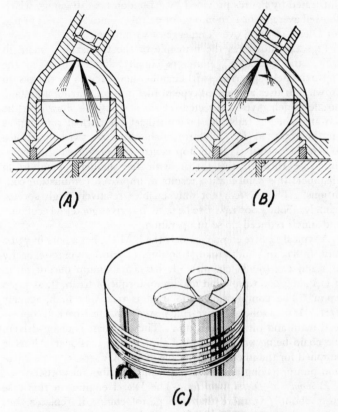

Fig. 248.—Showing Principle of the Ricardo Comet V Combustion System used on many Automobile and Commercial Vehicle Engines.

The Ricardo Comet V Combustion Chamber engines.— Recent engines for small cars, tradesmen's vehicles and taxicabs have been fitted with the later development of the Comet combustion system, originating with that indicated at *B* in Fig.

248. Typical engines using the later Comet V system, include the B.M.C. 1½ litre, Rover 2·052 litre, Fiat 1400 and British and French taxicab engines. The economy of these engines is illustrated by figures provided by a London taxicab garage which showed averages of 15·9 m.p.g. for petrol engines and 30·65 m.p.g. for Diesel engines, over a mileage of 5½ million.

Fig. 248 illustrates the principle of the Comet V system, in which the clearance volume is equally divided between the approximately spherical swirl chamber and cavities in the piston crown. It uses a modified type of fuel injector, having a Pintaux nozzle which gives a directed jet for starting, as shown on the left at *A*, in Fig. 248 and a modified jet for normal running, as on the right, at *B*. The arrows show the air swirl directions. When the piston is on its top centre the piston cavities communicate through the throat with the auxiliary combustion chamber; this combination results in improved combustion conditions. This system not only enables relatively high speeds, namely, about 4,000 r.p.m. to be used, but gives good fuel economy and much reduced noise in operation.

A typical 1½ litre engine, namely the B.M.C., has a bore of 73·02 mm. (2·875 in.) and stroke of 89 mm. (3·5 in.) giving a capacity of 1,489 c.c. (90·88 cu. in.). It has a maximum output of 40 B.H.P. at 4,000 r.p.m. and maximum torque of 64 lb. ft. at 1,900 r.p.m. The compression ratio used is relatively high, namely 23:1. It uses solid skirt five-ring pistons and steel-backed copper-lead main and big-end bearings. The camshaft is chain-driven, the chain being provided with a slipper-type tensioner. Fuel is supplied for the injectors by means of a C.A.V. rotary-type injection pump, having a main plunger and distribution ports.

Economy of Diesel Vehicles.—The Diesel engine, in this case costs about £85 more than the petrol engine it replaces, but accurate cost records show that with its fuel consumption of 38 m.p.g., as against 28 m.p.g. on a 12 cwt. van, for the petrol engine this cost is saved at the end of 10,400 miles, after which there is an annual saving, on a 20,000 mile basis, of over £20 per annum. There are, also, further savings, on account of the much longer major overhaul periods, e.g. 150,000 miles or longer and reduced regular maintenance attention.

The Dual Turbulence System.—The dual turbulence system shown at D (Fig. 247), utilizes a special cavity type piston. The air is given both horizontal and vertical movements during compression so that it has a high degree of turbulence when injection occurs. A wide angle spray is employed, the fuel injection nozzle having several holes for this purpose.

The Leyland cavity piston system is shown at E. This is really a combination of the turbulence and direct-injection systems, the heat losses being low and cold starting relatively easy. Later engines employ the toroidal-type cavity piston shown at D.

Another type of ante-chamber turbulence-head is indicated at F. This belongs to the spherical combustion chamber type, and is so arranged that the two halves of the sphere are formed in the cylinder block and cylinder head, respectively.

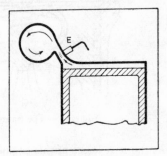

The original Perkins engines, used for cars and commercial vehicles, employed the combustion head shown in Fig. 249. In this case the fuel is injected into the air charge passing through the throat; the charge burning as the piston commences to to descend, for the air rushes out through the throat and meets the fuel spray.

Fig. 249.—The Perkins Combustion Chamber.

The Saurer Dual Turbulence Engine.—Originally devised by the Swiss Saurer Company, this form of combustion chamber, shown at D, in Fig. 247, has more recently been adopted in certain British engines on account of its high efficiency and performance.

The Saurer cylinder head is shown in Fig. 250, from which it will be seen that the combustion chamber forms a *toroidal* cavity of special shape in the piston head.

When the piston ascends on its compression stroke and is nearing the upper end of this stroke the compressed air is forced radially inwards towards the combustion chamber, as indicated by the arrows. The air, during the suction stroke, is given a certain amount of rotation about the cylinder axis, by means of *masks or deflectors on the inlet valves*.

This rotation persists throughout the compression stroke, so that the air that is forced radially into the piston cavity also has a rotational motion, as shown by the arrows in the cavity, as well as a certain vertical movement. The net effect is to give the heated air a high degree of turbulence at the commencement of, and during fuel injection, so that the fuel is burnt very efficiently.

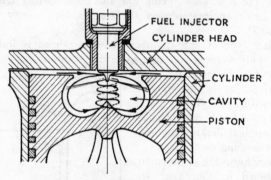

Fig. 250.—The Saurer Combustion System.

As the heat losses of this engine are relatively small, the thermal efficiency is high and, as a result, the fuel consumption per H.P. hour is about the smallest for any of the available C.I. engines. Thus minimum fuel consumptions of 0·34 to 0·37 are obtainable from commercial engines.

The Perkins Engines.—These engines have been made in various 4- and 6-cylinder models ranging from 1·62 litres (99 cu. in.) to 5·8 litres (354 cu. in.) with corresponding gross outputs of 43 B.H.P. at 4,000 r.p.m. to 112 B.H.P. at 2,800 r.p.m. Of these, from the car viewpoint the more recent Four-99 engine is the more widely used. It is the 1·62 litre four-cylinder vertical engine, which has been fitted not only to various British small cars, in place of the standard 1·5 litre petrol engines, e.g., Ford, B.M.C. and Vauxhall engines, but to taxicabs, here and in several European countries. Larger Perkins six-cylinder Diesel engines have been fitted to American cars and taxicabs. Perkins engines are fitted, as alternatives in two leading types of Russian cars. The principal advantages of the Four-99 engine are its marked fuel economy, reliability and very long periods before major

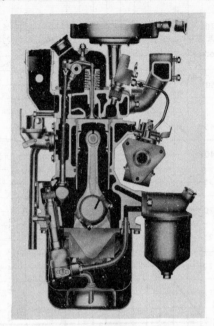

Fig. 251.—The Perkins Four-Cylinder
Four-99 Diesel Engine.

overhaul, e.g. over 100,000 miles. In regard to fuel consumption, the results of various tests under average road conditions, using cars fitted with standard petrol and alternative Four-99 Diesel engines show, conclusively, that with the Diesel engines, the mileages per gallon were from 36 to 42 per cent. greater than with the petrol engines.

Fig. 251 shows the Perkins Four-99 engine in side-sectional view. It is of the orthodox push-rod and rocker arm actuated overhead valve design with gear-driven camshaft. The B.H.P.

and torque curves for this engine, which is notable for its high maximum speed of 4,000 r.p.m. are shown in Fig. 252.

The new combustion system, which differs entirely from that of

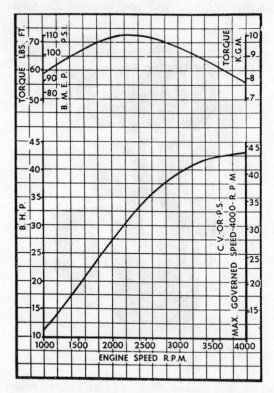

Fig. 252.—Power Curve of the Perkins Diesel Engine.

earlier engines using the system indicated in Fig. 249 can be seen on the right-hand side, just above the piston crown, in Fig. 251. In this, the upper part of the combustion chamber is hemispherical in shape and also partly machined in the cylinder head. The lower half is formed by an inserted machined plug of high heat-resisting alloy. It contains the throat connecting the auxiliary chamber to the cylinder. Fuel is injected into the combustion chamber through a pintle-type injection nozzle. Air compressed during the upward piston movement goes through the throat, producing an air swirl across the nozzle. In the last stages of the compression stroke the air movement changes from the vertical to

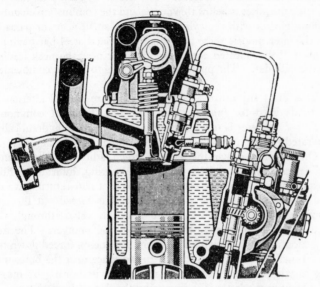

Fig. 253.—The Mercedes-Benz Car Diesel Engine.

horizontal, giving much turbulence and practically complete combustion to the fuel. The engine has a Lucas electric igniter based on a heated coil, with fuel supply from a small tank, to provide vaporized fuel to the air flowing from the induction manifold to the cylinders.

The Mercedes-Benz Engines.—Different sizes of Diesel engine have been fitted to cars of this make, a typical and recent one being the 190D four-cylinder engine which is an alternative to the petrol model. It has a bore of 3·43 in. and stroke of 3·29 in., giving a cylinder capacity of 121 cu. in. The compression ratio is 21:1 and the maximum output, 60 B.H.P. at 4,200 r.p.m. The maximum torque is 87 lb. ft. at 2,400 r.p.m.

The engines of this German company have a special design of auxiliary combustion chamber, which has been used on the various car models (Fig. 253). It has a hemispherical top, in which an inclined fuel nozzle is located. The heater, or glow plug for cold starting purposes is below this nozzle and the combustion chamber communicates with the cyliner head by a small neck, or passage. In the later models there is a small horizontal steel bar having a central "ball", located across the chamber near the neck leading to the cylinder. The purpose of this rod is to reduce combustion noise.

The Fuel Injection System.—There are several alternative methods used for injecting the fuel, but the most commonly favoured system is the mechanical pump and hydraulically-operated injection nozzle shown diagrammatically in Fig. 254.

On the left is shown the plunger which is lifted by the engine-driven cam and closed by the external spring, indicated. Fuel is sucked into the pump chamber through a non-return valve on the downward stroke of the plunger and is forced out through another non-return valve to the injection valve, through the delivery pipe, on the upward stroke of the plunger. The fuel injection nozzle is shown on the right, fuel being forced down the small inclined passage on the left to the space near the bottom of the nozzle plunger. The latter is kept on its seating by means of a spring; the latter is indicated by the dotted lines above.

Usually there is a separate pump unit for each of the engine cylinders, but all of the units are enclosed in a common casing. A four-cylinder engine will have a four-pump unit, a six-cylinder one, six, and so on.

Apart from its purpose of pumping the fuel to its appropriate injector, the fuel pump must be designed to deliver varying small quantities of fuel at exactly the right moment of injection.

To do this, each pump unit is provided with an adjustment for varying the quantity of fuel pumped per stroke of the plunger. All of these adjustments are connected to a single control, viz. the control rod shown in Fig. 255, so that by moving the latter all of the deliveries from the separate units are regulated together.

Most fuel pumps have a control rod (as shown in Fig. 255), in the form of a toothed rack, the teeth engaging with corresponding teeth cut on the cylindrical part of the lower end of each plunger.

When the rack is pulled endwise it rotates each plunger, so

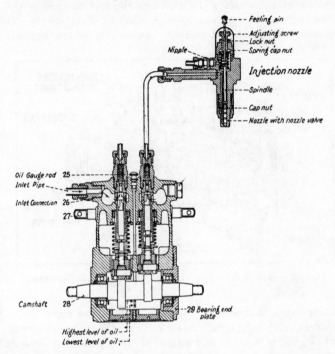

Fig. 254.—The C.A.V. Two-Cylinder Fuel Pump and Injector, showing the Principal Parts.

that spirally-cut portions at the upper end of each are moved up or down in order to control the quantity of fuel.

The cams operating the pump plungers are cut from the solid on a single camshaft—just as in the case of petrol engine camshafts; the cams are, therefore, arranged at their correct relative angles, these being the same as those of the cranks on the engine crankshaft.

Single Element Distributor Pump.—A more recent pump of special importance for small engines is the C.A.V. rotary pump, having a pair of opposed plungers actuated by a fixed external cam, which supplies fuel under high pressure to a single radial port member which rotates and distributes the fuel to each of the outlet ports supplying the respective injectors of the engine. It embodies the usual fuel quantity, slow-running, starting and

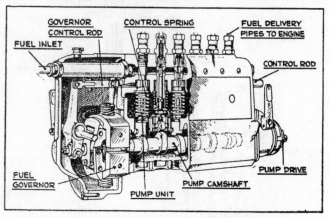

Fig. 255.—The C.A.V. Six-Cylinder Fuel Pump with Governor.

maximum speed controls. The advantages of this pump are its lightness, small bulk, suitability for vertical or horizontal mounting and the use of a single unit instead of the separate side-by-side pumping units of the previously described fuel injection pump. Moreover, the pump is lubricated by the fuel and, therefore, requires no external lubrication means.

Complete Fuel Injection System.—A complete fuel injection feed system is shown in Fig. 256 for the four-cylinder Dennis commercial engines. The arrows show the direction of travel of the fuel. Thus, from the main fuel tank below the fuel is drawn up the right-hand fuel pipe by the suction action of the mechanically-operated diaphragm fuel pump. Thence, the fuel is pumped upwards to the large capacity first fuel filter, shown on the upper right hand side. It then passes to the second fuel filter which has a pressure relief valve for by-passing the fuel back to the fuel tank should the pressure become excessive

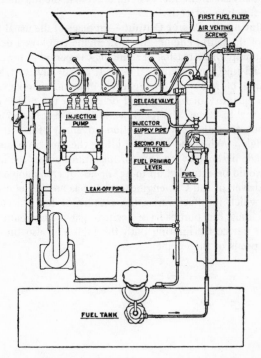

Fig. 256.—A Complete Fuel Injection System. (Dennis).

in the filter. The fuel from the second filter is drawn into the fuel injection pump on the left and thence is measured and distributed to each of the four injection nozzles, which are inside the cylinder head casing. Any leakage of fuel past the stems of the injection nozzles is taken away from leak-off unions, on the upper portions of the nozzle bodies, and thence through pipes to the common leak-off pipe shown in Fig. 256; this pipe leads the fuel back into the fuel tank.

The filters are provided with air venting screws which enable the air within the filters to escape when the fuel system is "primed", i.e., filled with fuel. In this connection it is very important to get rid of any air in the fuel system, otherwise the injection pump will not function correctly.

Regulating the Power Output.—In place of the usual throttle control of the petrol engine, the C.I. engine employs a device on the fuel pump for increasing or reducing the period of time during which fuel injection occurs. The C.I. engine always takes in a full charge of fresh air during each induction stroke, so that only the quantity of fuel injected determines the power output.

It is usual to "time" the fuel injection so that it has an advance (analogous to petrol engine ignition advance) of 20° to 30° of crank angle. When the engine is idling the fuel is injected for a crank angle period of 5° to 10°; for maximum power the injection period would be extended up to 25° or 30° of crank angle.

The driver of the C.I. engine vehicle has an accelerator pedal control which in effect varies the amount of fuel injected. In addition, most fuel pumps have a centrifugal or pneumatic type of governor device for limiting both the high and also the low, or idling, speeds of the engine.

CHAPTER VII

THE GAS TURBINE*

THE success of the modern gas turbine for aircraft, stationary and marine purposes has focused attention upon its possibilities for automobiles since it would appear to have certain advantages over the petrol engines in present use.

In this connection a considerable amount of research and development work has been carried out in England, on the Continent and in the U.S.A., which have shown that a practicable automobile gas turbine is possible, today, provided manufacturing costs can be reduced to petrol engine level and that the problems of low fuel consumption, smaller bulk and exhausting of the larger volume of exhaust gases can be solved. Already much progress has been made towards solution of these difficulties, so that the gas turbine will probably be first commercialized in larger vehicles, and subsequently in cars.

Working Principles.—Referring to Fig. 257, this shows the working principle of the more successful type of gas turbine, known as the continuous pressure or constant combustion one. It consists of three main units, namely, the rotary (axial or centrifugal) type of air compressor A, combustion chamber C and turbine or power unit B. The compressor is driven by the turbine rotor, both being fitted to a common shaft DD, with power output flange at D on the right. The axial compressor shown draws air into its casing at I and compresses this in a series of stages, in each of which the volume is reduced and the pressure increased, until the final stage, shown at the right of A is reached, when the compressed air is delivered through the duct F into the combustion chamber C, where the liquid fuel is sprayed through fuel injection nozzles S. When starting up from rest the air-fuel mixture is ignited by the electric resistance plug P.

* For fuller information see *Small Gas Turbines*, A W. Judge (Chapman and Hall Ltd., London).

In aircraft gas turbines there are several small combustion chambers with their spraying nozzles, but the chambers are all connected to a common outlet leading to the turbine unit; this is adopted to give a compact unit. Referring to Fig. 257, the burnt gases under about the same pressure as at F are forced along the outlet duct G into the turbine casing and there impinge at

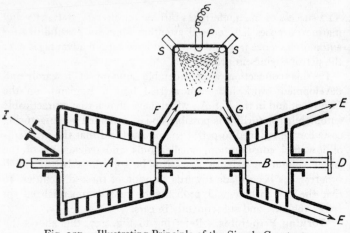

Fig. 257.—Illustrating Principle of the Simple Constant Pressure or Open Gas Turbine.

very high velocity upon one or more rows of turbine blades of increasing diameter until, finally, the gases—having lost a considerable amount of their energy to the rotor B—pass out into the exhaust E.

The energy given to the turbine unit by the combustion gases is greater than that required to compress the air in A, so that the difference is available for useful power output purposes at A. In applying the gas turbine to road vehicles, instead of the arrangement shown in Fig. 257, the exhaust gases from the turbine B would be supplied to a second independent turbine which would provide the power output for propulsion purposes. Thus, the turbine B would be employed, only, to drive the compressor A. The independent turbine arrangement enables the efficiency to

be maintained better at part loads and provides an easier method
of control.

The arrangement mentioned is illustrated in Fig. 258*, which
shows the separate turbine method, similar to that used in the
Rover, Austin and most American gas turbines. The air compressor A, of the axial flow type, delivers compressed air to the
combustion chamber B, where the gaseous products produced
by the combustion of this air and the fuel are directed on to the
vanes of the gas turbine B, which is used for driving the air
compressor A.

The rest of the energy in the hot gases is utilized in driving the
second gas turbine unit, shown at D, which drives the shaft E
gearing with the speed reduction gears shown on the right. The
reduced speed shaft F is employed to drive the automobile.

Proportion of Power Available.—It should, however, be
pointed out that of the power developed by the turbine unit B,
(in Fig. 257) only a small proportion—usually about one-fifth—is
available at the output coupling D. Thus, in the case of a Swiss
gas turbine locomotive the turbine unit developed about 12,000
B.H.P., of which about 10,000 B.H.P. was required to drive the
compressor, leaving the difference, 2,000 B.H.P., as useful
output. It will be thus evident that the turbine-compressor
shaft must be designed so as to be strong enough to transmit
considerably more power than the useful output.

Torque Curves.—With the separate gas turbine method, the
general form of the torque curve would be much more suitable
than that of the simple gas turbine arrangement, shown in Fig.
257, since in the latter method the maximum torque (*see* Fig. 259)
occurs at the maximum shaft speed, whereas for motor vehicle
purposes, it should be available for lower speed operation. The
upper curve, in Fig. 259, refers to the torque curve for the
separate turbine scheme and it shows that this torque-speed
relationship is suitable for traction purposes, since the greatest
torques occur at the lower speeds.

Advantages of Gas Turbines.—The gas turbine possesses
certain important advantages over the reciprocating-type petrol
engine, which may be summarized, as follows: (1) Smoother

* *Gas Turbines. Prospects in Road Transport.* R. H. H. Barr.

running, due to absence of reciprocating parts; (2) Absence of vibration; (3) Much higher mechanical efficiency; (4) More compact shape; (5) Silent operation due to absence of reciprocating members and to the continuous exhaust; (6) Low internal pressures (these seldom exceed about 75 lb. per sq. in.); (7) Reduced maintenance costs due to simplicity of design; (8) Very much lower oil consumption, namely, about $\frac{1}{50}$ that of equivalent petrol engine; (9) Much lower weight, namely about one-third

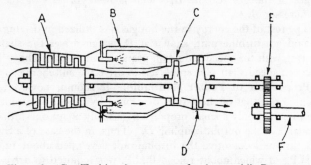

Fig. 258.—Gas Turbine Arrangement for Automobile Purposes.

that of the equivalent petrol engine; (10) Absence of clutch or gearbox, except for reversing and emergency low gear. (11) Ability to run efficiently on cheaper fuels, such as paraffin.

Some Disadvantages.—In its own sphere of application, namely, for aircraft engines and land and marine installations, the gas turbine has already made considerable progress. Thus, for aircraft use, where the backward momentum of the exhausted gases it utilized for propulsion application, the gas turbine* clearly established itself as a rival to the more expensive and complicated high output (2,000—3,000 h.p.) reciprocating petrol engine and has now replaced the latter engine in all of the larger airliners. In these applications the utilization of the ejected gases for propulsion purposes is the principal factor.

For large locomotives, power stations and certain marine

* Vide *Modern Gas Turbines*. A. W. Judge (Chapman and Hall, Ltd., London).

applications the gas turbine has proved a serious competitor to
the steam turbine, but in order to achieve its best efficiencies it
is necessary to depart from the simple type of gas turbine, i.e.
compressor-combustion chamber-turbine unit, and to introduce
multiple turbines, intercoolers, heat-exchangers, etc. These
complicate and increase the weight and bulk of the gas turbine
installation. As it stands the simple gas turbine (Fig. 257) is

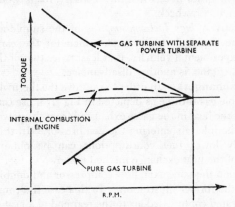

Fig. 259.—Torque Curves of Gas Turbines.

much less efficient than the petrol and Diesel engines and, in
the smaller sizes that would be necessary for automobile engines,
would be still less efficient, since there would be appreciable
losses due to the scaling down effect and limited room for such
refinements as efficient heat-exchangers. Moreover, to maintain
the efficient turbine and compressor blade speeds the smaller units
would require to operate at considerably higher rotational speeds,
namely, from 30,000 to 50,000 r.p.m. This would bring in
complications in regard to reduction gears and controls. In
order to minimise fuel costs, due to the lower efficiency of the gas
turbine, the aircraft types operate on paraffin fuels. If these
were used for road vehicles the large exhaust quantity might be
objectionable.

This serious objection to the use of gas turbines for automobiles is concerned with the *large quantity of exhaust gases* emitted, on account of the much larger quantity of air used per unit weight of fuel consumed. Thus, in most gas turbines, it is necessary to employ from $3\frac{1}{2}$ to $4\frac{1}{2}$ times the amount of air normally used to burn the same quantity of fuel in a petrol engine; thus, the volume of exhaust gases is, for this reason, much greater than for a petrol engine of similar output. The discharge of such quantities of gases over town and country roads would appear to be a serious drawback.

The relatively *high fuel consumption* of the simple automobile gas turbine depicted in Fig. 258, which for the earlier types used for experimental vehicles, is at least twice that of the equivalent petrol engine, is another disadvantage.

Heat Exchangers.—In order to improve the thermal efficiency of the simple gas turbine as depicted in Fig. 258, use can be made of some of the heat in the gases exhausted from the power turbine, to warm the inlet air entering the compressor. In this manner, appreciably lower fuel consumptions can be obtained. The principle of the heat exchange method is shown in Fig. 260 which is based upon the same axial flow compressor and double turbines shown in the former illustration. In this case the compressor A delivers heated compressed air to the rear end of a heat exchanger G—which may be likened to a kind of radiator unit but in which hot air from the exhaust flows through one tube system and cool air from the compressor passes around these tubes, thereby extracting heat from them. This heated air then passes into the combustion chambers B, from which the combustion gases pass rearwards past the blades of the compressor turbine C and then the power turbine D, after which the exhaust gases are discharged through the tubes, or matrices of the heat exchanger to the atmosphere.

It can be shown that the thermal efficiency increases, with the pressure ratio of the compressor, i.e. the ratio of outlet to inlet air pressures, and with the equivalent heating surface of the heat exchanger. Thus, in a specific example, if the efficiency of the simple gas turbine be represented by E, then with a given surface S sq. ft. of heat exchanger the efficiency was increased to $1 \cdot 2\ E$; with a surface $3\ S$, to $1 \cdot 4\ E$; with a surface $6\ S$ to $1 \cdot 5\ E$, while

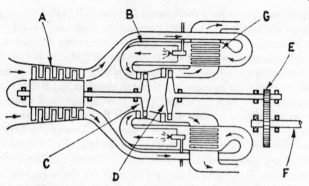

Fig. 260.—Illustrating the Principle of the Gas Turbine Heat
Exchanger, as applied to the Automobile-type Engine.

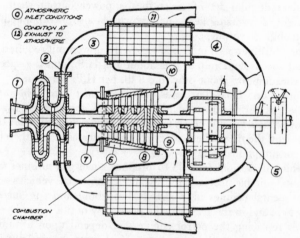

Fig. 261.—Schematic Arrangement of the Austin Automobile
Gas Turbine.

for a surface of infinite area it was calculated that the efficiency
would be $1.9 E$.   In each case the pressure ratio was about $3.5:1$.

The schematic arrangement of the Austin automobile gas
turbine, shown in Fig. 261 consists of a two-stage centrifugal air

compressor, driven by a three-stage turbine, followed by a single stage power turbine. A single combustion chamber is used. A heat exchanger of the crossflow type is also shown; it consists of two units located on either side of the engine. Air is drawn into the first-stage compressor at (1) and discharged from the second stage at a pressure ratio of 4:1 through the duct (2) and (3) and air spaces of the heat exchanger (11), through the twin ducts (4) and (5) and then reversing the heated air flow direction, forwards through the combustion chamber—indicated behind the turbine unit—and into the forward turbine intake (7). From there the combustion gases flow rearwards again through the three-stage turbine unit (8), finally exhausting through the ducts (9) and (10) through the heat exchanger and then rearwards into the outside atmosphere or the rear exhaust duct. It will be observed that the rear or power turbine, on the extreme right, is connected to a double reduction set of gears to the power output shaft, on the right; a reversing gear—not shown—is also incorporated for automobile drive purposes.

For a turbine speed of 21,000 r.p.m. the power unit required for a road speed of 92 m.p.h. was about 104 H.P., and the specific fuel consumption about 0·65 to 0·8 lb. per H.P. per hour.

Notes on Gas Turbines.—If the *inlet gas temperature* to the gas turbine rotor can be increased appreciably, by the use of *new alloys* with much *better high temperature strengths* than at present, the thermal efficiency could be increased and therefore fuel consumptions reduced.

Whilst it is quite feasible to produce small gas turbines with outputs suitable for larger cars and commercial vehicles—and experimental engines have already been built and run successfully here and in the U.S.A.—the possibility of this type of prime mover replacing the petrol engine will depend upon a number of factors, such as the comparative fuel costs, output per unit weight, solution of the mechanical problems of transmission, road speed and idling speed controls, response to fuel regulations, etc.

Using Cheaper Fuels.—Operating on cheaper fuels the lower thermal efficiency drawback may be overcome from the fuel cost viewpoint and with suitable combustion arrangements a free exhaust could be obtained, although the regulation of combustion over the wide range of operational speeds would require special consideration.

Starting Gas Turbines.—The starting of a small gas turbine with its purely rotational units is a relatively simple matter compared with that of the petrol or Diesel engine but a more powerful starting motor and larger battery would be needed. Moreover, if the gas turbine could be run at a few thousand r.p.m. with little power output it would be unnecessary to use a clutch for vehicle starting purposes, since the power developed would probably be less than the transmission resistance losses; otherwise a simple design of transmission might be effective.

Recent Progress.—The pioneer of modern gas turbines for automobiles is the Rover Company who, in 1950 demonstrated the world's first practical gas turbine car and has since developed the engine so that it has a much lower fuel consumption and better performance.

Later the Austin Motor Company produced a 120 H.P. model gas turbine, employing a two-stage centrifugal compressor and three-stage axial flow turbine, followed by a free turbine. A heat exchanger was also used and the experimental engine was installed in a car and demonstrated on the track.

In the U.S.A. appreciable sums of money have been spent by the General Motors Corporation, Chrysler and Ford companies in experimental work and development of the automobile gas turbine, with encouraging results. General Motors gas turbines have been fitted to specially-designed cars, of which the original GT-300 Whirlfire engine was tested in a Turbo-Cruiser car. Later engines include the GT-304 Whirlfire II, with a heat regenerator, to improve fuel economy. It had a rating of 200 H.P. —at a power turbine speed of 28,000 r.p.m., but would operate at 35,000 r.p.m. It was installed in a car and given various road and track tests with satisfactory results, including improved fuel economy and relatively low noise level. It had a fuel consumption of about 0·7 lb. per h.p. hour. The later GT-305 engine, which was installed in the Firebird III car developed 225 h.p. at 33,000 r.p.m. with a maximum turbine inlet temperature of 900° C. (1,650° F.). It used two separate can-type combustion chambers and had a (best) fuel consumption of 0·55 lb. per h.p. hour.　Later developments are taking place to reduce the consumption still further, namely, to that of comparable petrol engines.

The Chrysler Company has developed the automobile gas turbine, since the first 1947 model which embodied a heat exchanger.　Later engines have rotating heat regenerators.　A typical experimental engine, rated at 120 h.p. actually developed

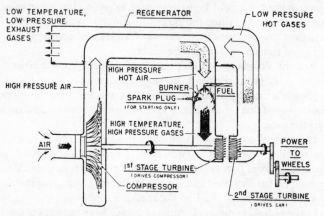

Fig. 262.—Layout of the Chrysler Automobile Gas Turbine.

160 B.H.P. at about 50,000 r.p.m., with an idling speed of 20,000 r.p.m. when fitted in a Plymouth car it had a fuel consumption of 15 to 17 miles per gallon, at 30–40 m.p.h.

One model Chrysler gas turbine is shown, schematically, in Fig. 262.

The operation is, as follows: Air from the atmosphere is drawn into the central suction side of the centrifugal compressor and delivered under pressure upwards and then along a horizontal duct which is surrounded by a concentric duct through which the exhaust gases from the turbine unit pass.　These gases heat the high pressure air on its way to the combustion chamber which is shown vertically and fitted with fuel sprayer and sparking plug— for starting purposes, only.　The heated and compressed air is used partly to burn the fuel and partly to cool the hot gases—which, otherwise would be too hot for the turbine wheel blades to withstand. The turbine unit has two stages, i.e., two turbine wheels, the hot gases passing through the left or 1st stage turbine to drive the compressor and through the 2nd stage turbine to deliver power *via* a speed reduction unit to drive the automobile.

This engine is rated at 120 B.H.P. and the actual compressor-turbine unit measures 32 ins. long, 33 ins. wide and 28 ins. deep. The engine is air-cooled and therefore requires neither a liquid cooling system nor radiator.

More Recent Progress.—More recently a Chrysler gas turbine installed in a Dodge car—named the Dodge Turbo Dart has been produced on a limited scale and between 50 and 75 such cars are being tested by various drivers on American roads. The CR2A engine develops about 140 B.H.P. and has a claimed fuel consumption under road conditions of about 20 m.p.g. (Imperial gallons).

The later Rover T4 gas turbine car, with heat exchanger is a marked improvement over the original car and a similar engine was installed in a B.R.M. racing car for the Le Mans 24-hour race, and completed the race at an average speed of 107·8 m.p.h., without experiencing any mechanical troubles. The Rover 25–150 engine develops 150 B.H.P. and weighs 200 lb. with its ancillary equipment.

CHAPTER VIII

LUBRICATION OF THE ENGINE

THE lubrication system is one of the most important items of the
high speed petrol engine since if the oil supply to the working parts
is stopped the engine will seize up after a short time.   The modern
engine contains a relatively large number of components rotating,
sliding or reciprocating at very high speeds, and in attaining these
speeds with engines which are light and compact, the bearing loads
and rubbing velocities used are much higher than those employed
in steam engine and machine practice.   It is therefore essential
for these parts to be kept well lubricated, and the lubricating
system must be absolutely reliable.   In passing, it may be mentioned that the majority of engine mechanical breakdowns may be
traced to failure of lubrication.

The essentials of an efficient lubrication system are: (1) reliability; (2) proper lubrication of all working parts; (3) minimum
oil consumption, otherwise much oil is by leakage past the
pistons, forming carbon deposit in the combustion chamber and
gumming up the valves; (4) efficient oil filters; (5) conspicuous
indicator or pressure gauge showing whether the oil is circulating
properly; (6) use of suitable lubricant; (7) accessible oil filler and
oil draining plug.

The Principles of Lubrication.—When two components
of a bearing, such as the shaft and its bearing, are adequately
lubricated with a suitable oil the two metallic surfaces are not in
metallic contact, but are separated by an oil film, and the resistance
to relative motion between the shaft and its bearing therefore
depends largely upon the physical properties of the oil, namely,
its viscosity; it is independent of the frictional coefficients of the
bearing metals.   This type of lubrication is termed the "*fluid*"
one.   There is, however, another form of lubrication liable to
occur in the case of bearings and sliding surfaces under inadequate
lubrication or high load conditions, or in instances where appreci-

able surface roughness of the components prevents the maintenance of "fluid" lubrication.   In such cases the oil film is insufficient in thickness to prevent occasional contact of the bearing surfaces, with the result that the frictional resistance is increased appreciably; usually the frictional coefficients under these "*boundary*" lubrication conditions are from ten to twenty-five times greater than for fluid lubrication, but are less than the "dry" coefficients.   It should here be explained that if the bearing surfaces were perfectly, or chemically, clean the coefficient of friction would be much higher, but a trace of oil on the surfaces reduces this value appreciably.   Thus, under the conditions of boundary lubrication the film of oil is infinitely thin, the actual value of the frictional coefficient will depend not only upon the nature of the oil, but upon the metals used for the shaft and bearing.   The friction effect in this case is proportional to the load on the bearing, so that the coefficient of friction is actually independent of the load.

There is a certain property associated with lubricating oils, which, for want of a better name, is termed "*oiliness*."   It is difficult to define, but it can be stated that it is a surface effect produced by the lubricant upon the metallic surface with which it is in contact.   In this connection fatty oil, i.e., oils saponifiable or those containing "fatty" ingredients, such as castor, rape or olive oil, exhibit a greater degree of "oiliness" than purely mineral oils, so that the frictional coefficient at any given temperature is lower under severe conditions of loading and slow speeds. Under extreme conditions of loading and high rubbing speeds, seizure between the shaft and its bearing would be less likely to occur with the lubricant of greater "oiliness".

In the case of pistons and cylinder walls it is very probable that the boundary conditions of lubrication, previously mentioned, exist, the most favourable circumstances for this condition being during the maximum cylinder pressure and initial expansion pressure period, i.e., in the upper part of the cylinder barrel.

Lubricating Oils.—A good lubricating oil should, at all temperatures, form an oil-film between the two sliding parts. If the oil becomes very thin at high temperatures, i.e., of low viscosity, the pressure between the sliding parts may cause it to break down, or be squeezed out, when the metal parts will make contact and may then seize up. The oil must therefore have sufficient viscosity at the engine working temperatures, but on the other hand it must not be too viscous (or thick) at amospheric temperatures, or it will be difficult to start the engine. The viscosity falls off fairly rapidly with temperature increase. Many reputable lubrication oil firms supply "winter" and "summer" grade oils.

These oils are obtained from petroleum by a process of distillation and refining, such oils from a base stock being known as *straight-run* lubricating oils. Modern lubricating oils to meet the needs of higher performance automobile engines, operating upon anti-knock fuels are often blends of oils prepared from distillates having relatively narrow boiling points, and certain additives, which are referred to, later.

It is well known that a *straight-run* oil has a low viscosity at low temperatures and a high one at high temperatures, so that hitherto it was necessary to use a separate viscosity-range oil for cold or hot weather operating conditions. Modern *multigrade* oils contain an additive, known as a *viscosity improver* or V.I., which reduce the effective viscosity at low temperatures and increases it at the higher engine temperatures. Tests have shown that these additives reduce combustion deposits and give improved fuel economy at the lower temperters and lighter engine loads. Multigrade oils can now be used to cover the full climate range of temperature.

S.A.E. Oil Ratings.—The Society of Automotive Engineers (U.S.A.) have standardized a system of classification for engine oils, in terms of their viscosities, which is also becoming used in this country. It specifies the Saybolt Universal viscosities of seven oils to cover the whole range of atmospheric temperatures. In this range, the thinnest oil is the 5W, while the 10W, and 20W would be used for Arctic temperatures (below 10° F. ($-12°$ C.) and extreme cold down to 10° F., respectively. The more widely used grades are the S.A.E. 20, 30 and 40 grades which cover normal air temperatures up to the tropical values. The other more viscuous oil is the S.A.E. 50.

Lubricating Oil Additives.—Lubricating oils produced by the usual distillation and refining processes have more recently been found liable to form undesirable deposits, such as moisture, sludge, varnish and acid in the crankcases, while also causing carbon deposits in the backs of the piston ring grooves, the piston tops and combustion chamber. The formation of these deposits can be prevented by adding specially selected agents to the lubricating oil. Thus, sludge and carbon deposits can be reduced by additives having *detergent dispersal* properties; varnish deposits on piston skirts and crankcase walls are reduced by *oxidation eliminators* and corrosion of the bearings, by special *corrosion prevention* additives.

It is possible for a high grade lubricating oil to contain as many as five different additives, to give the desirable properties of oiliness, viscosity over the engine temperature range, corrosion inhibition, sludge prevention and carbon and varnish coating dispersal.

High grade oils, which are both sludge-proof and free from deposit forming tendency are often prone to oxidise and become acid. While acids developed after prolonged usage of mineral oils in petrol engines will not usually attack cast iron or aluminium alloys, it has been shown that lead-bronze alloy main and big-end bearings are liable to corrosion—which may sometimes be severe—by the acid products of lubricating oils. The prevention of this bearing corrosion has now been effected by the use of additives to the oil.

Of the deposits which are liable to accumulate in engines, the following are the more important, namely, (1) carbon deposit on the top and underside of the pistons, in the piston ring grooves and on the combustion chamber walls. (2) Varnish-like deposits on the piston skirts, connecting rod surfaces and on the exposed parts of the crankshaft. (3) Black grease-like sludge in the crankcase, consisting of soot, oil and water, and (4) solid particles in the crankcase sump due to carbon deposit broken from inside the piston, siliceous matter from the air and broken metal burrs from the crankcase components—in the early stages of an engine's life.

In the case of engines that have been in service for longer periods, a detergent oil or a detergent added to the ordinary engine oil will in time get rid of the carbon deposits, gum and varnish, by solvent or loosening and flushing action, so that the ring grooves, piston skirts, connecting rod, crankcase walls and crankshaft surfaces become clean again.

Later Developments.—The usual periods for *changing the engine sump oil* are at the end of the first 500 miles, when new, and then at 3,000 miles intervals.

In recent American car engines, after presumably the initial new engine changing of the oil, it is then necessary, only, to change the oil at the same time as when changing the filter element, namely, at every 6,000 miles of road service. This improvement is due to the use of special "premium" engine oils containing long-life lubrication additives.

In certain Ford Mercury and Lincoln cars, in which the chassis greasing points employ specially designed bearings, using special lubricants and permanent rubber boots or seals, no lubrication attention is required until 30,000 to 36,000 miles have been covered. In order to cope with this condition for the engine lubrication, special oil filters are now fitted, which do not require replacement of their filtering elements, until 30,000 miles of service.

Lubricating Systems.—Automobile engine lubrication systems that have been or are in present use may be classified under definite headings, as follows: (1) *The Fly-wheel Splash Type*; (2) *The Assisted Splash and Constant Level Type*; (3) *The Low Pressure System*; and (4) *The High-Pressure System*. Before referring to these systems in detail let us consider for a moment the principal working parts which require lubrication. In order of importance these are as follows: (1) *The Cylinder Walls and Pistons*. The piston of a modern engine has an average bearing pressure on the cylinder wall due to piston thrust of 250 to 350 lbs. per sq. in. Its crown temperature is about 250° C. to 370° C. (2) *The Connecting-Rod Big End Bearings* and *Crank-Pin*. (3) *The Small End Connecting-Rod Bearing*. This rod end rocks a little only on its pin, but the temperature of the piston is relatively high, so that the lubrication of these parts is important. (4) *The Main (Crankshaft) Bearings*. (5) *The Camshaft, Cams and Bearings*. (6) *The Valve Tappets* and *Guides*. (7) *The Timing* and other *Gears*. (8) *The Overhead Valve Rocker Gear and Shaft*.

Most lubricating systems aim at oiling the main and crankpin bearings first, and allow the surplus oil to be splashed about so as to lubricate the cylinder walls, pistons and other parts. Where the former are liable to be over-lubricated, a special oil-scraper ring, or groove, is arranged in the lower part of the piston.

The Fly-wheel Splash System.—In this now historic method, once used on Ford engines, the fly-wheel has below it an oil-collecting sump, or trough, this being the lowest part of the enclosed crank-chamber, so that all the oil drains down to it. The

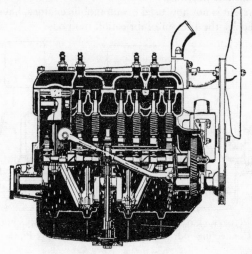

Fig. 263.—The Earlier Ford Engine Lubrication System.

fly-wheel rim runs in this oil, and the oil which is flung off the rim tangentially by centrifugal action is caught by specially shaped troughs or channels cast in the upper part of the crank-chamber, whence it gravitates to the main bearings and gearing. Below each connecting-rod, in its lowest position, an oil trough is arranged, into which an oil scoop, or projection, on the lowest part of the rod just dips; the oil thus caught is led through suitable holes to the big-end bearings. These troughs fill up automatically.

Fig. 263 shows the lubrication system employed in the earlier Ford engines, after the fly-wheel lubrication system previously described. In this case a submerged type of oil pump in the sump, driven by a vertical shaft, forces oil to the main bearings, whence the oil drains back into the sump. The connecting-rod big-ends dip into the oil troughs at the bottoms of their strokes, and thus lubricate the big-end bearings. The oil from the sump is first pumped up to a trough above the camshaft, whence it gravitates downwards along specially arranged passages to the main crankshaft bearings.

This system is not now used in automobile engines, having been superseded by the pressure lubrication methods.

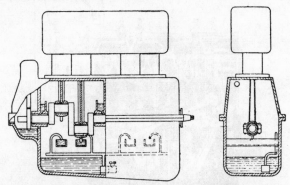

Fig. 264.—The Splash System of Lubrication, showing Oil Sump, Connecting-rod, Oil Troughs, and Oil Supply Pipes.

The Assisted Splash and Constant Level System.—In this method, here of historical interest only, an oil pump replaces the fly-wheel oil-flinger of the previously described system, and oil is pumped from an oil reservoir or sump, situated at the lowest part of the crank-chamber, to the main bearing and connecting-rod troughs (Fig. 264), whence it not only feeds the big-end bearings, but is also splashed, or flung about in the form of a spray to all the other parts, including the cylinder walls, to oil troughs above the main bearings, timing gears and camshaft bearings (whence the oil flows by gravity to these parts).

The splash system, was very efficient and reliable in the slower speed types of engine; it also possessed the merits of simplicity, durability and cheapness. The oil pump required for this system was a low-lift, non-pressure type, since it had only to raise the oil to the troughs.   Any surplus oil draining down the sides of the crank-chamber flowed through wire gauze gratings or sieves on a level with the bottoms of the troughs, whence it drained into the sump below; this enabled all solid matter to be trapped by the gauzes.

In this system, an oil level indicator, either of the two-tap type described, the float and wire, or the ordinary gauge glass

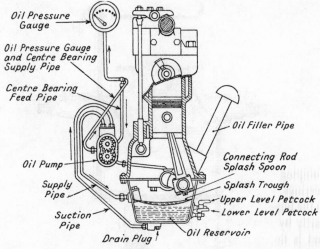

Fig. 265.—A Low-Pressure Lubrication System.

type, was employed.   It was also usual to fit some form of indicator in the system to show that the oil-pump was working. The simplest indicator was a small tap situated in the delivery side of the pump.   On opening this, oil flowed out if the pump was working.   Another indicator consisted of a brass fitting on the lower part of the dashboard, in which a small plunger and wire pulsated in synchronism with the strokes of the pump; a wire moved upwards and remained there when a rotary pump was used.   Another readily observable indicator was an ordinary oil gauge-glass on the dash, which showed the oil streaming down from a central jet inside.   Sometimes the delivery oil was made to impinge upon a vane, or small impeller, situated in a small brass box provided with a glass window; the movement of the vane or impeller at once showed that the pump was working.

The Low-Pressure System.—This system, which was the forerunner of the modern high-pressure system, consists in

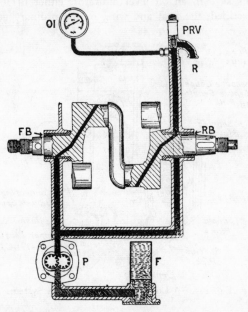

Fig. 266.—A Simple Pressure Oiling System.

pumping oil under a pressure (of 5 to 10 lb. sq. in.) to the main bearings, whence it passes through holes drilled in the crankshaft to the big-end bearings, impelled partly by the pump pressure and partly by centrifugal action. The oil escapes from the big ends and is flung out, whence it lubricates by splash the cylinders, gudgeon pins and other parts. Sometimes the oil was supplied under pressure (from 5 to 8 lbs. per sq. in.) to the main bearings only, and oil troughs are used for the connecting-rod big end lubrication (Fig. 265). It is essential in the former case that the oil ways be kept quite clear, otherwise any choking will cause the system to fail. Similarly any appreciable wear in the main bearings will cause the oil to leak away, and the big ends may become starved.

Fig. 266 illustrates a two-cylinder opposed engine lubrication system. This diagram will serve also to show the various items referred to in the preceding section. Oil is drawn by the gear-wheel type pump P from the sump, through a fine brass gauze filter F. From the pump it is delivered through holes or passages cast integral with the crankcase to the front main bearing FB, whence it flows through the drilled crankshaft to the left-hand crank-pin, and there lubricates the big-end bearings of the corresponding connecting-rod. The black lines show the paths of the oil under pressure. At the same time oil is delivered from the pump to the right-hand main bearing RB, and thence to the right crank-pin and its big-end bearing. From RB the oil is also led to the oil-pressure gauge OI, and to the pressure release valve PRV, situated at the rear of the engine in front of the fly-wheel. This valve is adjusted to prevent excessive oil pressure, any excess of oil returning by the pipe R via the crankshaft rear bearing plate, which is drilled for the purpose. The oil pressure gauge reads about 4 to 6 lb. sq. in. at idling speeds, and 14 to 16 lb. sq. in. at maximum engine speed.

In common with most oil pressure systems the gauge reads high (22 to 25 lb. sq. in.) when the engine is first started up with the oil cold.

The High-Pressure System.—In the case of the high-speed engines fitted in modern types of car, it is the practice to feed the oil positively, under a high pressure (35 to 70 lbs. per sq. in.) to all the friction surfaces. The oil is forced by means of a pump, usually of the plunger or gear wheel type, and submerged in the oil sump, direct to the main bearings, whence it travels through the drilled crankshaft to the big ends. Thence it is sometimes led to the gudgeon pins, usually by means of small copper tubes attached to the sides of the connecting-rods or by rifle-drilled holes through the connecting-rod webs themselves. The oil from the gudgeon pins is sometimes used to lubricate the pistons, but a more preferable method is to have separate oil ducts to the cylinder walls, or oil jets from the connecting-rod big-ends so that fresh cool oil is supplied. In addition to the main bearing oil supply, pipes are also arranged to supply oil under pressure to the camshaft bearings and oil-jets for the timing gearing. It is essential with this system to provide an oil strainer of large area around the oil pump supply, and to render this accessible for cleaning purposes. Further, an efficient oil filter unit must be fitted between the pump and main bearing supply pipe.

Lubrication of Overhead Valve Engines.—Overhead valve engines require a somewhat different method of lubrication to that of side-valve engines, for it is necessary in the former case to lubricate the bearings of the rocker-arm shaft; in most instances the valve stems and cam contacts with the valve stem ends are also lubricated.

The principle of the overhead valve engine lubrication is illustrated in Fig. 267 for the B.M.C. six-cylinder engine, in which the valves are operated by push-rods and rocker arms which have their bearings on the rocker hollow shaft.

The oil in the sump below is drawn upwards through a large area gauze filler by the oil pump, from which it is delivered under pressure through a full-flow oil filter to a longitudinal oil passage drilled in the cylinder block and running its whole length. From this oil gallery the oil flows down passages drilled in the main crankshaft bearing supports and thence to the main bearings themselves. Each crankshaft main journal, crankshaft web and crank journal is drilled with a slanting hole, so that the oil under pressure can reach the big end bearings from the main bearings.

Oil is also taken from the main oil gallery, through suitable drillings in the cylinder block, to the four camshaft bearings, shown in Fig. 267, almost in line with the oil gallery. Oil for the hollow rocker shaft, is taken through a restricted orifice on the main oil gallery, vertically upwards as indicated by the arrowed thick line, to a hole in the shaft, a little to the left of its centre. After lubricating the rocker arm bearings and tappet ends the oil drains down into the push-rod holes, to lubricate the tappets, which are operated by the camshaft cams. The used oil from the rocker gear, tappets, main and crankshaft bearings finally finds its way back into the oil sump.

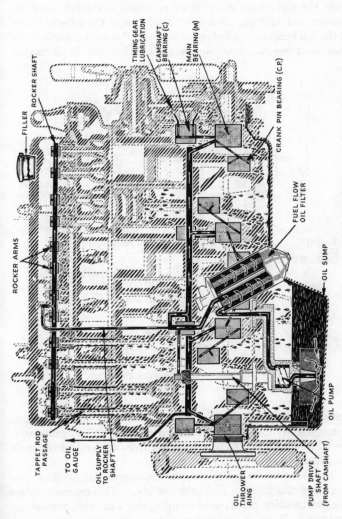

Fig. 267.—Lubrication System of Six-Cylinder Vertical B.M.C. Engine.

The valve timing gear is lubricated by a metered supply of oil from the right-hand main bearing, which has a felt insert to prevent end leakage. Similarly, as shown at the left-hand side of the left bearing an oil-thrower ring and threaded portion of the crankshaft beyond, stop any leakage of oil to the clutch casing,

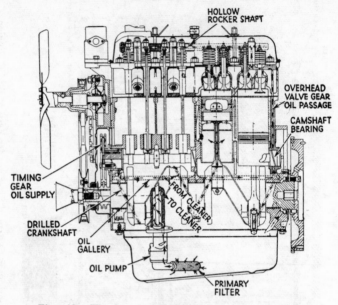

Fig. 268.—The Standard Vanguard Lubrication System, in front view.

surrounding the flywheel. This recovered oil is led downwards through a vertical oil tube to the oil sump. The oil gauge connection—shown on the left—is taken from the end of the main oil gallery.

The lubrication system of the Standard Vanguard 2 litre engine is illustrated in Figs. 268 and 269. This engine employs the two meshing gear type of pump, giving a delivery pressure up to 50 or 60 lb. sq. in., according to the engine speed. The pump draws the oil from the sump through a primary gauze filter and thence the oil passes through a channel in the pump casing to the annular space around the oil pump, as shown by the black arrows. The high pressure oil then passes through a passage in the cylinder block through an external oil filter; an oil pressure release valve is fitted between the pump and filter to limit the

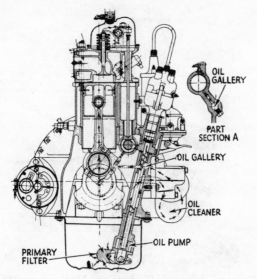

Fig. 269.—The Standard Vanguard Lubrication System, in side view.

maximum oil pressure to 50-60 lb. sq. in. The filtered oil then flows into a horizontal oil gallery, which extends the length of the cylinder block. It should here be mentioned that a portion only of the oil is filtered, the oil filter being of the *by-pass* type, described later.

The oil from the gallery goes, *via* drilled holes in the crankshaft to the three main bearings and to the crankshaft big-end journals to lubricate the bearings. Oil is also forced through drilled holes in the connecting-rods (shown in Fig. 269) to lubricate the small-end bearings.

Splash lubrication is further assisted by drilling into the oil passage between the small-end and big-end of the connecting-rod (shown by the side arrows at centre of the rod in Fig. 268); this oil lubricates the cylinder walls.

The three camshaft bearings are lubricated from by-pass passages from the main bearings. A by-pass from the rear

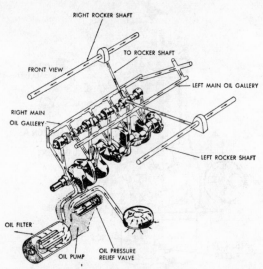

Fig. 270.—Schematic Illustration of Modern Vee-eight Engine Lubrication System

camshaft bearing conveys oil upwards into the cylinder head, whence it is led to the hollow rocker arm shaft of the overhead valve gear, to lubricate the rocker arm bearings, the oil leaving each rocker by a horizontal hole to lubricate the push-rod ball end. The timing chain is lubricated from the front camshaft bearing, there being four slots cut at 90° on the flange face; the oil is thrown out by centrifugal force on to the chain sprocket. Separate provision is made for lubricating the underside of the timing chain.

Vee-Eight Engine Lubrication.—In the 90° vee-eight engine there is a single four-throw crankshaft, a single camshaft located in the 90°-vee above the crankshaft and the overhead valve actuating gear on each cylinder block. This arrangement necessitates positive lubrication to the crankshaft main and big-end bearings, the camshaft bearings, the two rocker shafts (or tubes) and rocker and push-rod bearing faces.

Fig. 270 shows the layout of the Chrysler, De Soto engine lubrication system, in which a Hobourn-Eaton type oil pump draws oil from the sump through a floating-type filter and delivers it

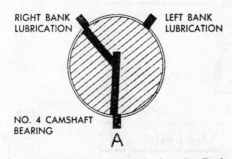

Fig. 271.—Method of Lubricating the Rocker Shaft Bearings, from the Drilled Camshaft, for Vee-Eight Engine, shown in Fig. 270.

first to a full-flow oil filter, from which it is delivered to the two main oil galleries, as depicted by the arrows. From the galleries the oil goes to each of the five crankshaft main bearings, whence it flows through drillings in the journals to the crank pin journals. Pressure oil is also taken direct to each of the five camshaft bearings from the right oil gallery. The No. 4 camshaft bearing and camshaft journal are drilled, as shown in Fig. 271, so that it receives oil from the right main oil gallery at A, *via* the supply to No. 4 main bearing and delivers "spurts" of this oil to the right and left hollow rocker shafts, alternately, as the drilled camshaft journal rotates. The oil from the rocker shaft lubricates the rocker arm contact surfaces with the valve stem ends and the push-rod ends.

In the 361 cu. in. Dodge engine, the crankcase oil capacity is 5 U.S. quarts, while that of the oil filter is 1 quart, making 6 quarts in all.   The oil pump pressure at 500 engine r.p.m. is 20 lb. sq. in., and at 1,500 r.p.m., 45 to 65 lb. sq. in.

The Dry-Sump Method.—In the previous system the main oil supply is kept in the sump below the engine crank-case.   There is another system (Fig. 272) which is sometimes used on racing engines, the more expensive car, and also on most aircraft engines in which there is no oil sump in the crank-chamber, but the supply

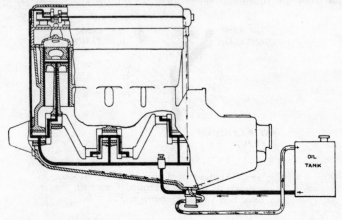

Fig. 272.—Illustrating the Dry Sump Lubrication Method.

is contained either in a separate tank, or in a separate reservoir, provided with air-cooling fins on the side or below the engine. Two pumps are employed in this case, one to force the oil under (high) pressure as in the preceding case, to the main, big-end and other bearings, directly, whilst the other, or "Scavenger," pump, which is the larger in capacity, sucks up the oil which drains down to the bottom of the crank-chamber, and returns it to the oil reservoir.   The principal advantage of this arrangement is that the oil is cooled in its circulation, and therefore has a better lubricating value.   The oil is also controlled in its circulation. Further, it is also more suitable for engines that may have to work in inclined positions.

Cylinder Wall Lubrication.—As mentioned earlier in this

chapter the usual method of lubricating the cylinder wall is by means of oil splashed from the rotating parts, e.g., the cranks and connecting-rod ends. The more recent tendency is to provide positive lubrication of the cylinder walls by means of an oil jet or jets from small holes drilled through the upper half of the big-end bearing and a corresponding hole in the crank pin. These holes come into line once every revolution and owing to their positions allow the high pressure oil supplied through the drilled crankshaft to spurt upwards on to the cylinder walls (Fig. 273). Usually in modern engines, a surplus of oil, above that actually required for lubricating the walls, is supplied to the walls and the oil control rings then scrape off most of this oil leaving the correct quantity for lubrication; this ensures adequate oiling of the walls, reduces cylinder wear and also tends to cool them more efficiently than would otherwise be the case. The oil consumption, however, is not increased owing to the efficient scraping action of the oil control rings. The drilled connecting rod method is used on Standard engines, as previously described.

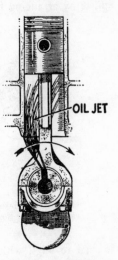

Fig. 273.—Method of Lubricating Cylinder Walls (Vauxhall).

In one or two instances a positive supply of oil has been led by means of a special pipe or oil passage to the cylinder walls, near their lower ends to ensure efficient lubrication. Sometimes this additional oil is supplied only whilst the engine is starting from the cold and warming up.

Lubricating the Rocker Arm Unit.—The rocker shaft bearings and those of the rocker arms of push-rod operated overhead valves, receive their oil supply by means of a drilled passage or pipe from the main or camshaft oil gallery pressure oil. This oil is reduced in pressure by either a pressure relief valve or a special smal hole metering plug, so that the oil arrives at the entrance to the hollow rocker shaft at a few pounds per sq. in.

Fig. 274 illustrates a typical rocker arm unit of a Chrysler Group engine, namely, the Dodge. The oil supply from the rocker arm tube, or hollow shaft, flows through drilled passages in the rocker arm to lubricate both the push rod end and the tip of the valve stem where it lubricates the rocker arm contact area. In this method the oil is supplied in frequent but small spurts by arranging for the lubricating oil hole in the camshaft

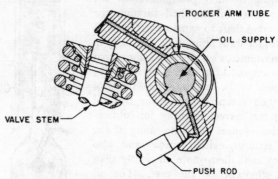

Fig. 274.—Method of Lubricating Push Rod End Bearing and Valve Stem Tip from Hollow Rocker Shaft Oil Supply (Dodge Engine).

journal to align with the passage leading to the rocker shaft, momentarily, as described earlier.

Crankcase Ventilation.—The crankcase compartments and also the overhead valve cover chamber are usually ventilated in modern engines. The purpose of this is to provide a direct air circulation through the crankcase, when the car is on the road, so as to cool the interior and also to get rid of any fumes, e.g., unburnt oil or fuel vapour or cylinder gases escaping past the pistons.

In some production engines crankcase ventilation fumes from the overhead rocker gear casing are conducted through a small pipe to the carburettor air inlet, or cleaner, to be burnt with the petrol charge in the engine. In recent *positive ventilation* systems the degree of ventilation can be controlled by a valve fitted for this purpose. Fig. 275 shows the ventilation system of the

Cadillac vee-eight engine, in which air from outside enters through the oil filler cap, which has a wire mesh air filter to keep out any dust, air is then drawn into the timing gear compartment whence it passes into the crankcase at the level of the crankshaft. The air then goes upward (on the right) and through the overhead

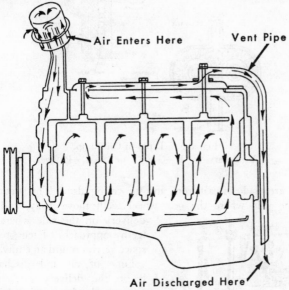

Fig. 275.—Crankcase Ventilation System used in Cadillac Vee-Eight Engine.

valve gear compartment, from which it is exhausted through the vent pipe shown, and discharged below the crankcase, causing a *partial vacuum* which creates the circulation of the air described.

Oil Pumps.—Except with dry sump systems, the oil pump can be made of quite simple design. For the high pressure and assisted splash systems the *gear-wheel type of pump* illustrated in Figs. 265, 276 and 277 has been widely used. It will run almost indefinitely with a very low rate of wear. The car engine gear type of oil pump will supply from 2 to 4 gallons of oil per min. at 40 to 60 lb. sq. in. when operating at 2,000 R.P.M. The

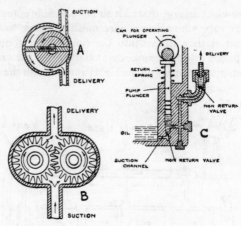

Fig. 276.—Types of Oil Pump.
A—Rotary.   B—Gear Type.   C—Plunger.

gears are made of cast iron; and the casing also. The suction is on the side where the gear-wheel teeth move away, and the

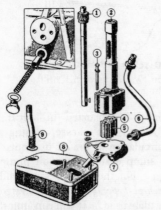

Fig. 277.—Components of Typical Gear Wheel Pump.

delivery on the side where the teeth approach. Each tooth-space carries round an equivalent volume of oil and discharges it on the delivery side. One gear-wheel only is engine-driven, the other meshing with it is thereby driven. It is usual to run this type of pump at one-half engine speed, and to submerge it in the oil sump so that the spaces are always full of oil, and no "priming" is therefore necessary. For a bigger delivery, three gear-wheels may be employed.

The Austin gear-wheel oil pump and its attachments are shown in Fig. 277. The driving gear (4) is driven by the vertical shaft (1) which has a helical gear at the upper end to mesh with a similar gear on the engine camshaft. The driving and driven gears (4) and (5) are housed in the oil pump body (lower end) (2) which is secured to the base-plate member (7) by four screws (3). The shaft (1) has its bearing in the tubular portion of (2). On the left under side of (7) there is an inclined oil suction pipe which is located within the fine metal gauze-faced oil strainer (8); the oil pump unit bolts on to the right top side of the strainer by the three studs and nut assemblies, partly shown, on the right. The oil delivery pipe that leads to the oil pressure release valve which is shown, in dismantled form, at the centre of the left-hand side in Fig. 277, and thence to the oil filter is shown at (6). The pipe (9) returns the circulated oil to the strainer.

This type of pump will deliver oil at 50 to 55 lb. sq. in., at normal engine running speeds.

The rotary eccentric type of pump (Fig. 276 (A)) occasionally used for low pressure lubrication, consists of a driving drum set eccentric relatively to the cylindrical pump casing, and provided with a pair of sliding vanes pressed outwards by means of a spring, so that the other ends of the vanes are always touching the casing. The oil is forced in the direction shown, the sliding vanes acting more positively than the gear-wheels. It is possible to attain pressures of 30 to 50 lbs. per sq. in. with suitably designed pumps of this type. The wear of the blades is the one disadvantageous feature of this pump.

The plunger type pump has been employed with medium and high-pressure lubricating systems. This pump is positive, and has a much higher suction lift than the preceding types, so that it can, if necessary, be placed above the sump. As a rule, however, the pump is driven either by one of the valve cams, or by a special eccentric cam, or eccentric sheath (Fig. 276 (C)) on the camshaft. On the upward stroke of the plunger oil is sucked from the sump past a non-return ball-type valve opening inwards; during the down stroke the oil cannot pass this valve and is therefore forced past another ball valve, opening outwards, into the delivery pipe to the bearings.

It is usual to surround the whole suction side of the pump with a cylinder of wire gauze, of ample dimensions, to trap any solid matter which might otherwise cause the valves to stick open. There is sometimes a gauze strainer on the top of the sump to strain the surplus oil which returns to the sump, and a gauze also in the oil filler. Sometimes the oil pump is made as a separate unit attached to the outside of the crank-chamber, where it is readily accessible.

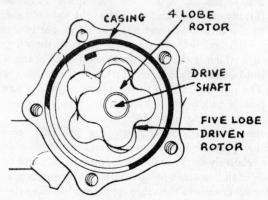

Fig. 278.—The Hobourn-Eaton Lubricating System Pump.

The Hobourn-Eaton Pump.—This more recent oil circulating pump, which is more compact than the gear-wheel pump consists of a four-lobe rotor which is driven by a shaft from the camshaft. It is located within a five-lobe internal casing which has a cylindrical outer shape and can rotate within the corresponding interior part of the pump casing. The four-lobe rotor shaft is eccentric to the centre of the pump casing so that when it rotates the five-lobe member is driven at a slightly lower speed. The "lobe tooth" spaces vary during each revolution of the drive rotor, so that oil drawn upwards into the larger spaces, through an inlet port below is forced into the smaller spaces which, in turn communicate with the upper side delivery port of the pump. In this way pressure is built up by the pump, such that maximum oil pressures up to about 75 lb. per sq. in. can be obtained. It will be seen, from Fig. 278 that the lobes of the driving and driven rotors are approximately of circular shape, but in another design the lobes are of a flat-pointed tooth form.   It may here be mentioned that the cover plate of the pump is a relatively close fit over the flat faces of the rotors.   This is necessary for the efficient operation of the pump for, when the pump is working this clearance space is filled with oil.

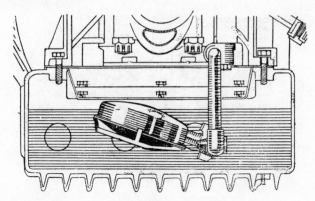

Fig. 279.—Constant Level Oil Intake Device.

Constant Level Oil Intake Device.—With the ordinary type of oil pump the suction pipe is arranged with its lower end near the bottom of the oil sump.   This arrangement is open to the objection that as the level of oil in the sump changes so does the flow of oil to the bearings, since the pump is working under a varying suction head.   Another disadvantage is that if there happen to be any very fine impurities in the oil these will be found near the bottom of the sump and tend, therefore, to be drawn into the oil pump.

A method of overcoming these difficulties, employed in many modern car engines, is to provide a float mounted on a hollow arm that can hinge about a horizontal axis, so that as the oil level in the sump changes the float moves up or down and the arm moves with it.   The lower face of the float carries a gauze filter inlet which allows clean oil to flow along the hollow hinged arm and up the vertical suction pipe to the oil pump.   With this method the oil flow is always constant and is unaffected by surging effects due to road inequalities, cornering or acceleration effects. The Wolseley oil intake is shown in Fig. 279, and it may be mentioned that a finned aluminium oil sump is employed for assisting in cooling the oil returned in the sump.

Pressure Indicators.—It is necessary, with pressure systems

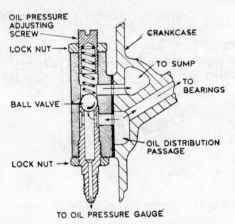

Fig. 280.—The Adjustable-type Oil Pressure
Release Valve.

of lubrication, whether semi- or full-types, to provide a pressure gauge (of the ordinary Bourdon type) on the dash-board to indicate the oil pressure on the delivery side, as shown in Figs. 265 and 266. In cold weather, with some systems, and at starting the oil pressure may be much lower than when the engine is thoroughly warmed up. In other cases the pressure is often higher for a short period after starting up, due to inertia and resistance of the oil in the system, but as the oil becomes warmer, the pressure will fall, as the oil experiences less resistance; this is often the case with plunger-pump systems. The pressure gauge is connected to the pump delivery side. It is the usual practice to mark, conspicuously, the normal working pressure graduation on the gauge so that abnormal readings can be quickly detected.

The Pressure Release Valve.—It is necessary in high-pressure systems to fit a release valve to obviate any excessive oil pressure due to an obstruction in the system. All that is required is a spring-loaded valve of the conical or ball type, opening outwards, and placed in the delivery side near to the pump. The spring is sometimes made adjustable by means of a screw cap, so that the valve can be arranged to open at any given

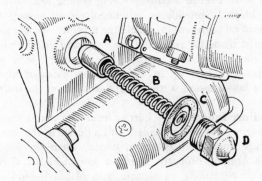

Fig. 281.—Components of Fixed Pressure type
of Pressure Release Valve (Morris)

A—Washer. *B*—Spring. *C*—Washer. *D*—Cap nut.

pressure. The oil thus discharged is led directly to the suction side or to the sump again. On many modern engines, however, the spring is not adjustable so that no regulation of oil pressure can be made, except by washers, or a change of spring.

Fig. 280 illustrates the principle of the adjustable-type pressure release valve unit which, in this case is attached to the side of the crankcase, where it is accessible for adjustment or cleaning purposes. The lower side of the ball valve is subjected to the oil pressure and, by adjusting the tension of the compression spring the pressure at which the ball lifts off its seating can be altered. When the valve opens oil flows past the ball through the passage shown and back to the oil sump. After long usage the seating of the ball may need to be re-trued. The valve unit can be removed from below for this purpose.

In the majority of recent engines as previously mentioned the pressure release valve is of the non-adjustable kind, thus giving a much simpler and cheaper design, a typical example of which is shown in Fig. 281, namely, for the Morris 1100 engine. This valve is on the side of the cylinder block and is held in position by the domed hexagon nut and sealed by a fibre washer. The tapered end of the left-hand component is the valve member and this is held on to a corresponding seating machined in the cylinder block to provide an additional return passage for the discharge of oil to the sump, should the pressure become too high. Provision is made for removing the valve and grinding it on its seating, should any wear occur. The pressure release valve is set to open at 60 lb. sq. in. No pressure gauge is fitted but, instead, a low-pressure warning light is fitted on the instrument panel.

Oil Failure Indicators.—Certain makes of car and commercial vehicle engines are now provided with a device in the form of an oil-pressure operated diaphragm, which, when the lubrication system is working properly, breaks a pair of electrical contacts in a circuit containing an electric bulb placed behind a green window on the dashboard. This may be a disc or cruciform frame. When the ignition is switched on and the engine is stopped or idling the circuit is "closed" by a spring so that the green window lights up. When the engine is speeded up, however, the diaphragm is deflected by the oil pressure so that the light goes out. Should the green light appear under ordinary working conditions, however, this is a sign of oil pressure failure and the engine should at once be stopped for investigation.

Upper Cylinder Lubrication.—In order to ensure adequate lubrication of the cylinder walls above the piston a special mineral lubricating oil may be mixed with the petrol; this procedure is recommended by some car manufacturers. In the case of new engines it is generally advisable to provide for this upper cylinder lubrication for running-in purposes.

The quantity of oil used for this purpose is usually in the proportion of about $\frac{1}{8}$-$\frac{1}{4}$ pint per gallon of petrol and it can either be mixed with the petrol or introduced into the induction manifold by means of a special device marketed for this purpose. If colloidal graphite is used the recommended amount is $\frac{1}{2}$ oz. per two gallons of petrol.

In connection with the running-in of new engines the use of a small quantity of a solid low-friction lubricant in the engine oil is recommended. The most widely used lubricants for this purpose are (1) *Colloidal graphite* and (2) *Molybdenum disulphide*. In each case the lubricant is in the form of particles so fine, namely, of the order of $\frac{1}{1000}$ millimetre that they can find their way into the pores of the siliding metal surfaces, to greatly reduce the frictional coefficient. These lubricants are supplied mixed to an emulsion form with a liquid lubricant ready to mix with the engine lubricating oil.

Oil Cleaners.—After the lubricating oil has been in use for some time it may become dark in colour, due principally to suspended carbon particles. Although gauze strainers are fitted to both the oil filler and oil pump suction parts, these are too coarse in mesh to deal with finely suspended matter, so that special oil cleaners are now fitted to get rid of the latter.

Two principal types of oil cleaners have been used, namely, (1) The Centrifugal, and (2) The Absorbent Type.

In the centrifugal types the hot oil, after its passage through the engine, is forced, under pressure, into a circular chamber whence, by means of spiral vanes, it is swirled around at a high speed, when all solid matter, being heavier than the liquid oil, is flung outwards by centrifugal action. This solid matter is in the form of a fine sludge and is caught in special channels, whence it it removed at regular intervals.

The centrifugal oil filter, is used on piston-type aircraft and on some C.I. engines, but now seldom for car engines.

Sometimes the crank-pins are made hollow and dirt-traps provided for solid matter thrown outwards by centrifugal action; in other cases the sludge has been caught in the flanged rim of the fly-wheel, whence it is removed every 20,000 miles or so.

Felt Type Oil Filters.—This type of lubricating oil-cleaning device is placed in the oil circuit so that the oil circulating around the engine is constantly filtered, thereby extracting all solid deposits from the oil.

A typical felt element type of oil filter has a vertical cylinder of star-section made of special felt material; the dirty oil in passing from the outside of the element to the inside has the solid matter removed by the felt. It is necessary to clean this type of filter every 3,000 to 4,000 miles, in a petrol bath, and to discard the felt element after 10,000 to 15,000 miles.

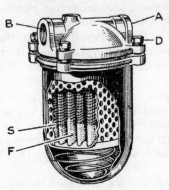

Fig. 282.—Typical Felt Element Type of Oil Cleaner.

The later types of felt oil filter employ metal supported fabric-covered elements of relatively large area to trap the fine solid matter, so that clean oil only passes into the central space, whence it is supplied to the engine.

Referring to Fig. 282, the dirty oil enters at A and passes into the space between the outside of the filter element and the inside of the outer metal container. The filter element consists of fabric F mounted on a gauze S and it is held against a cork joint in the head by the spiral spring shown. As long as the filter is working properly the spring retains the element in position, but should it become choked with solid matter the pressure developed on the inside of the element will cause the spring to depress when the oil is by-passed through the holes provided for this purpose, the engine will therefore not tend to become starved of oil.

The Paper Element Filter. Instead of using felt a special kind of paper filtering material is now becoming more widely used. Thus, in the Purolator filter a micronic paper element made in close pleated form and located between an outer perforated casing and an inner perforated central tube, is used. The pleated form gives a large filtering area, the dirty oil flowing from the outside and through the element, whence it is cleaned and returned through the central tube, past an oil pressure release valve to the delivery side.

The A.C.-Delco filter (Fig. 283) has a large number of contiguous impregnated paper discs, which form dirt traps between each pair of the discs. As before the oil flow is from the outside to the inside perforated tube. This filter which is bolted into a cylindrical recess on the cylinder block (previously shown in Fig.

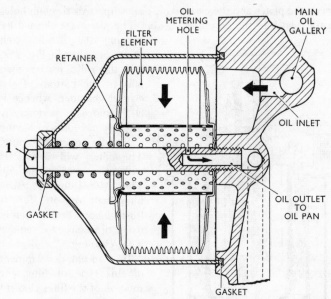

Fig. 283.—The AC-Delco Impregnated Paper Oil Filter used on Vauxhall Car Engines.

32) is readily taken apart, by unscrewing the bolt shown. Paper element filters cannot be cleaned when dirty but must be replaced by new ones about once every 6,000 miles of service.

Self-Cleaning Metal Filter.—Another class of oil filter is that known as the Autoklean self-cleaning one, illustrated in Fig. 284. In this filter the dirty oil is led to the annular space between the inside surface of the outer vertical cylindrical container and the Monel metal fine mesh gauze cloth cylinder within, which is carried in a perforated metal cage. This gauze element removes the larger dirt particles. The oil then passes between a series of thin metal discs arranged axially one above the other. The clearances between these discs are very small, but the total surface of entry for the oil is sufficiently great to prevent appreciable resistance to the oil flow. The finer particles are trapped between the plates and the cleaned oil passes upwards through holes in the plates to the oil outlet above. The plates are cleaned by rotating the handle seen above the filter in Fig. 284, and the solid matter then falls to the bottom of the central chamber, whence it is removed periodically through an orifice having a handle operated plug. The filter will also deal with any water in the oil. The Monel metal gauze filter will operate for at least 2,000 hours before it is necessary to remove it for cleaning.

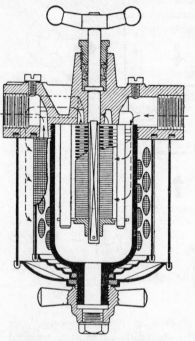

Fig. 284.—The Autoklean Oil Filter, with Handle Cleaning Device.

In a recent development of this type of filter the rotation of the filter element plates is performed automatically by means of a small oil-operated motor which utilizes the difference of pressure between the suction and delivery sides of the oil pump of the lubrication system.

External Types of Filter.—Oil filters have been much improved and it is now usual to provide external filters placed in accessible positions. These belong to one or other of two types, namely, the *Full-Flow* or *By-Pass* patterns. In the former

model (Fig. 285)* the full oil flow passes through the filter.   This
is satisfactory so long as the oil is hot and clean, but when dirty

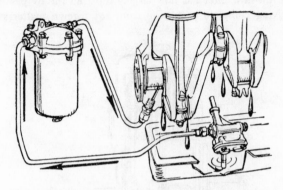

Fig. 285.—The Full-Flow Type Oil Filter.

and cold it may be too viscous to pass through the filtering element
and the oil flow in the system would stop, were it not for the
provision of a by-pass valve (Fig. 286) which opens and allows the

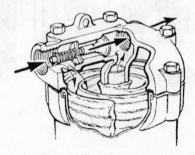

Fig. 286.—Showing By-Pass Valve in Full-Flow Oil Filter.

oil to by-pass the filter, by providing a direct connection between
the inlet and outlet orifices of the filter.   In Fig. 286 the thick
black arrows show the thick or dirty oil by-passing the filter.
One disadvantage of the full-flow type is that the large volume

* Courtesy, British Filters Ltd.

of oil and the speed of flow are not conducive to the most efficient filtering of the oil, unless large capacity filters are provided.

The full-flow filter has now largely replaced the by-pass type, thus giving complete protection from any foreign matter in the oil.

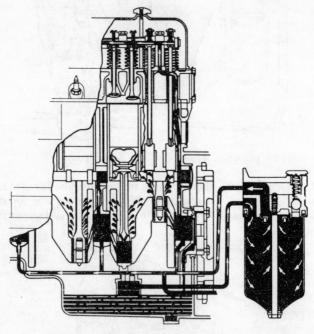

Fig. 287.—Full-flow type Oil Filter Connections to Engine Lubrication System.

Fig. 287 shows the application to a car engine, of the A.C.-Delco full-flow filter. The white arrows depict the directions of oil flow to and from and also within the filter itself. The spring-loaded ball valve which acts as the safety valve should the filter become choked with foreign matter.

In the by-pass type (Fig. 288) only a fraction, namely, about $\frac{1}{8}$ to $\frac{1}{10}$ of the full delivery is passed through the filter; this oil is cleaned and returned direct to the oil sump. In the course of time all of the oil in the system will have been filtered and the process repeated continuously. The felt or fabric elements are generally discarded after a stated number of miles, usually 8,000 to 10,000 miles, and new ones fitted. Felt-type filters are usually cleaned in petrol every 3,000-4,000 miles and replaced at the intervals previously stated.

An example of a by-pass filter is given in Fig. 289 for a Standdard Vanguard engine. This Fram filter has a restrictor by-pass to control the amount of oil that is filtered.

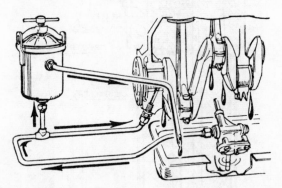

Fig. 288.—The By-Pass Type of Oil Filter.

The filter contains a removable cartridge element, which is filled with filtering media consisting of closely packed cotton yarn treated with triethanolomene, which latter substance increases the absorbing qualities of the fabric material contained by the element.

Referring to Fig. 289, oil from the pump, at about 60 lb. per sq. in. enters the inlet end of the oil gallery. A pre-determined proportion of this oil flows through the cleaner and is admitted to the "sump" gallery through the restrictor. When the speed of the engine is such as to deliver oil at a higher pressure than 60 lbs. per square inch, the release valve in the filter bracket casting above opens and a certain amount of oil is delivered to the sump through the same port as the filtered oil. The amount of oil so delivered is naturally a function of the pump delivery pressure.

The filter cartridge will clean the oil efficiently for 8,000 to 10,000 miles and should be changed at either of these mileages, according to its condition.

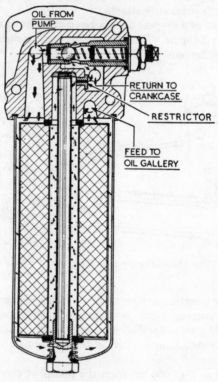

Fig. 289.—The Standard By-pass Oil Filter.

Oil Coolers.—It is well known that the higher the temperature of the engine oil, the less its viscosity and lubricating properties. Since in the majority of engines the oil is heated considerably, and is used to lubricate the working parts in this condition, it is advisable to arrange, in some convenient manner, to keep the average working temperature fairly low. For this reason some manufacturers make the crank-case sumps of aluminium alloy, (as shown in Fig. 279), and provide cooling ribs on the outside. In other cases oil from the engine is led to coolers of the radiator element type; these in one or two instances are neatly embodied in the water-cooling radiator block so as not to detract from the appearance of the car. With suitable means for oil cooling the engine will work more efficiently and wear better.

Oil Fillers, Indicators and Filters.—Automobile engines are now provided with oil filling tubes or openings situated in convenient positions and provided with dust-tight covers. All fresh oil added to an engine should be poured through a gauze filter; in most cases there is a cylindrical type of filter embodied in the oil filler orifice or tube.

Further, the oil used for lubricating the cylinder walls and the crankshaft bearings drains down the sides of the crank-case and passes into the sump below through gauze-covered openings, which extract most solid matter. These gauzes should be cleaned once every 8,000 to 10,000 miles.

In addition, the oil pump suction pipe is always provided with a gauze filter of generous proportions; this filter, also, should be removed and cleaned periodically.

The level of the oil in the crank-case sump must be maintained at a certain minimum level to insure correct lubrication and operation of the oil pump.

Various types of oil-level indicator are now fitted, the most widely used being the *dip-stick* one. In this case a graduated metal rod is used to ascertain the oil level.

Float-type gauges, glass gauges, and, more recently, electric oil level gauges, reading on a dial on the instrument board, are also included in automobile oil-level indicators.

Oil Consumptions.—An engine in good condition should show an oil consumption of about ·05 lb. per B.H.P. per hour, for air cooling, and about ·03 lb. for water cooling. For a $1\frac{1}{2}$-litre modern car engine, the oil consumption, when the engine is run-in, would be of the order of 3,500 to 4,500 miles per gallon.

CHAPTER IX

THE COOLING OF AUTOMOBILE ENGINES

It has been shown that about one-quarter of the heat of combustion of the petrol passes through the cylinder walls in an ordinary engine. Now this quantity of heat is considerable, and if the cylinder were a plain metal barrel, it would very soon become red hot, and the engine would cease to operate, since the incoming charge would ignite whilst the inlet valve was open; moreover, as the lubricating oil on the cylinder walls would all be burnt away, the piston would certainly seize up. When it is remembered that the temperature of the burning gases is about 1500° C. to 2000° C.,* and that of the exhaust gases from 600° C. to 800° C., it will be evident that the average temperature in the cylinder could be very high. It is therefore imperative to provide some means for keeping the temperature of the cylinder down to within reasonable limits.

In this respect the use of aluminium alloy for the cylinder and piston enables the temperatures to be kept down more readily than with cast-iron.

Aluminium is a much better heat conductor than cast-iron or steel. For the same temperature differences it will conduct away about three times the quantity of heat, whilst copper is better still, and will conduct about six times the amount of heat.

Applied to petrol engine construction, aluminium alloys enable the use of higher engine compressions, facilitate the engine cooling and lighten the engine parts, since aluminium has only about one-third of the weight (density) of cast-iron.

Concerning strength properties, one suitable aluminium alloy, which contains from 9 to 12 per cent. of copper, has a tensile breaking strength in the cast state of 9 to 11 tons per sq. in., and is thus as good as cast-iron; this alloy is now much used for

---

* Platinum, one of the highest melting point metals, melts at 1750° C., iron at 1530° C., and aluminium at 657° C.

cylinders and pistons.   Other much stronger casting and forging alloys used are R.R. and "Y" alloys.

Methods of Cooling the Engine.—In order to ensure that the cylinder temperature is not excessive, the heat conducted through its walls must be disposed of continuously.   This can be done in either of the following ways: (1) By providing cooling or radiating fins of sufficient area, and arranging for an air current to blow past these fins, in order to carry off the surplus heat—this process is termed *Air Cooling* (2) By providing a water jacket around the hotter parts of the cylinder barrel and the combustion head, and circulating water around same. (3) By arranging for the oil supplied for lubrication purposes to cool the cylinder barrels, as in the earlier Bradshaw engine; and (4) By injecting a water-spray into the cylinder.   Professor Hopkinson employed this

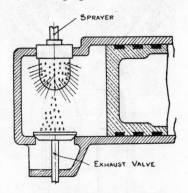

Fig. 290.—The Hopkinson System of Internal Cooling.

method (Fig. 290) for cooling gas engines.   The water-spray was arranged to occur during the combustion and expansion processes, the water being converted into steam; the heat required for this conversion was absorbed from the hot gases.   Part of the spray reached the exhaust valve and the combustion chamber walls, also, and tended to cool these parts directly.   This method whilst of interest, is not applicable to modern petrol engines.

In addition to the cooling methods mentioned, it is possible to combine these, and to employ, say, water cooling for the combustion head and valve passages and seatings, and air or oil cooling for the cylinder barrels.   Alternatively, as in the Bradshaw engine, the latter were oil-cooled, and the combustion chamber and valve seatings air-cooled.   Again, in some two-cycle engines the barrels and valve seatings are water-cooled and the detachable heads air-cooled.   The lubricating oil also assists in the cooling of most engines. Steam cooling of cylinders is another method.

Air Cooling.—The amount of heat carried off by the cooling air depends upon the following items: (1) The total area of the fin surfaces; (2) the velocity (and amount) of the cooling air; (3) the temperatures of the cylinder and fins and of the cooling air.

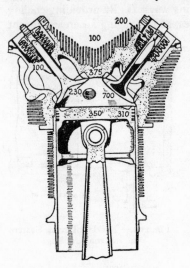

Fig. 291.—Air-cooled Cylinder Temperatures.

The greatest amount of heat will be carried off in the case of engines having a large area of cooling fin surface well disposed around the combustion head and valve surfaces, for which the speed, and cross-sectional area, of the cooling air stream is a maximum. The temperature of the air should be low for the best cooling effect.

It is, of course, quite possible to overdo the cooling, and to keep the cylinders *too cool*, so that the engine will lose in petrol economy and output. Most automobile engines have to run at light loads for the major part of their existence, and are apt to be over-cooled during these periods. It is a well-known fact that the petrol consumption of an engine is appreciably greater per B.H.P. hour (say) when running light than when on three-quarter to full load.

Cooling Fin Data.—The fins are usually made of about the cylinder-wall thickness at their roots (or junctions with the cylinder wall) tapering down to about one-half the root thickness. The length of the fins varies from one-quarter to one-third of the cylinder diameter; their pitch (i.e., the distance between the fin centres) is about one-quarter to one-third of their length; the total length of finned cylinder barrel is from 1·0 to 1½ times the cylinder bore, as a rule. Another rule based upon experimental considerations is to allow from $1\frac{1}{2}$ to $2\frac{1}{2}$ sq. ft. of cooling fin area per horse-power. This gives about the correct cylinder temperature at 30 to 40 m.p.h. air speed.

In the case of aircraft air-cooled engines, where much higher cooling air speeds are utilized the cooling areas for the high output per litre engines work out at about 17 to 20 sq. in. per B.H.P. or 75 to 85 sq. in. per litre; the cooling air speeds past the cylinder fins are 150 to 200 M.P.H.

The tendency in regard to cooling fins is to reduce the pitch and to increase the length of these. It is not, however, advantageous to reduce the pitch below about $\frac{1}{4}$ in. The usual fin lengths for high output radial engines are $1\frac{1}{2}$ to $2\frac{1}{2}$ in. for the aluminium alloy heads and $\frac{1}{2}$ to $\frac{3}{4}$ in. for the steel cylinder barrels.

A typical aircraft engine cylinder unit of the air-cooled pattern is shown in Fig. 291, which also gives the average working temperatures of the various parts. The inlet valve, on the left, has a relatively low temperature of 230° C. due to the cooling influence of the incoming mixture, whilst the exhaust valve on the right which is in the stream of outflowing hot gases has the highest temperature, namely, 700° C. The valve shown is of the hollow sodium-filled type. It is well known that one *limiting factor governing the maximum compression ratio that can be used* without detonation occurring is *the temperature of this valve*.

Applications of Air Cooling.—The simplest air-cooling system is that of the ordinary motor-cycle in which the air circulation, or draught, due to the forward motion of the machine is sufficient to carry off the surplus heat. The fins are now made much larger in area than hitherto, to ensure adequate cooling when running up long inclines on low gear, in following winds, and at low road speeds.

In the case of two-cycle engines, since these develop about 50 to 70 per cent. more heat, at the same engine speeds, it is necessary to provide a relatively larger cooling area, usually by deepening the fins and increasing their number.

In earlier light-car engines of the two cylinder opposed air-cooled types, similar to those of the Rover Eight and A.B.C., it was arranged for the combustion heads of the cylinders to project through the bonnet of the car, and to provide cup-shaped air-scoops facing the direction of motion of the car; in this way adequate cooling of the heads was ensured.

Fan Cooling.—This is employed on the larger air-cooled engines, more particularly on light and other cars. A two- or four-bladed fan is driven either at engine, or twice engine-speed, and the stream air is directed on to the cylinder heads. In fan-

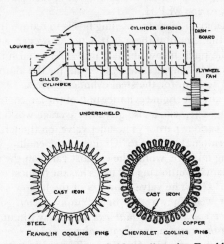

Fig. 292.—Showing Principle of Air-cooling the Franklin Engine and Method of arranging the cooling Fins.

cooled engines the cooling depends chiefly upon the engine-speed, and not upon the forward speed of the car. These fans usually absorb a small portion of the power output, namely, about 1 H.P. for every 15 to 20 H.P. output. In general, however, the power required to drive the fan increases as the cube of its rotational speed.* In the case of one well-known car engine of 12 H.P. (rated) the total output at 1,000 r.p.m. was 9·0 H.P., of which the fan absorbed about 0·5 H.P. The brake m.e.p. obtained, however, was 114 to 118 lbs. per sq. in. The cooling air speed was 20 m.p.h. and the maximum temperature of the cylinder head was 216° C.

* See also pages 456, 457.

In the case of small single-cylinder engines, running in enclosed situations, an excellent arrangement is that of a fan of about the same dimensions as the fly-wheel diameter (i.e., about twice the stroke) mounted on the main shaft, and enclosed in a metal casing so arranged that the air is drawn in at the centre and expelled peripherally through a bell-mouthed duct directing it on to the exhaust side of the cylinder.

For small air-cooled engines, the blower-type fan works quite

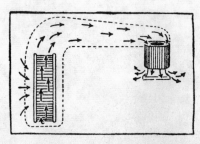

Fig. 293.—Showing principle of later model Franklin
Air-cooled Engine.

satisfactorily if suitable guides and ducts are provided for the air stream. The system usually adopted for larger engines is to shroud, or surround each cylinder with another barrel, and to *suck* the air by means of the fan, through these barrels, commencing at the crankshaft side. By placing the cooling system on the suction side of the fan, a more positive circulating effect is obtained. Sometimes the fly-wheel itself is designed to function as a cooling fan, and air is discharged backwards through it, after having been drawn past the cylinder barrels.

German Air-Cooled Engines.—The air-cooled engine appears to have received much attention in Germany in recent years, for there are several makes of air-cooled automobile petrol and C.I. engine, including 2, 4, and 8 cylinder models.

A good example of the modern air-cooled type is the Krupp four-cylinder opposed compression-ignition engine. This has a cooling fan fitted at the front end, and it is engine driven. It forces the cooling air through a casing around the front end of the crankcase and thence to the horizontal cylinder barrels which are ribbed and enclosed in rectangular casings.

Another more recent example is the Krupp eight-cylinder

Fig. 294.—The Krupp Air-Cooled Vee-Eight Petrol Engine.

Vee-type petrol engine, which has very similar cooling arrangements. (Fig. 294).

Modern Air-Cooled Engines.—The Volkswagen, Dutch D.A.F., Citroen 2-cylinder-opposed Chevrolet Corvair, 6-cylinder horizontally-opposed, Fiat 500D, 2-cylinder, in-line, and N.S.U. 2-cylinder, in-line, are typical examples of modern car air-cooled engines.

In most modern automobile engines provision is made to regulate the quantity of the cooling air thermostatically, so that when the air temperature of the discharge air from the cylinder increases above the normal value the thermostat actuates a large valve, vane or disc in the air outlet duct to allow a greater quantity of air to flow. Both the Fiat and Corvair engines have such air control devices.

General Considerations on Air Cooling.—Air-cooled engines in the past have been noted for their higher oil-consumption, due to some of the lubricating oil being burnt in the combustion chamber. They have not given the same petrol economy as water-cooled engines, and are still undoubtedly noisier in

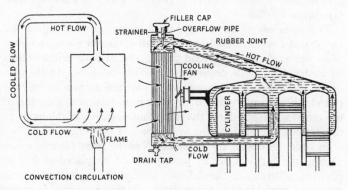

Fig. 295.—The Thermo-syphon System of Cooling.

action.* Their working temperatures are, on the average, higher than those of water-cooled engines (the cylinder heads being at 250° C. to 300° C.) and the maximum allowable compression ratios somewhat less.

The additional noise of the air-cooled engine, apart from the valves and silencer, may be due to the fact that there is no outer jacket and water content to damp down the noise as with the water-cooled types, so that the surface vibrations are of larger amplitude, and the emitted sounds therefore greater.

Water Cooling.—In the case of water-cooled engines, it is necessary to dissipate the heat conducted through the cylinder walls by circulating the water through a cooling device, known as the *Radiator*. Fig. 295 illustrates the principle of a typical car

* The single sleeve valve air-cooled engine is an exception.

engine water-cooling system. The water is made to circulate through a large number of very small section tubes in the radiator having large cooling surfaces. Air is forced past the outer surfaces of these tubes, and carries off the surplus heat. This system is therefore an air-cooling process, with the intermediary of water to carry the cylinder heat to the cooling device, or radiator. There are three principal methods of circulating the water in present use, namely; (1) *The Natural Convection*, or *Thermo-Syphon system*. (2) *Impeller-Thermo-Syphon*, (3) *Full Pump Ciculation*, and (4) *Steam Cooling*.

The principle of *thermo-syphon circulation* depends upon the fact that if a vessel of cool water is heated, the hot water, in virtue of its lighter density, will tend to rise, and its place will be taken by cooler water, thus causing a definite circulation. In the case of the automobile engine, the hottest part is the space around the cylinder head, and the hot water rises from this part to the top of the radiator, thence down the numerous passages in the latter to the bottom, and along to the lowest part of the water-jacket.

In order to obtain the maximum benefit from *thermo-syphon* cooling the engine cylinder jackets should be placed as low as possible, relatively to the radiator, so that the coolest water will always be available from the radiator. If the lower pipe from the radiator to the jacket be placed horizontal, as shown in Fig. 295, an excellent cooling system results. The advantage of *thermo-syphon* cooling lies in its simplicity and automaticity. It requires, however, a larger capacity radiator than when the water is circulated by pump, and the radiator must always be kept filled well above the level of the pipe from the cylinder to the top of the radiator.

It has been usual to arrange for a capacity of the complete *thermo-syphon* system, of 1 gallon of water to every 5 to 7 B.H.P. of the engine. The width of the water spaces around the cylinders is also often made from 20 to 40 per cent. greater than for pump circulation.

In order to assist cooling of the water in the radiator, a two- or four-bladed suction-type fan is mounted behind the radiator and is driven at 1 or $1\frac{1}{2}$ times engine speed, so that the amount of air drawn through is roughly proportional to the engine speed.

Heat Dissipation of Radiator.—The amount of heat that can be dissipated or got rid of by a radiator depends upon a number of factors of which the principal ones are as follows:—(1) The relative wind velocity. (2) The cooling surface. (3) The ratio

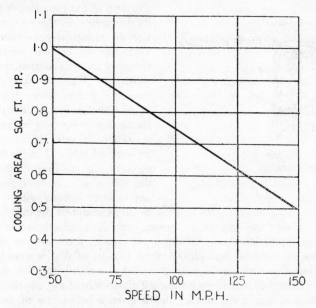

Fig. 296.—Radiator Cooling Areas and Speeds.

of tube depth to diameter. (4) The air density. (5) The water and air temperatures. (6) Design of radiator and disposition. (7) Conductivity of metals used.

The area of the cooling surface is inversely proportional to the air velocity; this is illustrated graphically in Fig. 296, from which it will be seen that at 150 M.P.H. the cooling area is 0·5 sq. ft. per B.H.P.; at 100 M.P.H., 0·75 sq. ft.; and at 50 M.P.H., 1·0 sq. ft.

A honeycomb radiator of 1 sq. ft. frontal area and 1 in. depth of tube gives about 8 sq. ft. of cooling surface, whilst one of 6 in. depth gives 48 sq. ft., although unless the ratio of length to diameter of tube is the same it will not necessarily give the same cooling efficiency.

The rate of heat transfer from the cooling water to the air through the radiator tubes has been shown to be dependent upon the ratio of the tube length to its diameter. In this connection the results of some tests* upon film or cellular matrix radiators of the construction shown in Fig. 297 are of interest. Tests were made upon radiator blocks of different depths up to 5 in. and at different velocities from 20 to 80 M.P.H. The results given in Fig. 298 indicate that the *rate of heat transfer* does not increase linearly with the depth as would be the case for free air flow through the air spaces, but falls away continuously as the depth increases. There is, of course, a progressive increase in the *total heat dissipated* on account of the increase in cooling area with core depth. It should be mentioned that the usual depths of radiators with the cell dimensions of Fig. 297 range from 2 to 4 in.; the smaller depth gives a better rate of heat transfer but requires a larger frontal area in order to obtain the same cooling surface.

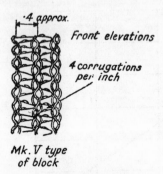

Front elevations

·4 approx.

4 corrugations per inch

Mk. V type of block

Fig. 297.—Radiator Cooling Elements.

Cooling of Supercharged Engines.—In the case of supercharged engines a larger cooling area is necessary since the heat loss to the cooling water increases in proportion to the supercharging pressure; the radiator area should therefore increase at about the same rate. Thus, increasing the H.P. of an engine of 3·5:1 compression ratio, 50 per cent. by supercharging, results in a 20 per cent. increase in losses to the cooling water; increasing the H.P. of an engine of 7·5:1 compression ratio by 50 per cent. results in an increase of 34 per cent. in losses to the cooling water.

* Automobile Cooling.—C. S. Steadman, *Proc. Inst. Autom. Engrs.*

Water Temperature.—Experience has shown that the hotter an engine can work without detonation, pre-ignition, or loss of power, the better. Not only is the thermal efficiency higher, but the engine's frictional losses, as a general rule, are lower with the result that the petrol consumption is reduced, and the power output is greater. For normal running, the temperature of the cooling water at the top of the radiator should lie between 75° C.

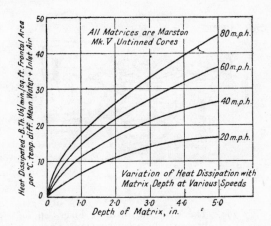

Fig. 298.—Radiator Cooling Data.

(167° F.) and 85° C. (185° F.). This allows a sufficient margin for temperature increase when climbing long hills on lower gears.

It is useful to have a radiator thermometer of the instrument panel type in order to keep a check on the water temperature. These instruments if accurately graduated are useful in detecting "over-cooling" in the winter, and over-heating in the summer. It will sometimes be found, more particularly with American cars used in this country, that they are over-cooled in winter, the running temperature being from 55° C. to 65° C. Where other means, such as radiator-shutters, or thermostats, are not provided for regulating the temperature, it is possible to blank off a portion of the radiator with a panel of metal placed in front of the radiator inside the bonnet. Alternatively the fan can sometimes be disconnected during winter.

Improved Liquid Cooling.—In order to reduce the weight of cooling liquid carried and also that of the radiator, it has been proposed to employ some liquid other than water, which has a higher boiling point. Examples of such liquids are glycerol

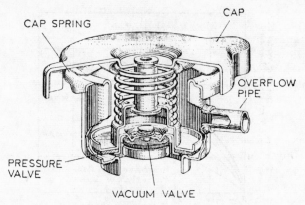

Fig. 299.—The A.C. Delco Radiator Filler Cap, for Pressurized Cooling Systems.

(glycerine) and ethylene glycol (195° C.). The latter liquid was used on modern aircraft piston engines. It is thus possible to use a radiator of less than one-half the size of a water-cooled type. The engine operates at about 130° C., but loses about $2\frac{1}{2}$ per cent. of its maximum power through working at this temperature; with water-cooling the temperature is 80° to 85° C.

Pressure Cooling System.—In the case of ordinary water cooling systems where the cooling water is subject to atmospheric pressure, only, the water boils at 212° F. In the pressure system of cooling, which is now widely used on American car engines and on most British engines, the cooling system radiator filler cap is sealed and fitted with a valve that opens after a certain pressure in the cooling system is reached. Usually this pressure is from about 4 to 7 lbs. per sq. in., although on some American cars it is as high as 12 to 15 lbs. per sq. in.

The result of raising the pressure is to increase the temperature of the boiling point of the water. Thus a pressure of 4 lbs. per sq. in. will raise the boiling point from 212° F. to 223° F. The higher water temperature gives a more efficient engine operation and affords additional protection under high altitude and tropical conditions; also for long hard driving periods.

The widely-used A.C.-Delco radiator cap is fitted with two valves, namely, a vacuum valve, as shown in Fig. 299 and a pressure valve. The vacuum valve is to prevent excessive pressure—due to the atmosphere—on the system as the water cools down and tends to form a partial vacuum. The pressure valve is adjusted to open the system to the atmosphere when the designed excess pressure is attained, so as to protect the system against excessive pressures that would otherwise damage it. The radiator filler cap is provided with a special bayonet joint and gasket to ensure pressure tightness.

Sealed Cooling Systems.—In the endeavour to reduce maintenance attention to the minimum the cooling systems of some recent cars have been sealed, so that it is unnecessary to top up the cooling water at regular intervals, as in the past. Instead, the radiator, pump hoses and cylinder coolant spaces are filled with a special cooling solution consisting of an anti-freeze and corrosion-inhibitor, so that once the radiator cap is screwed down the system remains sealed until such time as the cylinder head needs removal. It is, of course, necessary to ensure that all the rubber hose and other connections are fully water-tight.

The usual arrangment for the sealed system is to provide the radiator with a thermostatic relief valve which will operate when the temperature of the system exceeds a given value. The system (Fig. 300) also incorporates a separate expansion tank, having a pipe which is connected to the top of the radiator. Thus when the water in the radiator and cylinder block, and their connections, reaches a predetermined temperature, e.g., in one particular case, 190°–200° F., the coolant expands through the relief valve into the tank and, when the coolant temperature falls, it flows back into the ordinary cooling system. It is usual to fit a pressure relief valve in the expansion tank, to operate just above the given pressure employed in the system for its normal working condition. In regard to the expansion tank this is usually made about one-eighth of the capacity of the cooling system, i.e. about $1\frac{1}{2}$ pints, where the total coolant volume is $1\frac{1}{2}$ gallons.

Pump Circulation.—In the case of earlier small cars, the additional size of the radiator and the extra weight of cooling water

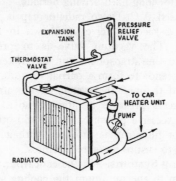

Fig. 300.—Typical Sealed Cooling System used on Certain Modern Cars.

required with the *thermo-syphon* system does not matter much, but for medium and large car engines, *thermo-syphon* cooling would usually necessitate large radiators and an appreciably greater weight of cooling water. It therefore became the rule to circulate the water positively, by means of a centrifugal type of pump, and at a much greater rate than is possible with the previous method. The water capacity of the system and the size of the radiator can thus be reduced appreciably.

The pump draws water from the lower part of the radiator tank and ejects it by centrifugal action at its periphery into the lowest part of the cylinder jacket, whence it is forced through into the top of the radiator. It is driven at 0·9 to 1·3 times engine speed. A water-tight gland was required where the driving shaft enters the pump in the earlier type of separate pumps.

This was usually packed with asbestos or hemp string soaked in tallow, a union nut being provided for tightening purposes.

Greasers were supplied to both the pump spindle and to the fan bearing; this method is sometimes employed on modern engines.

Fig. 301 illustrates a typical centrifugal pump that was used in certain Austin engines. The rotor B is driven by means of

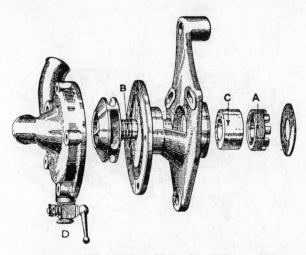

Fig. 301.—The Separate Water Circulating Pump.

a shaft from the engine and leakage of water along the shaft is prevented by a special packing ring C, which is held in place, and maintained so as to provide a water-tight gland by means of the nut A. In later designs of pump the packing ring C is a *graphite disc* held in position by means of a compression spring.

The pump casing is in two parts bolted together by means of studs and nuts. A drain cock D was provided for letting out the water in the pump casing when the radiator was drained in frosty weather.

Impeller Thermo-Syphon Circulation.—A method which has found much favour in modern cars is a combination of the two previous ones. A small straight-bladed impeller fitted usually on the fan-shaft spindle in a circular casing is connected at its centre, or suction, side with the water inlet pipe from the lower part of the radiator, and discharges the water into the cylinder jackets. It will be evident that as there is a free passage for the water past the straight blades of the impeller, the natural, or

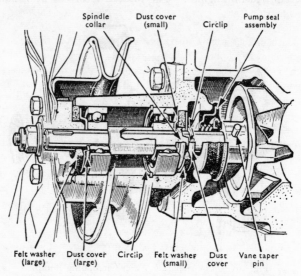

Fig. 302.—Combined Fan and Water Impeller Drive Unit.

convectional circulation of the water can occur, but the impeller assists this circulation, although not so rapidly as in the case of the centrifugal pump.

The fan-pulley-driven water impeller used on B.M.C. engines is shown in section, in Fig. 302. This unit is mounted on the cylinder block above the crankshaft pulley and behind the radiator. The same vee-belt drives the fan on the left and impeller on the right; these have a common drive shaft mounted on a pair of ball bearings, which are provided with efficient felt washer seals, for the fan end and a special water-tight graphite disc pump seal unit between the bearings housing and the pump body. For a typical four cylinder engine of 1,476·5 c.c. capacity, the cooling system capacity is about 15 pints, i.e., 1·875 gallons.

Plastic Impellers.—In order to obtain a more accurately-shaped, better balanced and lighter impeller, the moulded plastic (phenolic) impeller has been introduced, more recently. It is accurately moulded and has an excellent surface finish, thus reducing the water frictional loss. It withstands, indefinitely, the highest operating water temperatures.

A Complete Water-Cooling System.—Fig. 303 illustrates the complete water-cooling system of the Austin six-cylinder engine, which employs the fan-pulley driven water impeller-assisted *thermo-syphon* circulation method. The arrows show the directions of the water flow.

Commencing at the radiator lower tank E, the water flows upwards on the right to the impeller casing G, whence the water is expelled into the cylinder water jackets, as shown by the arrows on the left. From the upper part of the cylinder head water space, the water flows back to the radiator top tank and thence down the water spaces of the radiator into the lower tank E. The thermostat D is mounted in the hot water outlet at the front of the cylinder head and when the engine is cold restricts the water flow to the radiator, as explained, later. F denotes the system drain tap.

Modern cooling systems include a supply of heated water for the interior heating of the car. In Fig. 303 is shown a water cock J which, when opened, allows hot water to flow to the heating unit A, where heat is extracted from the unit by a motor-driven fan and circulated into the car interior. The return pipe B from the unit A conveys the cooler water to the impeller casing.

The Austin engine has a cylinder capacity of 3,995 c.c. and the capacity of the cooling system is 24 pints, i.e. 3 gallons.

The Radiator Fan.—This fan, is usually driven by a vee-belt from the crankshaft vee-pulley, at speeds, varying in different engines from 0·90 to 1·30 times crankshaft speed. While this constant drive method for the fan is simple, and inexpensive to make, it has one *serious drawback*. Thus, while the fan will draw sufficient air through the radiator air passages when the engine is at rest or operating at speeds up to 15 to 20 M.P.H., to cool the water in the system, at higher road speeds there is a sufficient flow of air through the radiator matrix to keep the water at about its correct temperature; the cooling fan is then not necessary.

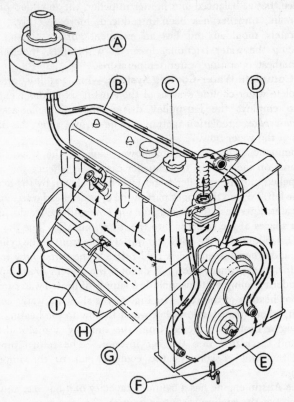

Fig. 303.—The Austin Engine Cooling and Interior Heating System.

Theoretically the *power absorbed*, from the engine, in driving the fan increases as the *cube of the fan speed*, i.e. if the speed is doubled, eight times the power is needed to drive the fan. The power needed depends upon several factors, including the blade design and blanking effect of the engine block, but the results of numerous tests have shown that the power absorbed by the fan in recent engines is between 3 and 7 per cent of the B.H.P. Thus, for a 50 B.H.P. engine, the fan may absorb between 1·5 to 3·5 H.P.

To obviate *the wastage of fan power* at the medium to maximum road speeds several methods are available commercially. Of these those mostly favoured include:

(1) The electric-motor driven fan which is switched on and off by a thermostatically-controlled switch. Thus, should the water temperature increase above its normal value the thermostat, in the cooling system, acts through an electric relay to switch on the fan motor. The fan then cools the water to its normal value, after which the motor is switched off, automatically.

(2) By a temperature-controlled clutch which disengages the belt drive to the fan when the temperature of the water increases above its normal value. A typical drive is the Eaton Tempatrol*. The coupling consists of two circular members, whose surfaces are close together. Between them is a film of silicone fluid which transmits the drive—just as in a fluid coupling or torque converter. Referring to Fig. 304, a slide valve is opened or closed by a thermostat, actuated by the engine bonnet inside temperature. When this temperature is below the normal one, the slide-valve is closed, so that the fluid chamber is empty and the fan is disconnected from its drive. When the bonnet temperature rises above the normal thermostat setting the slide-valve is opened by the thermostat, to allow silicone fluid to enter the space between the two circular members, thus connecting up the drive from the fan driving pulley to the fan itself. With the drive engaged adequate cooling is obtained at low engine speeds. The coupling is about $5\frac{1}{2}$ in. diameter and $1\frac{3}{4}$ in. wide.

(3) By a magnetic powder clutch, similar in principle to the Eaton (or Smith) clutch used in the automatic transmission system. The unit contains an electric coil which is energized when the thermostat actuates a switch. The coupling then acts as a solid member to drive the fan.

(4) By an hydraulic (oil) type clutch between the fan drive and the fan. This method is based on the principle of a piston in a

* Made in this country by Smiths Motor Accessory Division, Witney, Oxford.

cylinder filled with oil, such that when the inside bonnet temperature increases the oil expands and allows the piston to operate the presser plate of the friction-type clutch, so that the latter is

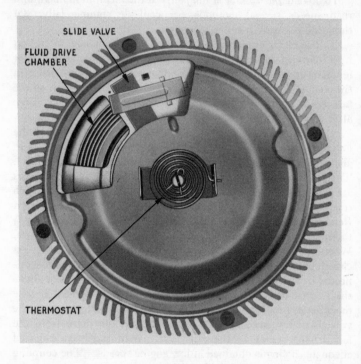

Fig. 304.—The Tempatrol Thermostatically-controlled Fan Unit.

engaged and drives the fan unit positively. In the Clark automatic fan clutch the axially-movable pressure plate is held out of engagement by a permanent magnet. When the piston is actuated it compresses a spring which releases the pressure plate from its magnet.

The Fan Blades.—The number of blades varies, in different cars from two to as many as sixteen; in this connection the **more**

recent Jaguar engine fans have twelve blades. With the increased number of blades, for a given cooling volume and air speed, the overall diameter of the fan can be reduced. The blades should not be too close to the engine unit, otherwise it operates less efficiently and, at the higher fan speeds becomes noisy. In some instances *fan blades* are made from *nylon* or a strong *plastic material*, accurately modelled to shape.

Plastic Fans.—The all-metal fan with its stamped and riveted blades is being replaced in the U.S.A. by a single-element plastic fan. The advantage of this fan is its relative lightness, lower first cost and ease of fitting. A typical example is the four-bladed fan made of a high-strength polycarbonate resin plastic, plain glass fibre reinforced, a material retaining the required tensile value up to the highest engine bonnet (hood) temperatures and of high impact property. Known as Lexan, is a General Electric product.

Fan and Impeller Drive.—The combined fan and impeller unit are driven by means of a vee-belt from the crankshaft end, which has a belt pulley of approximately the same size as the fan pulley. Usually, the dynamo is driven by the same belt, so that a simple form of common drive is obtained.

Fig. 305 illustrates the fan and dynamo drive of the Armstrong-Siddeley "Sapphire" engine, and also the usual provision made for adjusting the belt tension. For this purpose the dynamo is mounted on a pair of steel plates—one at each end. These plates are held on hinges (4), but are prevented from moving by means of a pair of steel links, clamped to the cylinder block at their ends (1) and to the dynamo plate-lugs (2) by bolts and nuts. When the nuts (1), (2) and (3) are slackened the dynamo can be partly rotated, the lugs (2) moving relatively to the slots in the link members (1)-(2). When the belt is properly adjusted, it should give a lateral deflection of $\frac{1}{2}$ in. on its longest section. The nuts (1), (2) and (3) of each end-plate are then tightened.

The same illustration (Fig. 305) also shows how the crankshaft pulley is marked with a notch, or vee, at (6), which is set against a fixed timing pointer (5) on the crankcase, to show the *top dead centre position* of the piston in the front, or No. 1 cylinder. This setting is used for ignition and valve timing purposes.

Quantity of Water in Cooling System.—With the modern fan-impeller system and efficient radiators, the quantity of water provided in the system varies, according to the different designs of engines, from about 0·15 to 0·20 pint per B.H.P. (maximum).

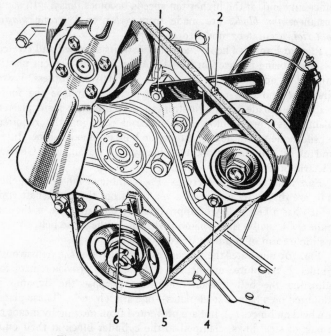

Fig. 305.—The Fan and Dynamo Drive of the Armstrong-Siddeley Sapphire Engine.

In some instances, e.g. the larger 250 to 400 B.H.P. American engines, from 0·1 to 0·15 pint per B.H.P. are required for efficient engine cooling, in most cars. The effect of using a car interior heater is to increase the total capacity of the cooling system by 1 to 2 pints for cars up to 2 litres engine capacity and 2 to 4 pints for the larger cars, including American models.

Improved Cooling Methods.—A good deal of progress has been made in recent years in connection with the cooling of water-cooled cylinders and exhaust valves by improvements in the design of the water jackets and other means.   One notable result of these developments is the provision of water jackets for the full length of the cylinder barrels instead of the previous method of partial jacketing.

Another development is that of the exhaust valve seating which

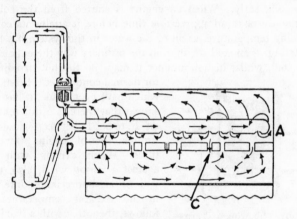

Fig. 306.—Cooling System of the Vauxhall
Six-cylinder Engine.

is now liberally jacketed for cooling purposes. A recent improvement in exhaust valve seat cooling, consists in the provision of nozzles within the water jacket for directing streams of cool water from the radiator on to the valve seatings and also the sparking plug seating.

Modern In-line Engine Cooling System.—A later improvement in cooling systems, illustrated in Fig. 306 in the case of Vauxhall four- and six-cylinder engines employs a water-distributor tube located lengthwise in the cylinder head, as shown at A. This delivers streams of cooler water from the impeller (pump) P to the areas of the exhaust valve seats and sparking plug bosses. The water passes from the cylinder head through cored holes C into the cylinder block where it circulates, as shown by the arrows before leaving the head *via* the thermostat T to the radiator top tank. From there it flows down the matrix water passages to the bottom tank. When the engine is cold the thermostat is closed, to shut off the cylinder water jackets, so that the water circulates through a by-pass drilling in the pump body.

Thermostats.—It has been stated that there is a definite mean temperature for the cooling water, at which an engine works most efficiently. When an engine is started from the cold, it frequently takes an appreciable time before it attains its correct working temperature, as all of the water in the system has to be heated up. In cold weather, in the ordinary way the engine will run cool, whilst in hot summer it may run too hot, if adjusted for normal conditions. From tests which have been made, it has been shown that there is a difference of 40° F. to 60° F. between winter and summer petrol engine temperatures. When driving with and against a strong wind, also, there will be an appreciable difference in the radiator temperature. It follows, then, that with a fixed area radiator, it is impossible to keep the mean circulating water temperature constant under all circumstances.

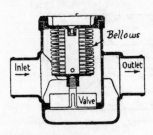

Fig. 307.—Showing a Thermostat Cooling Water Temperature Regulator.

For these reasons, it is necessary to employ a device, known as a *Thermostat*, the object of which is always to keep the temperature of the jacket water fairly constant. This is usually effected by cutting out the radiator when the engine is cold and by varying an opening in the outlet water pipe to the radiator according to the temperature of the engine. The jacket water temperature variations in this case, operate a thermostatic element, which opens or closes a valve in the upper outlet from the cylinder to the top of the radiator.

Thermostat Opening Temperatures.—In modern car engine cooling systems it is usually arranged for the thermostat valve to begin to open when the water temperature reaches 155° to 165° F. in cooling systems having low pressurized values, e.g. 3 to 6 lb.

per sq. in., while in systems using pressures of 12 to 15 lb. sq. in. the thermostat opening temperatures are usually from 175° F. to 183° F.; this is chiefly on account of the higher coolant boiling points, with such pressures.

Typical Thermostats.—Fig. 307 illustrates one type of thermostat used on automobile engines. It consists of a series of thin copper discs, made in the form of bellows, somewhat like the pressure element of the ordinary aneroid barometer. The interior of the bellows is sealed and contains a volatile liquid, such as ether, and its vapour; methyl-alcohol also is used in many modern thermostats. When such a bellows is heated up, the vapour pressure increases and causes the bellows to expand in the direction of its axis. One end of the bellows is anchored rigidly, and the other is attached to a flat or cone-seated valve. As the water temperature on the inlet or jacket side of the bellows increases, the bellows expands and lowers the valve from its seating, thus allowing the water to flow from the jackets to the radiator. Thus, when the engine is heated up, the radiator is fully connected up, but whilst the engine is cool, it is disconnected and can therefore warm up quickly. It is usual to fit the thermostat between the top of the cylinder head and the radiator as shown in Fig. 306.

The Wax Pellet Element Thermostat.—A more recent thermostat originating in the U.S.A. makes use of the expansion of a melting wax element to open the valve to allow the cooling water to circulate from the radiator to the cylinder water jackets. Fig. 308 illustrates the wax pellet type used on General Motors car engines. It has a wax pellet which is connected through a piston to a water flow valve, such that when the pellet is heated to the predetermined temperature the wax melts and pressure is exerted against a rubber diaphragm which forces the valve to open downwards and allow the hot water from the radiator to pass through the valve opening to the cylinders. When the water temperature falls the wax hardens and contracts; this allows a coil spring to close the valve so to shut off the water in the radiator from the cylinders.

This type of thermostat is made in different models, to open at specific temperatures and one of its special features is that it enables the engine to warm up quickly in cold weather and gives efficient regulation of the water temperatures.

Radiator Shutters.—In some cases the thermostat has been placed in the top of the radiator, and its movement then operates an elbow lever, which in turn works a series of Venetian blind type radiator shutters. When the engine is cold the shutters are closed,

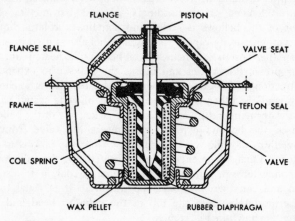

Fig. 308.—The Wax Pellet Element-type Thermostat.

thus blanking off the radiator. As the engine warms up the hot water flowing into the radiator causes the thermostat to expand, and to operate the lever connected to the hinged members of the shutters, causing these to rotate so as to admit more and more air to the radiator.

It is an advantage to employ one of the cooling water thermometers now available, in conjunction with the thermostat, as a check on its working, and for adjustment purposes.

Shutter Control Thermostat.—The principle of one type of radiator shutter control is shown in Fig. 309. In this case provision is made not only for operating the shutters as the water temperature changes, but also when the engine stops. The advantage of this scheme is that the heat of the engine is conserved by closing the shutters when the engine stops, whereas with the ordinary thermostat the shutters would remain open until the temperature of the water fell to the air temperature.

In addition to the thermostat there is a bellows which is connected to the main lubrication system.

When the engine is running the oil pressure expands the bellows and the gap between the plunger and thermostat is immediately taken up. When the engine stops the bellows contract through the loss of oil pressure and the gap is re-established with the result that although the thermostat may be expanded owing to the heat

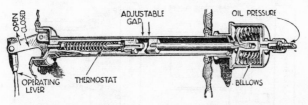

Fig. 309.—Thermostat for Controlling Radiator Shutters.

of the engine the expansion is compensated for by the gap, and the radiator shutters are at once closed by their springs.

Preventing Cooling System Corrosion.—As the metal of the cooling system—in particular the cast iron water jackets—is exposed to corrosive action causing solid deposits in the system, tending to block the radiator water passages, it is now customary to add a soluble oil corrosion inhibitor to the cooling water. This forms a protective coating on the metal surfaces. In the case of most car cooling systems about 1 volume of soluble oil is used to every 120 to 150 volumes of water.

Radiator Requirements.—As previously stated the purpose of the radiator is to dissipate, or get rid of the heat of the cooling water as rapidly as possible, and with the lightest and least bulky form of construction. Modern radiators are now very efficient and light, yet they will stand up to their work for long periods.

The principle of most radiator designs lies in the provision of a large number of small sectioned water spaces extending from top to bottom, and a maximum external cooling surface area exposed to the cooling air. There is a water reservoir at both the top and bottom of the radiator, into which the outlet pipe from, and the inlet pipe to, the cylinder jacket are led, respectively. The radiator tube system connects these two water reservoirs, or headers, so that when the engine is working there is a continuous series of water streams flowing from top to bottom. In addition the radiator is provided with a filling cap at the top, an emptying tap or plug at the bottom, and an overflow, or steam pipe inside. The latter is a small copper or brass tube (of about $\frac{1}{4}$ to $\frac{1}{2}$ in.

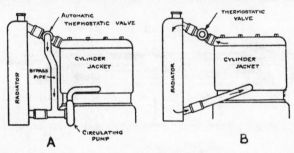

Fig. 310.—Showing Alternative Thermostat Arrangements. In *A* the cooling water is by-passed, and in *B* the flow is stopped when engine is cold.

diameter) extending from near the top of the radiator, or filler, to the bottom of the radiator, either outside or inside, and to the air. If the radiator is filled to too high a level, the surplus flows out of this tube. Similarly when the engine is started up, the water as it warms up expands, and any surplus flows out through this overflow pipe. If the water boils under any running circumstances, the steam also escapes through this pipe.

The Cold-Filling Radiator.—When engines are used under very cold weather conditions, unless anti-freeze coolant is used, it is necessary to empty the water from the cooling system when the vehicle has to stand for any appreciable period. In order to prevent the water from freezing in the radiator block whilst filling the system, in the cold-filling radiator method, illustrated for the Austin vehicles in Fig. 311, the majority of the water poured in goes directly to the bottom tank, then across the tank to the

outlet pipe where it is circulated by the pump on the closed
thermostat system whilst the radiator is still filling. The engine
is generally started *before re-filling* with water. The water
entering the radiator during the filling process, *via* the direct

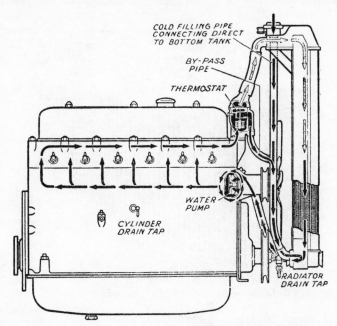

Fig. 311.—The Austin Commercial Vehicle Type Cold
Filling Radiator.

filling pipe is warmed in contact with the bottom tank before
reaching the radiator, so that there is no risk of freezing in the
latter. The cold filling pipe therefore obviates the necessity for
any water to be passed down the cooling tubes or water spaces of
the radiator. Thus, when the engine is first started this system
allows warm water to be by-passed across the bottom tank instead
of it going direct from the cylinder head to the water-pump as
in previous types; it also fills the radiator with warm water. This

system has no disadvantages when used in ordinary or warm climates.

Referring again to Fig. 311, the black arrows refer to the water circuit before the thermostat opens and the grey arrows to the isolated portion of the system whilst warming up.

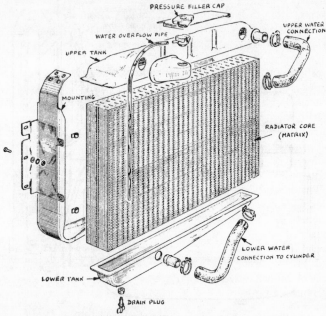

Fig. 312.—Components of Typical Radiator Assembly.

A Typical Radiator Assembly.—Fig. 312 illustrates the various components of a modern car radiator, namely, that used on the Vauxhall 162 cu. in. Velox and Cresta cars. The radiator element is of the film-type with the top and bottom tanks reinforced by means of a strap having support brackets bolted to a panel behind. A filler and inlet pipe are located at opposite sides in the rear of the top tank, with a central fan shield. The bottom tank has an outlet pipe and drain tap. An external plastic overflow pipe connects with a pipe in the filler neck. The top tank has a pressure-type filler cap. This provides for a pressure of $3\frac{1}{2}$ to $4\frac{1}{2}$ lb. sq. in. for the hand-operated gearbox models and $6\frac{1}{4}$ to $7\frac{1}{4}$ lb. sq. in. for the automatic transmission model.

The radiator has a nominal capacity of 7 gallons (4·2 U.S. quarts), while the whole cooling system contains $18\frac{1}{4}$ pints (10 U.S. quarts), including the car heating system. The thermostat is set to open at 170° to 180° F. and is fully open at 200° F. When *testing this type of radiator for leakages* in the water elements, air at a pressure of 7 to 10 lb. sq. in. is applied to the radiator overflow pipe, after sealing the radiator main water inlet and outlet pipes with wooden plugs. The radiator is submerged in a water bath and any leakages are indicated by rising bubbles.

Alternative Radiator Construction Methods.—There are three common methods of constructing radiators (and also a number of special methods which space prevents mention of here) as follows: (1) The Honeycomb Tube, (2) The Film or Corrugated Strip, and (3) The Gilled Tube Systems.

The honeycomb tube is perhaps the most efficient type. It resembles externally the ordinary honeycomb of the bee. It is made up of a large number of thin brass tubes, having their ends expanded outwards, as shown in Fig. 313 (*A*). The tubes are assembled with their expanded ends all in the same planes and the ends are soldered (usually by dipping the surface into a molten bath of solder), so as to fill up the end interstices; the soldered portions are shown in Fig. 313 (top). In this way the ends are sealed, while leaving narrow water-spaces about 1 mm. wide between the tubes, but also leaving the interior surfaces of the tubes quite open and free for the cooling air to flow through. The blocks of tubes thus formed are soldered to the top and bottom tanks, and at the two sides, to form the complete radiator. The thin brass used for (1) and (2) constructions is only from $\frac{5}{1000}$ to $\frac{10}{1000}$ in. thick.

The film type construction (Fig. 313 (*B*)) consists in building up the tubes by means of corrugated strips, with their edges (or ends) expanded as in the former case. These strips extend from top to bottom of the radiator, and a pair of strips, with their expanded ends, form a long zig-zagged narrow water-space from top to bottom of the radiator. Their external surfaces form hexagonally-shaped air-spaces. This cheaper mode of construction gives a radiator element resembling the true honeycomb type.

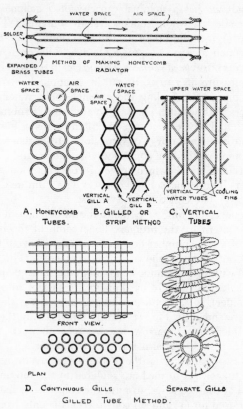

Fig. 313.—Illustrating Different Systems of Radiator Construction.

The gilled tube system (Fig. 313 (*D*)) is perhaps the oldest type of radiator construction, although it is still used, principally on commercial vehicles, on account of its strength and

freedom from leakages. In this method the top and bottom
tanks are connected by means of a number of straight vertical
copper or brass tubes, usually of about $\frac{5}{16}$ to $\frac{7}{16}$ in. external
diameter, thus providing a corresponding number of vertical
paths for the water; in this case the water flows *inside* the tubes.
Each tube has a large number of annular rings or fins pressed
firmly over its outside surface, and spaced at uniform intervals,
somewhat on the lines of the cooling fins of an air-cooled engine.
These fins conduct and radiate the heat from the water inside the

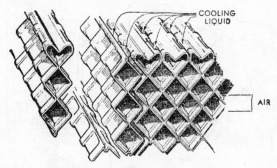

COOLING
LIQUID

AIR

Fig. 314.—Galley Radiator Construction.

tube, the air current flowing through the spaces between the gilled
tubes.

The Galley radiator construction shown in Fig. 314 represents
for an automobile radiator a method of providing the maximum
cooling effect from a given air velocity. The air spaces are
roughly of square form and the water film spaces between the air
orifice portions run in serrated fashion from top to bottom of the
radiator. The materials used for the radiator cooling fins include
brass, copper, steel and nickel-copper alloys.

The ratio of the fore-and-aft length of the air orifices or tubes
(when the honeycomb radiator construction method is used)
to diameter ranges from 30:1 to 80:1, the modern tendency
being to use higher ratios of length to diameter for minimum
frontal areas. The thickness of the metal used for honeycomb
radiator tubes varies from 0·005 in. to 0·010 in.

The Cross-Flow Radiator.—The construction of this type of radiator is similar, in principle, to that shown at C in Fig. 313, but instead of the water tubes being vertical they are arranged horizontally, between vertical water tanks, which have their water pipes to the cylinder block hoses, at the top of one tank and the bottom of the other. The top header tank contains the usual filler cap.

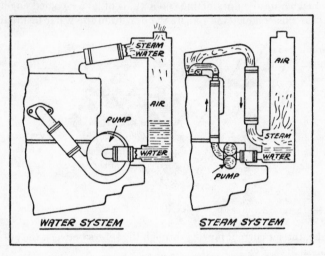

Fig. 315.—Illustrating Steam Cooling Methods.

The Steam Cooling Method.—In the ordinary pump and and thermo-syphon methods of cooling an engine the water passing through the cylinder jackets has its temperature increased by some 20° to 30° F. Each 1 lb. of water then deals with 20 to 30 British Thermal Units. If however, the same weight of water were converted into steam it would be able to deal with no less than 966 heat units, that is, about 33 to 48 times the quantity.

It will be evident, therefore, that if the cooling system is so arranged that instead of merely raising the temperature of the water in the jackets, it is converted into steam, a very much smaller quantity of water will be required in the system. It will, of course, be necessary to employ a "radiator" to act as a condenser for this steam, the condensed water being returned to the cylinder jackets.

This method which has been used with satisfactory results on motor-car and aircraft engines has resulted in an appreciable saving of weight.

As compared with the cooling water method it gives a rather higher cylinder temperature, and in most cases a gain in efficiency.

Fig. 315 illustrates the ordinary steam and the Rushmore steam cooling systems that have been used on some American motor vehicles. In this case the top water outlet on the engine, instead of being connected to the top of the radiator, is carried down and enters the lower tank a few inches from the point where water enters the line leading to the pump. A header-pipe lies in the bottom of the lower tank and has holes drilled through it to distribute the steam through the width of the radiator.

A gearwheel type of water pump in the Rushmore system draws the condensed water from the base of the radiator and delivers it to the bottom of the cylinder jackets. In this system of cooling it is necessary to fit a safety-valve in the top of the radiator to prevent steam pressure occurring.

Radiator Casings.—The earlier type radiator block was soldered into a pressed brass casing, and a back cover plate then soldered into place, a space being left at the front and back for the air to flow through. The casing was usually provided with cast brass holding down brackets, for attaching it to the chassis members. Sometimes ordinary angle brackets were used, and a block of fibre, canvas or rubber inserted between the bracket and frame to insulate the radiator from shocks. In some designs a hinged fork or ball joint is provided at each side, thus relieving the radiator of any stresses due to the two side members of the frame distorting differently. In other cases, notably in commercial vehicle radiators, horizontal trunnions are provided, one at each lower side of the radiator. These trunnions are housed in capped bearings bolted to the chassis side members, so that the radiator is free to tilt fore and aft; sufficient end play is allowed for endwise movement. Each bearing is provided with a greaser.

A type of radiator construction now widely used consists in

making the radiator block, complete with its headers and water pipes, and inserting the whole within another separate chromium-plated casing, designed for appearance purpose only. This method enables the radiator block to be removed without its casing, and cheapens the cost of replacement.

In some recent cars the radiator casings are dispensed with and louvres made in the inside portions of the mudguards, or the lower front parts of the engine cowling or bonnet, allow the cooling air to reach the radiator block within the cowling.

Care of the Radiator System.—As the water passages in the honeycomb and film type radiators are of very fine sections, it is necessary to prevent any solid matter in the water from reaching them. All water should therefore be strained before filling the radiator, and "hard" water, which is likely to deposit magnesia or chalky matter on the surfaces, should not be used. It is best to use rain-water, or to "soften" the water with one of the boiler water-softening compounds sold for the purpose. Should any deposit have formed in the tubes, it can usually be removed by filling the radiator with a strong, hot solution of washing soda unless there is any aluminium alloy in the cooling system components when such a solution should not be used, since it attacks aluminium.

In frosty weather the water should be drained off completely to prevent freezing of the radiator and cylinder jacket waters, otherwise a leaky radiator or cracked cylinder will result. Where pump circulation is used, the pump casing must be drained of water; a tap is usually provided for this purpose on the casing. (*See* Fig. 311).

Anti-Freezing Solutions.—To obviate emptying the water system in cold weather, an anti-freezing solution can be used in place of pure water. Glycerine mixed with water will lower its freezing-point several degrees. Thus 10 per cent. of glycerine will freeze at 2° F. below freezing-point, 20 per cent. at 7° F. below, and 40 per cent. at 32° F. below (i.e., at 0° F.). It is not advisable, however, to use more than 20 per cent. solution, or the circulation will tend to become choked up, and the rubber hose connections attacked. Alcohol, in a 15 to 20 per cent. solution (using methylated spirits), has a freezing-point several degrees below that of pure water. In the case of all anti-freeze solutions any loss by evaporation should be made up for by the addition of mixture, *not* water alone. Any loss of solution by leakage in the cooling system must be made up by more of the original solution.

Ethylene Glycol Solutions.—Glycerine has been replaced as an anti-freeze, by solutions of ethylene-glycol in water, but with the addition of a chemical, known as the corrosion inhibitor, i.e., it prevents corrosion within the cooling system.

Ethylene glycol anti-freeze liquid is available under various trade names, but the majority of these are made up to the British Standard No. 3150 Specification. The following are the proportions of ethylene-glycol solution (as purchased), to soft water, and the corresponding freezing points:

Percentage of Anti-freeze	10	15	20
Pints of Anti-freeze per gallon of mixed coolant	$\frac{3}{4}$	$1\frac{1}{4}$	$1\frac{3}{4}$
Freezing Point	17° F.(-8° C.)	7° F.(-14° C.)	0° F.(-18° C.)
Degrees of Frost F.°	15	25	32
C.°	8	14	18

When using an anti-freeze, the cooling system is emptied and flushed out. Then, soft water to about one-third the cooling system capacity is poured into the radiator. Next, the measured amount of anti-freeze is added and, finally, the balance of the water. The engine is then started and run until the coolant becomes warm; this ensures adequate mixing of the anti-freeze with the water.

CHAPTER X

TESTING OF AUTOMOBILE ENGINES*

Practical Tests.—Without the use of special testing apparatus it is possible only to make rough qualitative tests, but nevertheless an experienced engineer can usually tune up his engines satisfactorily. Assuming that an engine has been overhauled completely, the crankshaft re-ground and main bearings re-fitted, or adjusted, cylinders re-bored and new pistons fitted, valves skimmed and ground, new valve springs, big-end bearings replaced, new small-end bearing and other minor repairs, the first item is to check the clearances of the working parts, such as the valve tappets and contact-breaker. The engine will usually feel quite "stiff," and must be run in for a period at light loads.

The modern garage will have a special running-in electric motor or belt drive from the machine-shop shafting. A temporary wooden pulley is fixed to the crankshaft, or the fly-wheel itself, can be used in some cases, and the engine is motored around with the petrol supply and ignition switch off. The engine should be given an excess of lubricating oil, and the speed of running-in should correspond to a road speed of 20 to 25 m.p.h. (engine speed 1,000 to 1,500 r.p.m.); the running-in period is from 2 to 4 hours.

The engine should then be run light under its own power for about an hour, and then, if satisfactory, accelerated gradually to a speed corresponding to a top gear speed on the road of about 35 to 40 m.p.h. The engine is then throttled down and note is made of any points requiring attention, such as oil leakage, exhaust leaks, excessive noise in the tappets or gearing, and similar items. It will usually be found necessary to re-adjust the engine tappets and to check the cylinder head and manifold securing nuts after these tests; this should be done with the engine warm.

Before running-in a new or reconditioned engine it is a good

* A full account will be found in *Testing of High Speed Combustion Engines and Automobiles*, A. W. Judge. (Chapman and Hall).

plan to add a measured amount of a lubricant additive, known as a *running-in fluid*, such as colloidal graphite in suspension in oil, or molybdenum di-sulphide. These compounds are available commercially, in tins of suitable capacities for automobile engines.

Running in the Engine on the Road.—In the case of new engines the machined surfaces of the cylinder walls, pistons and the various bearings are usually left with very fine tool or hone feed marks on them; even with ground bearing surfaces there are fine irregularities. It is, therefore, necessary to remove the crests of these machining marks before one can say that the surfaces are properly bedded down for normal duty purposes.

The engine, after a relatively short running-in period, therefore has to complete its bedding-down process by a period of road running, namely, over distances of 500 to 1,000 miles, at the manufacturer's specified maximum speeds. For this purpose, the new owner is usually instructed to drive his car on top gear at speeds not exceeding 35 to 45 m.p.h. for the first 500 miles or so, with correspondingly reduced speeds on the lower gears; most mass-produced engines are somewhat "stiff", when new, for this reason. To ensure that the driver did not over-run his engine many manufacturers used to fit restriction washers in the carburettor-inlet pipe joint. These were removed after 500 miles.

Road Tests.—When the engine is fully run in, i.e. after about 1,000 miles with production engines the ignition timing, sparking plug gaps and valve clearances should be checked and the nuts on the cylinder heads, inlet and exhaust manifolds tested for tightness. The carburettor settings (if adjustable, e.g., the S.U. model) should also be checked and the slow-running adjustment examined. Then, the car should be taken on the road and tested for: (1) Petrol consumption on the level at normal running speed, say 30 to 40 m.p.h. (2) Maximum speed on the level, but, for short intervals, only, at this stage. (3) Hill-climbing capabilities. (4) Acceleration, to various top speeds.

The petrol consumption may be checked by filling the main tank up to a given fixed mark, with the car on level ground, and then running the car for a known distance (by speedometer or milestones) and noting how much petrol must be poured into the tank in order to fill it to the same fixed mark on level ground.

Sometimes it is more convenient to rig up a small auxiliary tank of a few pints capacity on the dash-board, and to run the engine off this. If a glass or float gauge be fitted, the quantity of petrol consumed for a given distance of running is readily deducible.

Petrol Consumption.—Thus if x pints of petrol be used whilst the car covers y miles the consumption $= \dfrac{8y}{x}$ miles per gallon. For example, if $1\frac{3}{4}$ pints are consumed for a road distance of $7 \cdot 4$ miles, the consumption will be:—

$$\frac{8 \times 7 \cdot 4}{1\frac{3}{4}} = \frac{8 \times 7 \cdot 4 \times 4}{7} = 33 \cdot 8 \text{ m.p.g.}$$

Instruments known as mileage-per-gallon or consumption meters, are now on the market, which read the consumption direct. Of these the Milegal, Gallometer, and Zenith flowmeter are typical examples. Fig. 316 shows the Zenith Mileage Tester, which incorporates a graduated glass vessel of $\frac{1}{10}$ gallon capacity, electric fuel pump, and a three-way cock. The tester is provided with padded hooks so that it can be hooked over the door window. There are two rubber tubes, one of which gives from the three-way cock to the carburettor and the other to the most convenient part of the fuel supply line to the carburettor; the normal carburettor feed pump is disconnected. When the handle of the cock is *vertical* petrol is pumped direct from the main fuel tank to the carburettor. When the handle is turned to the *right* petrol is pumped both to the carburettor and the glass vessel. When filled to the graduated top mark the glass vessel is filled ready for test.

When the engine is at its working temperature, the car is run at a given speed on a flat road, the cock handle is turned to the "*Test*" position and the speedometer mileage reading taken. When the petrol has fallen to the lower graduated mark the speedometer reading is again taken.

The difference between the two mileage readings gives the distance that has been covered on $\frac{1}{10}$ gallon of petrol, so that by multiplying this mileage difference by 10, the result will give the mileage per gallon.

Flowmeters.—Although more elaborate than the graduated vessel or tank method, the flowmeter instrument will give the actual rate of petrol consumption of the engine *all the time*, so that fuel consumption tests can be made more quickly and accurately: These instruments are generally used for laboratory tests on engines, but special models have been made for road testing purposes and are used for road tests made by the staffs of motor publications. A typical flowmeter is shown in Fig. 316.

The petrol consumptions of average modern cars normally loaded and when operated at speeds not exceeding about 50 M.P.H. should be approximately as follows:—

TABLE 10

Engine Capacity. cu. cm.			Miles per Gallon
700 to 1,000	..	..	45 to 40
1,000 to 1,500	..	..	40 to 33
1,500 to 2,000	..	..	33 to 27
2,000 to 3,000	..	..	27 to 22
3,000 to 4,000	..	..	22 to 17

If it is possible to ascertain the B.H.P. of the engine, the fuel consumption can be checked from the fact that a good design of water-cooled petrol engine at three-quarters full load uses only 0·50 to 0·55 lb. of petrol per B.H.P. per hour.

Ton Mileages.—For accurate comparisons of car performance, the weight of the car, as well as the fuel consumption, should be taken into account. Most of the fuel consumption tests results on cars made by the R.A.C. are expressed in this way.

If the weight of a car be W tons, and $C =$ its fuel consumption in m.p.g., when $W \times C =$ ton mileage per gallon.

The highest values recorded to date are those for heavy commercial vehicles, namely, 120 to 130 ton miles per gallon. Most cars range from 40 to 60.

Laboratory Testing of Engines.—Complete equipment is available for laboratory and routine tests on petrol and Diesel engines. Typical self-contained equipment includes an hydraulic or electric dynamometer, air measuring tank, fuel flow meter, thermometers for water inlet and exhaust gas temperatures, and, when required pressure-volume type high-speed indicators, which will produce indicator diagrams. The available indicators include the optical or minor types, the Dobbie-McInnes "Farn-boro" indicator, micro-indicators and the cathode ray oscilloscope.

The recording-type optical and "Farnboro" indicators actually draw pressure-volume diagrams, so that by measuring the areas

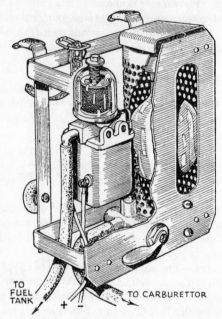

TO FUEL TANK

+ −

TO CARBURETTOR

Fig. 316.—The Zenith Flowmeter for Testing Fuel Consumption During Road Tests.

of these, the indicated horse-power can be calculated from the formula given in Chapter I. With the aid of the equipment outlined before estimates or measurements can be made of the I.H.P., B.H.P., I.M.E.P., B.M.E.P., fuel consumption for H.P. per hour, heat balance, thermal, volumetric and mechanical efficiencies, etc.

Works Tests of New Engines.—In the case of mass-produced engines, each engine after assembly is flexibly coupled to one of a group of special block testing D.C. electrical dynamometers which is arranged to operate both as a motor for running-in the engine and then as a generator or dynamo for measuring the engine output. After the initial period of working under its own power, the engine is speeded up until it attains its normal temperature. Next, the speed is reduced and any adjustments necessary are made, e.g., carburettor or ignition. The engine is then tested over its whole working speed range at full throttle and the power and engine speeds measured. The observed maximum B.H.P. and speed must attain certain minimum values before the engine is passed for instalment into its chassis.

In this method of "block" testing new engines, water, lubricating oil and petrol are fed to the engines from a central supply, while the electric current for ignition is provided by special generators. The cooling water is circulated by pumps and the temperatures regulated automatically; heat exchangers are used to cool the water as required.

The cooling air is supplied by an external fan or blower and not by the usual fan of the engine. The exhaust pipe (or pipes) is lead to a special exhaust system of large capacity with good silencing properties. As explained earlier in this volume, the horse power thus measured is the gross B.H.P.

"Road" Tests in the Laboratory.—It is also possible to test the engine, gears, and transmission of a car in the test house by arranging a large diameter metal drum under each of the rear wheels, and providing a spring balance from the rear axle to a fixed post at the back. The engine is started up, clutch engaged with the gear in the gear-box in mesh, when of course the rear wheels will revolve, turning the drums. The car cannot move forward owing to the spring balance tension cable which measures the tractive effort. The power delivered to the drums can readily be measured, electrically or otherwise, and the efficiencies of the gears and transmission computed. Messrs. Heenan and Froude provide a special car (and also a motor-cycle) testing plant on these lines; by using drums having irregular surfaces, road inequalities can be reproduced in the tests. Arrangements can be made to attach the two rear axles each to a separate dynamometer, such as the Froude hydraulic type, and to measure the power delivered at the rear axles, on all the gears. Much useful information can be obtained without actual road tests.   The Sprague Electric Branch of the G.E.C. in America have supplied a large number of chassis testing plants of this type, with swinging field dynamometers for absorbing and measuring the power at the wheels.

Engine Pressure Indicators.—The pressure-volume indicators are usually expensive and complicated; some types require the services of a highly-trained operator.   In many cases all that is necessary, for the test purposes, is a knowledge of the *compression* and the *combustion pressures*, for checking the performance of an engine.

It is not practicable to use an ordinary pressure gauge of the Bourdon type, for the pressures in the cylinder vary so rapidly that the mechanism of the gauge would quickly be rendered unreliable and inaccurate.

The compression pressure should, of course, be taken when the engine is hot, and the cylinder in question not firing.   It should be remembered also, that the compression value will depend upon the throttle opening, the maximum value being obtained at low-running speeds and at full throttle opening.   At higher speeds, owing to wire-drawing of the charge, the compression pressure, for the same throttle opening, falls off somewhat.

Special compression-type gauges are available for measurements of the compression pressure with the engine cold.   A feature of these gauges is a maximum reading hand which remains stationary when the pressure is released, so that accurate readings can be made at leisure.   Other types of barrel and dial gauges are made to read the maximum cylinder pressures while the engines are operating.   Typical examples of such gauges are the Dunedin dial gauge that can be used on petrol and Diesel engines in their chassis and can read pressures up to 2,000 lb. per sq. in. and the Dobbie-McInnes and Okill barrel-type gauges.   These gauges are all supplied with adapters and connections, to fit into the sparking plug (or fuel injection nozzle) holes.

Brake Horse-Power.—The horse power obtained at the crankshaft is less than that given in the cylinder (i.e., the I.H.P.) owing to the power lost, or absorbed, by friction at the working surfaces, namely, the piston and its rings, cylinder walls, connecting-rod, camshaft and main bearings, and in other ways.   Also, there is a loss, known as the *Pumping Loss* due to the power absorbed in moving the charge and exhaust gases in and out of their cylinders. The horse power lost due to friction is termed the *Friction H.P.* The available horse power at the flywheel or clutch is termed the *Brake Horse Power* (B.H.P.), and we have the following simple relations:

Brake Horse Power = I.H.P. — Friction H.P.*

$$\text{and Mechanical Efficiency} = \frac{\text{B.H.P.}}{\text{I.H.P.}} = \frac{\text{I.H.P.—Friction H.P.}}{\text{I.H.P.}}$$

The mechanical efficiency of an engine should be as high as possible; the maximum efficiency is ensured by minimum friction losses. The friction losses can be reduced to a minimum by the use of well-lubricated surfaces, of ample, but not too large, a bearing area, the lightening of reciprocating parts by the use of high tensile steels and aluminium alloys, by the use of ball and roller bearings wherever possible in place of plain gun-metal or white-metal bearings, and by the correct balancing of the crank webs and the reciprocating parts, etc.

The mechanical efficiency of a well-designed engine ranges from about 90 per cent. at low speeds to 80 per cent. at high speeds. The falling off in efficiency is due partly to increased friction losses on account of increased friction with speed or rubbing, and partly to increased losses due to drawing in and expelling the charge, i.e. pumping losses.

Measuring the B.H.P.—It is better to measure the B.H.P. directly, however, with one of the dynamometers or power brakes now on the market for this purpose.

A Prony brake, for giving approximate results, is not difficult to rig up for small engines. In its simplest form it consists of two blocks of wood held together with spring tensioning bolts, around the fly-wheel rim, or pulley fixed temporarily to the crankshaft. The frictional drag of these blocks opposes the fly-wheel's motion and absorbs the power output. The torque arm (Fig. 317) is loaded with weights W until there is no tendency for it to rotate with the flywheel. Stops are usually provided to

* In this term we here include the *Pumping Losses*.

limit its motion, and a pointer on the torque arm moving over a fixed scale enables the balance to be maintained.

Using the notation shown on Fig. 317, the Brake Horse Power is given by the relation

$$\text{B.H.P.} = \frac{2\pi N\,W\,d}{33,000}$$

where $N = $ r.p.m. d is measured in feet, and W in lbs. $\pi = 3\cdot14159$.

With a weight of 25 lb. at the end of a 3 ft. torque arm it is possible to absorb about $14\cdot3$ H.P. at 1,000 r.p.m.

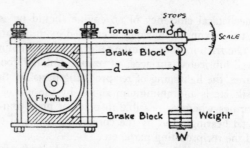

Fig. 317.—The Prony Brake for Horse Power Measurements.

This type of brake is simple and inexpensive to make; the fly-wheel, however, becomes fairly hot, and there is a tendency on the part of the friction surfaces to "snatch." A water-cooled rim can, however, be used.

The air, or fan, brake shown in Fig. 318 has been extensively used for tests on small engines. It consists of a pair of arms which bolt rigidly on to the crankshaft. Each arm has a rectangular metal plate bolted at the same radius, the plates being normal to the direction of motion. The air resistance of these blades opposes the engine's torque, and absorbs the horse power. The plates can be bolted at different radii on the arms, and plates of different area can be fitted readily. The size of plate and radius are chosen by trial so that the engine on open throttle attains the selected speed. Tables or curves are supplied with the commercial brake which give the corresponding horse power for this speed, blade area and radius and air density.

The advantage of this type is simplicity and low cost. The disadvantages lie in having to alter the blades or radii of fixing to vary the load at any speed. The brake's accuracy is also liable to be affected by draughts and air-currents due to extraneous causes. By enclosing it in a square or circular casing, provided with a variable central inlet area, or external outlet, a much wider range of control is obtained.

Absorption Brakes.—The majority of production-type automobile and also aircraft engines in this country are now tested for B.H.P. at various speeds by means of a water or hydraulic

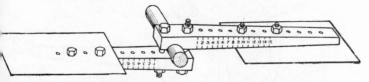

Fig. 318.—The Walker Fan Brake.

brake which absorbs the power given out by the engine by churning the water within the brake casing and thus heating it. This and the two types of brake previously described are termed *Absorption Brakes*. The alternative kind kind of brake, known as the *Transmission* type, transmits the engine power but enables the latter to be measured; the electric generator brake mentioned later is an example of this type.

The principle of the hydraulic brake or dynamometer is illustrated in Fig. 319. The crankshaft extension or coupling of the engine is connected to the main shaft of the brake by means of a flexible coupling—which allows for small errors of alignment. The brake shaft contains a rotor in the form of a kind of paddle which rotates within a closed casing containing projecting vanes, just clear of the paddle blade tips, and filled with water. Glands are provided between the rotor shaft and casing to prevent water leakage. The whole of the casing is mounted on low friction bearings, so that if free it could rotate around the rotor shaft. When power is transmitted to the rotor the paddle churns the water and in doing so tends to drag the casing round in the same direction as its own rotation. Above a certain minimum speed of several hundred r.p.m., below which slipping occurs, the torque or drag effect on the casing is equal to the torque applied to the rotor. The casing is prevented from rotating by means of a weighted arm, and if the casing is initially balanced, when at rest, the torque applied to the casing when the rotor is working is equal to the product of W and l in lbs. ft., assuming that the arm of the casing is just floating between the stops shown.

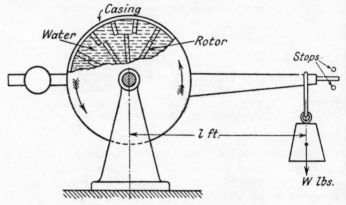

Fig. 319.—Principle of Hydraulic Brake.

It is then a simple matter to work out the horse power absorbed, for this is proportional to the product of the torque and speed of rotation as in the previous example on page 484.

$$\text{Thus B.H.P.} = \frac{2\pi\, W\, l.\, N}{33,000}$$

where W = weight on torque arm in lbs. = distance between C.G. of weight and shaft in feet, N = r.p.m.

For any particular machine the length l is constant, so that the expression $\dfrac{2\pi l}{33,000}$ is also constant and can be denoted by the constant quantity k, so that the formula may then be simplified as follows:—

$$\text{B.H.P.} = k\, W.\, N.$$

It is, therefore, only necessary to measure the weight and speed values in order to find the horse power of the engine. To obviate the use of a number of weights *a spring balance is usually employed*, so that the value of W in the previous formula can be read off directly. In the case of a hydraulic brake having an arm length of 5 ft. $3\frac{1}{40}$ ins. long the value of $k = \frac{1}{1000}$ whilst for a metric machine having an arm of 1432·4 mm. length $k = \frac{1}{500}$.

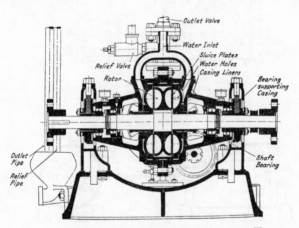

Fig. 320.—The Froude Hydraulic Absorption-type, Torque-measuring Dynamometer.

The crude paddle arrangement shown in Fig. 319 is replaced in modern dynamometers by cup like depressions both on the rotor and inside the casing, to enable much higher power absorptions to be obtained for relatively small overall diameters of casings.

Typical Dynamometers.—One Froude type hydraulic brake shown in Fig. 320 is now largely employed for automobile engine tests. In this case the power is absorbed by hydraulic resistance. The engine shaft is connected to the shaft of this brake by means of a flexible coupling. To this latter shaft is keyed a rotor having a number of semi-elliptical pockets divided from one another by oblique vanes. The internal faces of the water-tight casing have similar pockets. The resistance offered by the water to the motion of the rotor reacts on the casing, which latter is free to rotate within certain limits, or stops in its anti-friction bearings, but is constrained by weights hung at the end of a long torque arm attached to the casing. The product of the weight by the

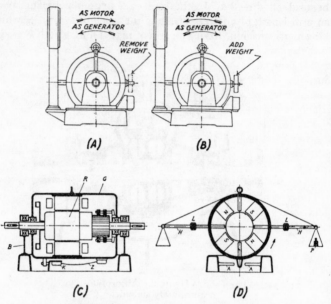

Fig. 321.—Illustrating the Principle of the Electric
Swinging Field Dynamometer.

A—Stops.   B—Pedestal Bearing.   G—Casing.   H—Torque Arm.
L—Movable Weights.   P—Scale Pan Weights.   R—Armature (Rotor).
(K-Z)—Electric Cables.

length of the torque arm, to the centre of the rotor, is equal to the mean torque of the engine. The friction of the bearings is included with the hydraulic resistance, so does not affect the results.

Since the water would get very heated if it were left in the casing, it is necessary to arrange for a continuous circulation from an outside source. It is necessary to provide an independent electric motor or belt pulley drive to start up the engine with this type of brake.

Another type of dynamometer of the transmission type much used for accurate engine tests is the *electric swinging field type*. Here the armature of a motor is connected to the engine under test by means of a flexible coupling, and the field magnets and their casings are so arranged that they are capable of rotating about the same axis as the armature. The torque required to prevent this rotation is meaured exactly as in the previous case, and is equal to the engine torque. This form of dynamometer has the advantage that it generates electric power which can be utilized for other purposes while the dynamo can be converted into a motor by suitable switches, for starting the engine, and for measuring its frictional losses, by making suitable switch connections.

The principle of the electrical swinging field dynamometer is shown, diagrammatically, in Fig. 321 (A) to (D). The electrical unit can, by suitable switches be used as either a motor to rotate the engine crankshaft for running-in or cold starting purposes, as indicated at (A) and (B).

The electrical rotor R is mounted on ball-bearings in the outer field casing G, which can swing in ball-bearings in the fixed pedestal B, as shown at (C). Seen in end view in (D), the casing is provided with two torque arms H along which weights L can slide, while for smaller measurements, weights P in scale pans are used. The swinging movements of the casings are limited by a short projector below and fixed stops A. Instead of using weights L, a spring balance can be employed at the end of one arm H, so that direct readings of the torque can be made. This type of dynamometer can deal with both directions of rotation of engines under test.

CHAPTER XI

LATER DEVELOPMENTS IN AUTOMOBILE ENGINES

SINCE the last edition of this book was published the previously noted steady improvements in engine design, materials and construction methods have continued so that the modern car engine contains certain new features, has a much improved power output, is generally quieter, rather more reliable and requires reduced maintenance attention. In consequence the more recent cars have much better performances, e.g., appreciably improved acceleration with higher cruising and maximum road speeds, but not always with lower fuel consumptions. In this connection the basic principles and methods used for increasing the power outputs of production types of car engines, described in Chapter I are equally applicable today. Typical examples of such improvements include increased compression ratios, higher volumetric or charge efficiencies, increased maximum speeds, adoption of overhead-type camshafts, improved combustion chambers, larger and better located valve ports, extended use of aluminium in engines, cooling systems at higher pressures, automatic radiator fan controls, improved exhaust systems, special measures for reducing exhaust pollutants, etc.

The accompanying Table gives some performance figures for a limited selection of British and foreign cars, fitted with engines of different capacities.

By comparison with the values given in the Table on page 51 the results will be seen to emphasize the improvements mentioned earlier; in particular, it will be seen that the B.H.P. per litre values have gone up appreciably.

Sports Car Engines.—This engine group includes the G.T. versions of certain production engines and those designed especially for this type of car. In some cases the manufacturers fit a larger size of production engine as used in a larger car model. A typical example is that of the sports car which uses the 1·5 or 1·6

PERFORMANCES OF MORE RECENT CAR ENGINES

Country of Origin	Compression ratio		Engine capacity (litres)	Maximum B.H.P. (gross)	at r.p.m.	B.H.P. Litre	Mean piston speed at Max. B.H.P. in ft. per min.
	Range	Mean					
Great Britain	8·5 – 9·5	8·8	1·1 to 3·5	50 to 150	4500–5500	45–65	2400–3150
France and Italy	8·5 –10·0	9·2	1·1 to 2·7	60 to 160	5250–6500	50–73	2500–3050
U.S.A.	8·8 –10·8	10·2	3·5 to 6·7	150 to 400	4400–4800	45–55	2300–2950
Japan	8·3 – 9·0	8·6	1·2 to 2·25	70 to 115	4200–6000	50–63	2400–3150

litre engine instead of the standard 1·3 litre one, but with certain chassis component modifications. However, the usual G.T. models have specially tuned standard engines; in some cases the manufacturers supply special tuning kits for their standard production engines.

In general, sports car engines have higher compression ratios and give greater power outputs, but at higher engine speeds. The compression ratios vary from about 9·0:1 to 11·0:1, and the B.H.P. per litre values are from about 60 to 90 for engines of 1·3 to 3·5 litres. The average for twelve different engine makes was 60. In these engines the maximum B.M.E.P. values were about 150 lb per sq. in at speeds of 3,200 to 4,000 r.p.m.

Racing Car Engines.—In order to obtain the maximum outputs specially designed engines have to be used. These have higher compression ratios, engine speeds and use special racing grade fuels. While many cars employ multiple carburettors, e.g. one per cylinder for vertical in-line engines or multi-barrel carburettors for groups of two or three cylinders, more recently the fuel injection systems—with their known advantages—have been employed. Electronic ignition systems are used. Another interesting feature of these engines is the use of engines having a smaller number of cylinders, e.g. the vee-six, vee-eight and vee-twelve types; also horizontal engines of the opposed cylinder kind with six, eight or twelve cylinders.

It has been shown theoretically and also experimentally that a smaller bore cylinder will give a greater output per litre than one of larger bore. In this connection, mention may be made of a highly-developed motor cycle air-cooled engine of 0·4 litre capacity, which gave a maximum output of 100 B.H.P. per litre, while two, four, six and eight cylinder engines of the same capacity gave 125, 160, 183 and 200 B.H.P. per litre, respectively. Mention should also be made of a highly developed Austin Morris 998 c.c. engine, designed for drag-racing which gave over 200 B.H.P. Its dragster captured several world records.

According to their cylinder capacities and number of cylinders, racing engines have given maximum outputs equivalent of 130 to 170 B.H.P. per litre at speeds ranging from about 7,000 to 12,000 r.p.m. Compression ratios of between 10:1 to 12·5:1 are employed and single or twin overhead camshafts are used on these engines. It should here be mentioned that the outputs per litre and also the maximum outputs of racing car engines appear to be steadily increasing year by year. Another type of engine for racing purposes is the much modified American vee-eight of about 5 to 7·5 litres engine capacity from which maximum outputs of from 500 to over 750 B.H.P. at about 7,000 to 8,000 r.p.m. have been attained. In some cases the original engines are supercharged to give the required maximum outputs.

Horse Power Ratings.—As mentioned on page 40 there is a difference in the horse power of an engine as tested on a dynamometer, or "brake" and the actual horse power available at the crankshaft, or flywheel, for transmission eventually to the road wheels. In such tests on production-line engines a separate exhaust system and cooling arrangements are employed. The horse power measured under these circumstances is termed the *Gross* horse power. Since, as stated earlier, the engine as installed in the chassis of the car, has to drive its own accessories, the power available for useful external purposes, i.e. at the crankshaft, is always less than the Gross value, and this is the horse power figure usually stated in manufacturer's literature and unless stated otherwise in this volume it is assumed.

The DIN Rating Method.—The German method of engine power rating refers to the results of B.H.P. tests upon an engine using all of its engine-fitted accessories, e.g. the exhaust system, cooling fan and water circulating pump, electrical generator, the carburettor, air cleaner, etc., this rating method also contains corrections for the air temperature at the time of the test and also the barometric pressure. It is also used to express the actual maximum torque values.

A good example of the relative values of the Gross and Din. ratings is that shown in Fig. 322 for the recent Jaguar 5·345 litre vee-twelve engine, described later in this Chapter. It will be seen that these graphs show the values, over the speed range, of the B.H.P., Torque, B.M.E.P. and Specific Fuel Consumption.

The S.A.E. Rating Method.—This method as specified by the American Society of Automotive Engineers gives the B.H.P. of an engine as shown by the dynamometer, but after the various

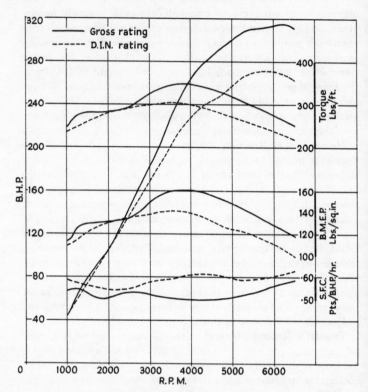

Fig. 322.—Gross and Din Rating Graphs for Jaguar Vee-Twelve Engine.

engine accessories have been removed. These include the cooling fan, water circulating pump, electrical generator, air cleaner, etc., but allows the car's exhaust system to be replaced by separate pipes leading to a common exhaust box with a single outlet to the atmosphere. In effect this method gives a gross rating and therefore over-rates the available horse power at the crankshaft.

Petrol Engine Pollution Products.—More recently much attention has been given to the problem concerning pollution of the atmosphere due to certain of the exhaust products, notably carbon monoxide, nitrogen oxides, hydrocarbons and sulphur oxides. These products have been shown to be a serious health hazard under continuous or dense traffic conditions so that engines in the U.S.A. must conform to certain rather stringent legislations regarding the allowable amounts of the pollutants in the emitted exhaust gases. Further, all engines from other countries imported into the U.S.A. are covered by these regulations. Since this subject is dealt with in some detail in the 8th Edition of these Manuals, viz. "Carburettors and Fuel Injection Systems" it is unnecessary to dwell at length with it here, but the following brief notes should give a general idea of the principles employed.

Notes on Exhaust Problems

The following are among the more important items which affect the constitution of the exhaust gases. (1) The mixture strength, (2) the combustion chamber design, (3) the crankcase blow-by, (4) nature of the fuel used and (5) petrol vapour.

(1) *The Mixture Strength.* The modern carburettor can supply a mixture with the proportions of air and fuel which will give theoretically perfect combustion of the fuel, in which case the exhaust gases would consist of carbon dioxide, nitrogen and water vapour and there would be no exhaust pollution problem. If, however, the mixture is richer in fuel than the one for perfect combustion then the exhaust gases will also contain the deadly gas, carbon monoxide; the richer the mixture the greater will be the proportion of this gas. Unfortunately, it is necessary to use rich mixtures in car operation, as when starting from the cold and accelerating; most carburettors include an acceleration pump to give an extra amount of fuel for this condition. If the carburettor supplies a mixture weaker than the correct one, then there is no carbon monoxide in the exhaust gases, but the power output of the engine is reduced continuously, as the mixture becomes weaker.

(2) *The Combustion Chamber Design.* In earlier considerations of some modern combustion chambers, mention was made of the "squish" effects and quench volumes. In this connection it can be stated that by reducing quench volume in a combustion chamber better conditions for fuel burning are obtained and the hydrocarbons which otherwise would form are reduced. The

piston bore-stroke ratio, and the design of the inlet induction and exhaust systems also are of importance in reducing pollutants.

(3) *Crankcase Blow-by.* This refers to the escape of the air-fuel mixture past the pistons into the crankcase chamber during the compression and combustion processes. This results in a mixture of oil and petrol vapour which escapes *via* the oil filler cap into the atmosphere. In the case of Austin and Morris engines, over a long period, this problem has been solved by means of a rubber tube from the sealed overhead valve cover to the air supply side of the air cleaner, where the mixture is burnt with the fuel mixture from the carburettor to the engine.

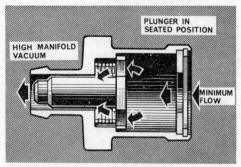

Idling or Low Speed

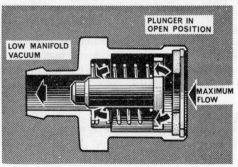

High Speed.

Fig. 323.—The AC-Delco Crankcase Ventilation System.

Fig. 323 shows a crankcase ventilation system to meter the flow of crankcase gases direct to the inlet manifold. It is fitted to the overhead valve cover and connected by rubber tubing to the inlet manifold. The valve has two parts, namely a tapered plunger and stainless steel spring. When the engine is running slowly, the manifold vacuum is at a maximum and the plunger is then pulled forward to reduce the gas flow. At higher engine speeds, with their reduced manifold vacuum, the plunger is moved to the fully-open or middle position to give maximum gas flow. With the engine switched off, the manifold vacuum ceases and the valve closes. Another example of a closed system as applied to a vee-eight American engine is depicted in Fig. 324; it has been

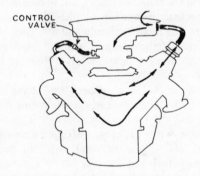

CONTROL
VALVE

Fig. 324.—Closed Crankshaft Ventilation System for Vee-type Engine.

approved by the State of California which first initiated serious research into pollution problem solutions. The system uses one pipe which is fitted with a control valve connecting the inlet manifold to the crankcase, through which the oil and petrol vapour are drawn into the manifold. There is also a second pipe connecting the air cleaner to the overhead casing. Air is thus supplied through this tube to ventilate the crankcase. If the volume of these vapours is too great for the valve, the vapours pass through the ventilation tube into the air cleaner, so that there is no escape to the atmosphere.

(4) *Lead Additive Petrols.* When used in car engines of the high compression types such fuels can give rise to exhaust products which form a serious health hazard; it has also been shown that their presence in the combustion chamber increases the hydrocarbons in the exhaust gas.

Reducing Exhaust Pollution. Various methods to reduce the amounts of the constituent products are described in the Manual No. 2, mentioned on page 495. One favoured method is to supply air into the exhaust manifold, from an engine-driven compressor, to provide the oxygen necessary to oxidize the carbon monoxide and hydrocarbons into relatively harmless products.

Another method would be to use a larger maximum power output engine to the standard model, but to operate this on rather weaker mixtures than the one for complete combustion. The loss of power due to this mixture should still not bring the maximum output below that of the standard model. Moreover, it is well known that such weaker mixtures give appreciably lower fuel consumptions.

(5) *Petrol Vapour Pollution.* Petrol loss, as vapour, can occur from the fuel tank filler cap, carburettor petrol feed pipes and float chamber and the feed pump system. This source of loss can be from about 6 to 9 per cent. Means are now being used to minimize this loss. Thus the tank cap can be sealed after replenishing with fuel and any vapour in the tank allowed to escape through the crankcase ventilation system to the inlet manifold. Again, the carburettor float chamber can be vented, by tubing to the air cleaner or inlet manifold. Another system used in the U.S.A. leads the fuel vapours to an activated charcoal-filled chamber. When the engine is running there is an air flow through the chamber which takes up any fuel vapour and directs it to the inlet manifold. When at rest the charcoal absorbs the petrol vapour.

Combustion Chambers.—More recently, the demand for better performances from cars led to a study of the various means possible to attain the increased power outputs from conventional production engines. As mentioned elsewhere in this Chapter one of the most important methods is to use higher compression ratios, but in order to employ these the actual volumes of the combustion chambers have had to be reduced; this, in turn has introduced problems relating to valve port areas and also better combustion efficiency. A further necessity is to employ a combustion system which minimizes exhaust pollutants which arise during the reactions, e.g. nitrous oxides and hydrocarbons.

An examination of the various combustion chamber shapes available has resulted in the adoption of certain now standard types, namely, (1) the Wedge shape, (2) the Bath Tub shape, (3) the Hemispherical shape, (4) the Recessed or Cavity Piston shape, (5) the Disc shape and (6) the combined Wedge and Cavity Piston shape.

(1) *Wedge Shape Chambers.* This is now more widely used than any other since it provides very good combustion conditions, but valves of the overhead engine have to be inclined at an angle of about 15 to 25 deg. to the cylinder axis and the sparking plug located at the thicker end of the wedge, so as to give the maximum flame area. Fig. 325 shows a typical chamber for a compression

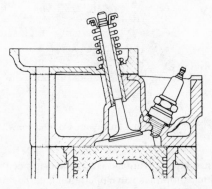

Fig. 325.—Wedge-type Combustion Chamber.

ratio of about 10·5 : 1. It will be seen that the piston at the end of its stroke comes very close to the cylinder head. The object of this is to give the desired mixture "squish" effect and also reduce exhaust pollution.

(2) *Bath Tub Chambers.* In this case the combustion chamber, as shown in Fig. 326, consists of an upper portion having a limited flat shape with its ends curved down to the piston crown end. This type has been used on several British and American

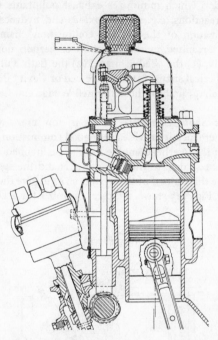

Fig. 326.—Bath Tub-type Combustion Chamber (Buick).

engines. Its principal advantage is that of using central vertical axis valves, which simplifies valve operation and machining costs; it also gives a more compact engine, but is used for engines in the rather lower compression ratios. In this design the sparking plug is readily accessible although its location results in a long flame travel path, but there is no "squish" effect. However, turbulence is given to the compressed mixture by means of the inclined inlet ports.

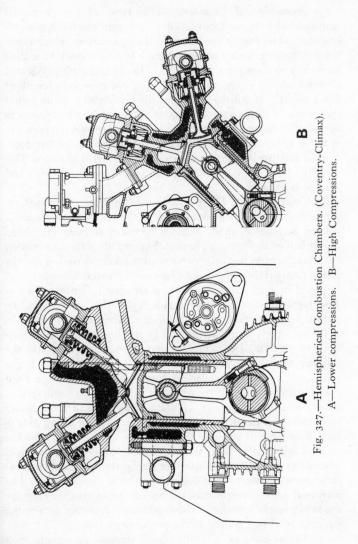

Fig. 327.—Hemispherical Combustion Chambers. (Coventry-Climax).
A—Lower compressions. B—High Compressions.

(3) *Hemispherical Chambers.* This type, of full or partial hemispherical shape, lends itself to accurate machining, so all chambers of an engine can be made of equal volume, which is not always the case with cast chambers. It is necessary to incline the valves, as shown in Fig. 327 (*A*). The sparking plug is located almost centrally in the chamber, to give the minimum length of flame travel. It will be seen that the outer parts of the piston crown comes close to the outer part of the cylinder head so as to obtain the "squish" effect. In high compression engines, in order to obtain the smaller volume before ignition of the compressed mixture, the piston crown is less domed as shown in Fig. 327 (*B*). This type of chamber is used on some British and American engines.

(4) *Heron or Cavity Piston Chambers.* For engines having high compression ratios, e.g. over 10·5:1, the combustion chamber volume becomes so small that it is necessary to make the piston crown hollow to form part or as in Fig. 328 all of the combustion chamber. It has been the practice over a long period to use the piston cavity thus in the much higher compression Diesel engine. In petrol engines, examples of which are certain Ford models and The Jaguar vee-twelve type engine, cylindrical cavities in the pistons form the combustion chambers. Apart from combustion advantages the piston cavities can be machined to exact dimensions to give equal combustion chamber volumes and also the cylinder head lower face can be left flat. This enables larger valves to be used and cheapens costs. Fig. 328 shows a Ford cylinder unit with a cavity piston at the top of its compression stroke, when the piston crown comes very close to the cylinder face; this gives the desired "squish" effect. The sparking plug is located as near as possible to the centre of the cavity and the vertical inlet valve close to the cylinder axis. The principal advantages of this type of chamber were given earlier, but against these must be offset the following drawbacks; namely:– (1) The increased piston weight and (2) Special arrangements must be made to prevent transfer of the combustion chamber heat to the gudgeon pin bearings and to the piston rings.

(5) *Disc Chambers.* As the name implies the combustion chamber consists of a flat-topped piston and a cylindrical recess in

the otherwise flat cylinder lower face. It is used in the Hillman Avenger engine and has some of the advantages of the cavity piston chamber. Further, compression ratios can readily be changed merely by altering the depth of the head cavity. It does not have the two drawbacks of the cavity piston mentioned earlier.

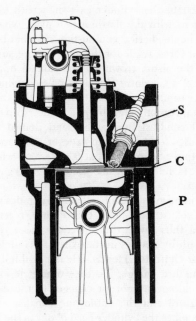

Fig. 328.—Cavity-piston, or Heron Combustion Chamber (Ford).
S—Sparking plug. *C*—Piston cavity.
P—Piston.

(6) *Combined Wedge and Cavity Chambers.* In this arrangement some of the advantages of both types can be obtained, but with cast cylinder heads it is more difficult to obtain equal combustion chamber volumes and as smooth surfaces as with all-machined heads and pistons. Further, the costs are higher than in types (3), (4) and (5).

Some General Notes on Combustion Chambers.—Apart from the principal types mentioned earlier there are other ones which have been used and some special ones. In some instances there have been certain combinations of two different types of chamber to give the desired combustion conditions. In engines having two camshafts per cylinder the valve stems are more widely inclined to the cylinder axis than with single camshaft engines. Then, with the flat-headed valves, what was initially intended as a hemispherical combustion chamber becomes one of almost angular form, while for the high compression ratio used in such engines the piston crown may be domed or conical. The cylinder unit of a Coventry Climax 2·5 litre engine, having twin overhead valve camshafts shown in Fig. 327 (B), is a good example of a high compression, high output model with approximately hemispherical combustion chamber and domed piston crown. This engine used a compression ratio of 11:1 and developed a maximum output of 240 B.H.P. at 6,800 r.p.m. with a maximum B.M.E.P. of 211 lb. per sq. in. Later engines gave appreciably greater outputs.

Modern combustion chambers are being designed so as to reduce the exhaust pollutants before they reach the exhaust manifold, and thus lower the obnoxious gases quantity before being dealt with by devices in the exhaust manifold.

Cross-Flow Cylinder Heads.—It is possible to increase the amount of mixture charge, i.e. the volumetric efficiency of an engine by avoiding any abrupt changes in the direction of the inflowing mixture and the outflowing exhaust gases. In more recent Ford engines the cylinder head of one is shown in Fig. 329. The combustion system is of the piston cavity type and mixture from the carburettor enters the cylinder on the suction stroke through the inlet port as shown in (A), thus giving the necessary degree of turbulence. During the ensuing compression stroke (B) the mixture at the top of the stroke, is ignited; the sparking plug being shown on the right. When the exhaust valve opens, towards the end of the exhaust stroke (C), the gases are ejected through an exhaust gas passage, on the right, so that the complete flow of the mixture and gases is across the combustion chamber with the minimum change of direction.

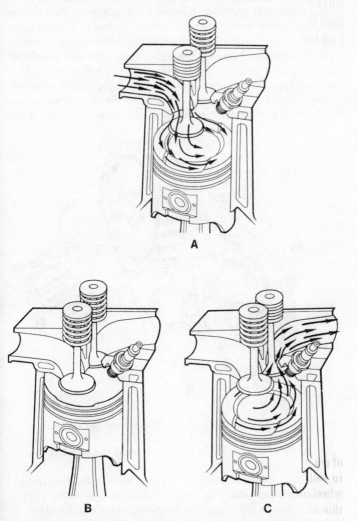

Fig. 329.—Principle of the Ford Cross-flow Cylinder Head.
A—Suction stroke. *B*—Compression and Ignition. *C*—Exhaust stroke.

Toothed-Belt Drives.—This type of camshaft drive from the engine crankshaft sprocket, referred to, briefly on page 177, has since been developed further and is now used as a standard drive on many production car engines with satisfactory effects over long running periods. Initially, this type of vee-belt was reinforced against tension stresses by spirally-wound high-strength steel wires. More recently continuously wound glass fibre cords are used in neoprene-impregnated nylon fabric material to give a very long belt life without stretch Fig. 330. In a typical design, teeth

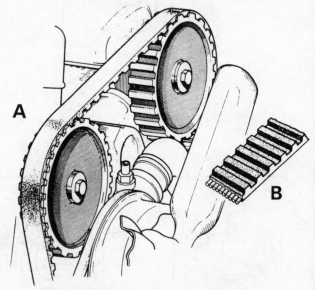

Fig. 330.—Toothed-Belt Drive on Vauxhall Overhead Camshaft Engine.
A—Drive to sprocket wheels. *B*—Construction of the belt.

of 0·5 in. pitch are moulded into a 1 in. wide belt and are arranged to roll in contact with the steel or face-hardened cast iron gear wheels. Such a belt has a tensile strength of about 3,500 lb. In this connection, when the belt is under full engine power conditions in engines of 150 to 200 B.H.P. its maximum tensile pull

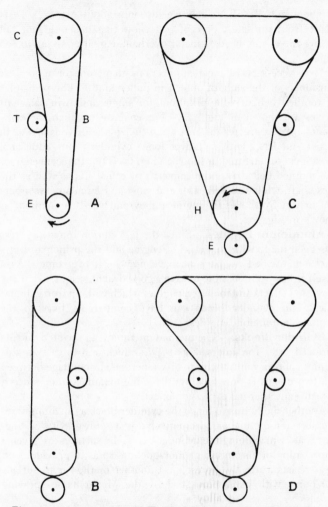

Fig. 331.—Arrangements of Toothed-Belt Drives to Camshafts.

A—Single camshaft, direct drive. *B*—Single camshaft drive from engine half-speed shaft. *C*—Twin camshafts drive from engine half-speed shaft. *D*—Twin camshaft drive with two belt-tensioners.

B—belt. *C*—camshaft. *E*—crankshaft. *H*—half speed shaft. *T*—tensioner.

is about 330 lb. The belt weighs only about 9·5 oz. Tests made from the latest type of belt show that it should not fail before the engine needs a major overhaul, e.g. after about 50,000 miles.

It is necessary in most applications to fit one or more belt tensioners of the smooth outer rim pulley kind. The principles of toothed belt drive applications for single and twin camshaft drives are shown in Fig. 331. For engines in which the camshafts are at a greater distance from the engine drive gear—as in larger models, with relative long cylinders—an additional tensioner is often fitted, as shown in Fig. 331 (D). In recent engines the toothed belt drives the camshaft or camshafts as well as the jack shaft, while the alternator and fan-water circulator members are driven by the usual engine pulley and vee-belt method, as shown in Fig. 332.

Aluminium Engines.—Since the last edition much progress has been made with this class of engine and the principal drawbacks mentioned on page 99 have now been overcome. As a result it has been possible to save a good deal of weight, avoid any corrosion effects and to design engines which without iron cylinder liners and cylinder heads can have comparable lives to the orthodox automobile engines.

Cylinder Blocks.—The method of employing serrated outside surfaced cast iron liners, about $\frac{1}{8}$ in. thick and to cast an aluminium alloy around them, to give a permanent bondage between the two metals is now favoured by some manufacturers, since it obviates any thermal expansion effect.

Another method is to make the cylinder block as an aluminium pressure die-casting or, alternatively, as a semi-gravity casting. Then, after machining the cylinder bores, the surfaces are given a thin coating of chromium or molybdenum so as to provide good wear-resistance to the bores. Yet another method is to surface the bores with hard alloy steel by the Metco flame spraying process.

The special alloys used for cylinder blocks and heads include the high silicon and silicon-copper ones. A typical American aluminium alloy used for car engines is the Reynolds 390 alloy which contains about 17 per cent silicon and 4·5 per cent copper.

It has good casting qualities and a hard surface, giving cylinder-wear results comparable to those of alloy cast iron. It is now usual to coat the piston skirts with a hard wearing metal, e.g.,

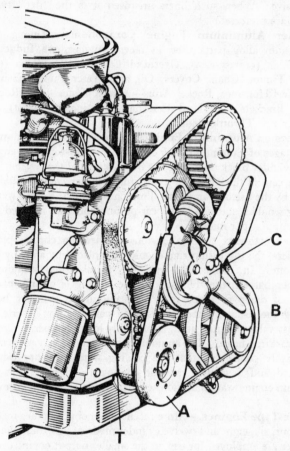

Fig. 332.—Application of Toothed-Belt and Vee-Belt Drives to Vauxhall Victor Engine.

A—Engine pulley and gearwheel (behind). *B*—Vee-belt drive to alternator and fan-impeller. *C*—Cooling fan. *T*—Belt tensioner.

chromium or alloy steel applied by the plasma arc spraying process. There are still manufacturers who prefer aluminium blocks with cast iron liners, but the engines are a little heavier and more expensive. When such liners are used it is the "dry" types which are preferred.

Other Aluminium Engine Components.—Among the aluminium alloy parts used in recent engines, are Induction Manifolds (gravity-cast), Overhead Camshaft Housings (die-cast), Timing Chain Covers, Oil and Water Pump Casings, Flywheel Housings, Rocker Arms—instead of steel ones, Rocker Shaft Brackets, Overhead Camshaft Housings (die-cast) and Carburettor Throttle Bodies.

Notes on Aluminium Engines.—One of the most important advantages of aluminium engines is their weight-saving over cast iron cylinder block ones. It has been estimated that as much as 30 per cent of the weight of a cast iron cylinder engine could be saved by the use of aluminium alloy components. Such a weight saving would result in a better car performance and reduced fuel consumption per mile.

When orthodox types of pistons are used in chrome-plated cylinders, the top piston ring should be replaced by an alloy cast iron one. In connection with the metal-coated aluminium cylinder barrels, the thickness of the coating is usually of the order of 0·005 to 0·010 in. With aluminium cylinder heads having the usual bronze or cast iron valve seating inserts the exhaust valves operate at an appreciably lower temperature than for a cast iron head. The aluminium block and head can be more effectively cooled, so that the cooling water capacity can be reduced and a smaller radiator employed. Further, the aluminium engine warms up quicker than the cast iron one from the cold.

Vee-Type Engines.—More recent automobile engines include the four, six, eight and twelve cylinder models. Of these the two former are employed for use on the smaller output engine cars of about 70 to 110 B.H.P., although certain larger cars and occasional racing cars have employed six vee-type engines from about 120 to over 200 B.H.P.; the earlier American Buick engine had an output of 135 B.H.P. at 4600 r.p.m.

Vee-Four Engines.—Later examples to those mentioned on pages 251 to 254 include the Ford engines made in two sizes, namely with engine capacities of 1·664 and 1·996 litres with compression ratios of 9:1 and 8·9:1, respectively, and maximum outputs of 81·5 and 93 B.H.P. both at maximum speeds of 4,750 r.p.m.

These engines have their two-bank cylinder blocks at 60 deg. and employ high camshafts, gear-driven from the crankshaft, with short push rods and rocker arms to the overhead valves. A special feature of these engines is the use of cavity piston combustion chambers and the Ford "cross-flow" gas directional method, described earlier.

The advantages of this type of engine are, as follows:– (1) it gives an appreciably shorter engine, (2) has a stiffer and lighter crankshaft, (3) enables the carburettor and ignition distributor to be located in the vee-space, (4) the camshaft, also, is in the vee-space. Some drawbacks include (1) the necessity for a separate balancing shaft with counterweights, driven at engine speed, (2) greater engine width than the four-cylinder in-line engine, (3) rather heavier than the latter type and (4) usually more costly to manufacture.

Vee-Six Engines.—These are more popular than the vee-fours and have been adopted by several European manufacturers. It enables higher outputs to be obtained and does not need the vee-four's balancing shaft, since the unbalanced forces can be balanced by means of counterweights on the front crankshaft flange, the flywheel rim and on the crank webs. It may be mentioned that it is possible to have an angle of 90 deg. for the cylinder axes as in the Maserati C 114–1 2·67 litre engine which has a maximum output of 170 B.H.P., net at 5,500 r.p.m. This engine can be balanced properly but in doing so the firing intervals are uneven.

The Vee-Eight Engine.—While this type of engine has become almost universal for most American engines it has been used to a much smaller extent on European ones; the British vee-eight engines now include the Rolls-Royce, Jaguar, Rover and Stag models. All of these engines belong to the higher output class, ranging from about 120 to over 200 B.H.P., in touring cars as distinct from sports and racing ones.

The American engines range from about 7 to 8 litres with outputs of between about 340 to over 400 B.H.P. at 4,400 to 5,000 r.p.m. A typical example of a European vee-eight engine is that of the Mercedes-Benz 90 deg. one, illustrated in Fig. 333. It has a bore of 92 mm. and stroke of 65 mm., giving a bore-stroke ratio of 1·42. This greater bore-stroke feature is now common to

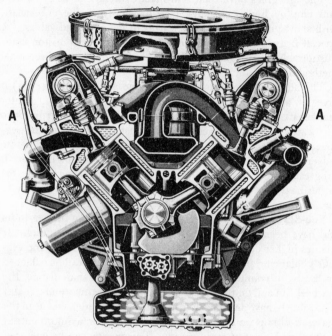

Fig. 333.—Mercedes-Benz Vee-Eight Engine, showing at *A*—Valve clearance adjuster.

most high output engines since, as stated earlier it allows greater valve diameters for higher volumetric or charge efficiency and also lower maximum piston speeds. The cylinder capacity of the engine shown in Fig. 333 is 3·5 litres and it has a maximum output of 200 B.H.P., net, at 5,800 r.p.m. Special features of this engine include a wedge-type combustion chamber, single overhead camshafts, a fuel injection system and hydraulic coupling driven cooling fan for the radiator.

The camshafts are chain-driven from the crankshaft sprocket with an hydraulic chain tensioner. The engine has its induction ports arranged in the vee-space between the cylinder blocks, while the exhaust ports are on the outsides of the blocks.

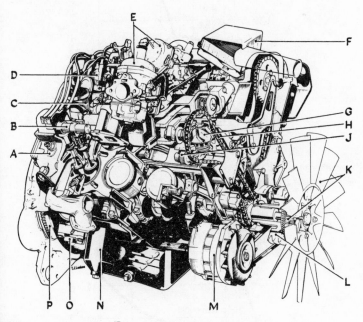

Fig. 334.—Triumph Stag Vee-Eight Engine, showing some of its Components.

A—Cylinder head. *B*—Camshaft. *C*—Ignition coil. *D*—Distributor. *E*—Carburettors. *F*—Air Cleaner. *G*—Water pump. *H*—Timing chain. *J*—Jackshaft. *K*—Viscous-type cooling fan. *L*—Torque damper and belt pulley. *M*—Alternator. *N*—Oil filter. *O*—Oil pump. *P*—Flywheel.

A good example of a recent British vee-eight engine is that of the Triumph Stag, shown in Fig. 334 which has several interesting design features. The cylinders have bores of 86 mm. and strokes of 64·5 mm., thus giving a bore-stroke ratio of 1·33. The

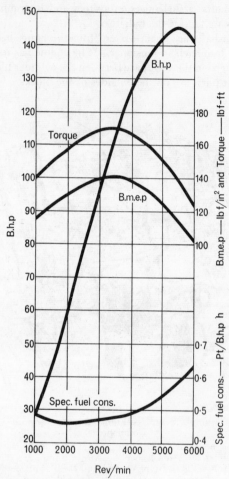

Fig. 335.—Performance Graphs for Triumph
Stag Engine.

cylinder capacity is very nearly 3 litres. With a compression
ratio of 8·8 : 1, the engine has a maximum output of 145 B.H.P
at 5,500 r.p.m. and maximum torque of 170 lbs. ft. at 3,500

r.p.m. The maximum B.M.E.P. is 140 lbs. per sq. in. at the same speed. Fig. 335 shows the performance graphs for the engine. The cylinder block is of cast chromium iron and the head of a suitable aluminium alloy with separate valve insert rings. Single overhead camshafts are employed, the cams of which act directly on the valve inverted bucket members. The two camshafts have separate chain drives from crankshaft sprockets. The left drive is direct to the camshaft and it has a chain tensioner on its left side. The right chain drive also operates the jackshaft and it too has a chain tensioner between the latter and the crankshaft sprocket. The combustion chambers are wedge type with the valve stems inclined to the flat piston crowns. The crankshaft is made from an alloy steel forging, with integral balance weights and it runs in five main steel-backed lead bronze bearings with indium coatings—a now popular feature of many British engines. Each camshaft runs in five bearings. The flywheel is of cast iron with a shrunk-on hardened starting gear wheel rim. A combined torsion vibration rubber-type damper, with combined pulley—for driving the cooling fan and water impeller—is fitted at the front end of the crankshaft...Another pulley on this unit is for driving the alternator and power steering oil pump, using vee-type belts in each case. Twin side draught Stromberg carburettors are fitted, the fuel for same being supplied from an electric diaphragm-type fuel pump at the rear of the car; the fuel pressure is 2·7 lb. per sq. in.

A hot water-heated four branch aluminium alloy inlet manifold, between the castings is used and a three-branch cast iron exhaust manifold is located at the outside of each cylinder unit.

The cooling system is of the "No loss" kind using a radiator pressure of 13 lb. per sq. in. The cooling fan has 13 blades and employs a viscous type of coupling.

Exhaust pollution is reduced mainly by the type of carburettors fitted but also by a valve-less closed circuit breathing system from the depression side of the carburettor to the inside of the valve rocker cover. To meet with the U.S.A. regulations an air inlet hot spot and a thermostatically-controlled air inlet valve can be fitted.

The Twelve-Cylinder Engine.—The two British car engines

of this type described on pages 283–285 were eventually super-ceded by vee-eight engines, but recently Jaguar Cars Ltd. Coventry have developed a new engine, illustrated in Fig. 336, which has a number of advanced design features.

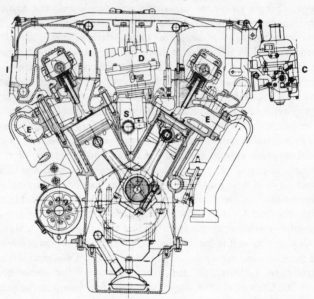

Fig. 336.—Jaguar Vee-Twelve Engine in Cross-sectional View.
I—Inlet passages. D—Distributor. E—Exhaust passages. S—Jack-shaft drive to distributor.

The 60 deg. vee-twelve engine has a bore and stroke of 90 mm. and 70 mm. giving a bore-stroke ratio of 1·28 and total cylinder capacity of 5·343 litres. It has a compression ratio of 9:1 and develops a gross B.H.P. of 314 at 6,000 r.p.m. The maximum torque is 349 lb. ft. at 3,600 r.p.m. The maximum B.M.E.P. is 161 lb. per sq. in.

The performance graphs for this engine are given in Fig. 322. In these graphs the full lines refer to the gross rating and the dotted Din ones to net values.

Referring to Fig. 336 it will be seen that the combustion chambers are formed in the piston crowns, so that a flat faced head can be used to give larger valves and to control exhaust emissions more effectively. The cylinder blocks are made from sand-cast aluminium alloy and are fitted with wet-type cast iron liners. The cylinder heads are of aluminium alloy and have sintered Brico metal valve seating inserts. Special inserts are also used for the sparking plugs. The engine has single overhead camshafts to

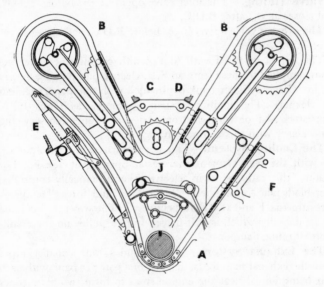

Fig. 337.—Arrangement of Jaguar Engine Timing Chain Drives.
A—Crankshaft sprocket. *B*—Camshaft sprockets. *C*, *D* and *F*
—Rubber-faced dampers. *E*—Morse automatic chain tensioner.
J—Jackshaft.

each cylinder block, made of alloy cast iron with chill-hardened cam faces. Each camshaft runs in seven bearings which are machined in a die-cast aluminium alloy detachable tappet block. The two camshafts are driven from the crankshaft sprocket by a single stage duplex-type timing chain, shown in Fig. 337. The chain, which also drives the jackshaft, is located in the vee space of the engine and has a Morse automatic tensioner between the crankshaft and the first camshaft sprocket and three other rubber-faced dampers located as shown in Fig. 337. The jackshaft also drives the ignition distributor unit shaft. Each forged alloy steel crankshaft runs in five main bearings of the copper-lead kind. The inlet valves, of 1·625 in. diameter, are made of silicon chrome steel while the smaller 1·360 in. exhaust valves are of austenitic steel.

Valve Timing.—The inlet valve opens at 17° before T.D.C. and closes at 59° after B.D.C.

The exhaust valve opens at 59° before B.D.C. and closes at 17° after T.D.C.

The engine is fitted with four Zenith carburettors, the fuel being supplied to these from an S.U. electric pump at 1·5 lb. per sq. in. Each cylinder bank is provided with one paper element air cleaner. The ignition system is the Lucas Opus fully-transistorized type with electronic trigger action. No servicing is necessary with this component.

The Cooling System.—This is pressurized at 13 lb. per sq. in. with the wax element thermostat opening at 180°F. For cooling the radiator twin electric thermostatically-controlled four-blade fans are supplied. The "make" and "break" temperatures are 194°F and 185°F, respectively. The water is circulated by an impeller which is belt-driven from a pulley on the crankshaft vibration damper.

The Exhaust System.—This includes four exhaust pipes from the exhaust manifolds, namely, one pair per bank with each pair being joined near the engine so as to form two main pipes leading to two silencers, located midway between the engine and rear of the car. From these two further pipes lead to a single rear mounted silencer with four tail pipes.

The Cooling System.—More recent cars employ the "No-loss" or sealed cooling system, mentioned briefly in Chapter IX, using water containing a corrosion inhibitor and, for winter driving conditions, an anti-freeze solution.

The problem of fan blade noise, including that of resonant vibrations, has been solved by the use of plastic moulded fans of Nylon or Polypropylene materials, having more blades, ranging from about 6 to 13 and in overall diameters from about

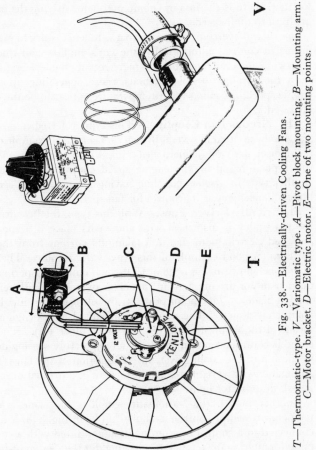

Fig. 338.—Electrically-driven Cooling Fans.

T—Thermomatic-type. V—Variomatic type. A—Pivot block mounting. B—Mounting arm. C—Motor bracket. D—Electric motor. E—One of two mounting points.

7 in. to 13 in. according to the engine size.

The principal drawback of the fixed blade engine-driven fan has been its inability to supply the correct amount of cooling air, through the radiator, over the wide range of engine speeds, including cold starting, idling and the air flow through the radiator, due to the car's speed and its atmospheric temperature. Further, there is the loss of engine power for driving the fans at medium to high speeds.

In order to overcome these drawbacks it is necessary to provide means to control, automatically, the fan's air flow and this has been done in some fans fitted on more recent engines.

Of the various possible methods, those depending upon the engine speed or cooling water temperature appear to be the more favoured.

The Viscous Fan Coupling.—This type, referred to on page 457, has been fitted in several British cars, a typical example being the Holset used on Jaguar, Triumph, Volvo cars, etc. While this type is based on the engine speed control effect, there is another type of viscous coupling having a temperature sensing bimetallic element as its control for fan speed.

Electrically Driven Fans.—With this type, the belt drive to the fan pulley is dispensed with and in its place is a specially designed electric motor-fan unit taking its current from the car battery. A good example of this type is the Kenlowe Thermomatic one shown in Fig. 338 at (T) which employs a permanent magnet motor driving a 10 in. diameter ducted fan with aerofoil shaped blades. The motor bearings are lubricated for life, being completely sealed. The unit is readily fitted to car engines, by means of the mounting arm (B) and pivot block (A). The fan speed is controlled automatically by a temperature sensing device in the top of the cooling system. The action of the fan is fully automatic, with variable temperature setting, a manual over-ride and dash indicator light.

Another model, the Variomatic, employs the same type of temperature sensor control, as shown at (V) in Fig. 338, which is connected to a control unit on the dash and fitted with a control knob for altering the temperature setting, for the motor cut-in and out, for temperatures between 20°C and 120°C; the setting can be locked at any position. Further, there is a safety bridge member to prevent the fan being cut off manually if the temperature needs reduction. Road tests made on different cars show that the reduction in petrol consumption per mile varies from 8 to 14 per cent. Other advantages include very quiet running at all speeds; quick warm-up of the engine from cold; automatic control action independent of any engine thermostat setting; readily fitted to various cars and operation equally well under hot summer and cold winter conditions.

Variable Pitch Blades Fan.—Another type of variable output fan is one having four separate blades, each of which can be rotated about a radial shaft by torsional twist created from centrifugal force action. A typical example is the Aerofan, shown in Fig. 339 in which the fan blades are given a coarse pitch to provide a good flow of air for cold starting and idling conditions. When the engine is accelerated from idling the effect on the fan is to provide increasing centrifugal force to gradually reduce the blade pitch until at a certain road speed, namely 30 to 40 m.p.h., the blades assume a fine pitch with practically no flow, thus leaving the cooling of the radiator to the relatively high vehicle speed only.

Other Devices.—Among the alternatives to the automatic fans mentioned there are fans controlled by centrifugal clutches which bring the fans into action when a certain speed is attained by their drive; of the possible types the electromagnetic one is generally favoured. Another control method is a thermostatic switch for sensing the cooling water temperature, using the clutch to bring the fan into or out of action.

Fig. 339.—The Auto-fan Variable-pitch Blade Fan.

Lubrication Oil Cooling.—In some of the larger car engines, the heated engine oil in the bottom of the crankcase is cooled by passing it through a heat-exchange unit in which water from the bottom part of the radiator—which is the coolest in the system—circulates through the heat exchanger and is then returned to the inlet housing of the water circulating pump.

Overhead Valve Clearance Adjustment.—The overhead valves of more recent engines are provided with simpler means for adjusting valve stem end clearances, thus dispensing with the adjusting screw and locknut at one end of the rocker arm. One favoured method used on Vauxhall engines is illustrated in Fig. 340. Here, the push-rod has a rounded top which engages with a curved depression in a special design of rocker arm, made as a shell-moulded casting, having a lower hemispherical seating, which is held in contact with a fixed rocker ball. The bearing is lubricated direct from an oil gallery *via* a hollow part of the stem on which the rocker ball is mounted. Adjustment of the valve stem clearance is effected by the adjusting nut above the rocker ball. In a later design, oil is delivered through hollow push-rods having special metering valves.

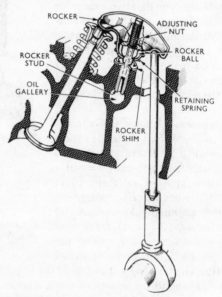

Fig. 340.—Method of adjusting Push-rod to
Valve Stem Clearance (Vauxhall).

In the case of overhead camshaft engines one method for adjusting the valve stem end clearance is by means of shims, placed between the bucket-type tappets and the valve stem ends, as shown in the right hand valve unit of the Jaguar engine in Fig. 336.

Fig. 341 illustrates a simple method used in certain Vauxhall camshaft engines, in which the cast bucket-type tappet is provided with a screw, the hole for which is inclined at 5·5°. The screw, which is turned by an Allen hexagon screwdriver, has a flat portion which engages with the valve end. One complete rotation of the screw alters the clearance by 0·003 in.

In the case of the Mercedes-Benz engines the valve stem clearance adjuster is that shown at (A) in Fig. 333, in which a short arm having a ball-and-socket bearing is interposed between the cam and valve stem end. The ball member has a threaded support in the cylinder head with provision for altering the clearance as required.

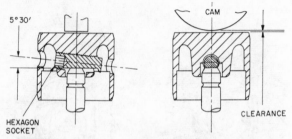

Fig. 341 —Overhead Camshaft Valve Clearance adjusting Method.

The Wankel Engine.—Since its successful test bench engine at Neckarsulm in 1958, the Wankel engine has been in constant development by four major automobile manufacturing countries in Germany, Japan and France and for other applications by Curtiss-Wright in the U.S.A. In its automobile designs it has been fitted to a number of limited production cars. As a result of its intense development the earlier drawbacks have now been largely overcome and a wide range of engines from small air-cooled single rotor ones, of about 6 to 12 B.H.P., up to four-rotor automobile models, with maximum outputs exceeding about 360 B.H.P., have been made.

Operation of the Later Wankel Engine.—Instead of the arrangement shown in Fig. 205 for the earlier layout and cycle

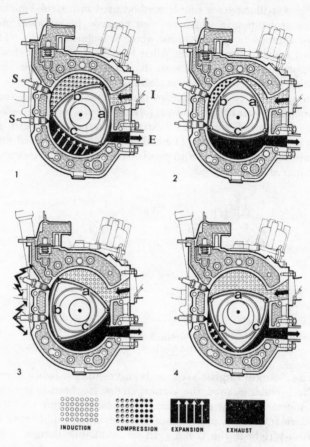

Fig. 342.—Illustrating the Operation of Later Type Wankel
Engine.

of operations the later one illustrated in Fig. 342 is now in general
use.   From this it will be seen that the outer casing of the epi-
troichoidal chamber casing is water-cooled, its temperature being
thermostatically controlled by a unit located near to the outlet at
the top.   The air inlet and exhaust outlet ports are depicted by

the I and E, respectively, on the right-hand side of the casing. Instead of the single sparking plug, used on the earlier engines, the two separate ones shown at S, S are now used; the reasons for these are explained later.

The cycles of operation are, as follows:–

(1) The chamber (a) is at minimum volume, denoting the end of the exhaust phase and the start of induction. The chamber (b) mixture is under compression and that at (c) is in the expansion stage.

(2) Here, the chamber (a) continues to grow larger, while the mixture induction process continues. The mixture in chamber (b) is being further compressed while that in (c) has attained its maximum volume, i.e., at the end of the expansion stage and the exhaust port E is uncovered.

(3) At this stage the chamber (a) continues to grow larger as the induction process continues, while the chamber (b) is at its minimum volume, corresponding to maximum compression, so that the mixture is about to be ignited. Also, the chamber (c) is getting smaller as the exhaust gas escapes through the port (E).

(4) Here the chamber (a) has approached its maximum volume and the inlet port (I) is about to be closed. In chamber (b) the heated combustion products are acting on the rotor flank thus producing torque on the eccentric shaft, while in (c) the exhaust operation is proceeding. During the four operations described three complete cycles have occurred for one revolution of the eccentric shaft. As shown on page 309, the rotor rotates about its own centre at 2/3 of the angular velocity of the main shaft, so that a single rotor engine is equivalent to a two cylinder four stroke engine of the same displacement.

The Rotor Member.—During its development period the rotor of the Wankel engine has suffered from the troubles connected with sealing of the rotor tips, corners and its two sides. Any appreciable wear in the seals could result in leakage of mixture or combustion gases between the chambers and also result in excessive oil loss within the chambers. In this connection an extensive research has since been carried out, resulting in a considerable improvement in the rotor sealing methods, with notable reduction in the wear of the chamber rubbing surfaces and the seals, themselves. Initially, chamber wear was reduced by chrome plating the surface of each chamber and induction hardening of the side surfaces. Later, this expensive method was replaced by one in which the surfaces are coated, by a special process, with silicon carbide embedded in a nickel matrix; this gives a much longer wear life. Alternatively, the wearing surfaces of the chamber can be molybdenum-sprayed. While the side faces, gas and oil seals have been much improved, it is the seals at the three rotor tips which have provided the more difficult sealing problem.

Fig. 343 illustrates one method for sealing the rotor tips and also the corners. The latter are provided with cylindrical slotted link blocks, as shown at (A), while the slots along the whole length of each apex are provided with apex seal members, as shown at (B). For a period these members were of carbon, but later they were made of piston ring-type cast iron. To ensure adequate sealing at starting and low engine speeds these members were provided with light apex seal springs. For medium to maximum speeds centrifugal force provided the required pressure. Gas pressure during the compression and firing stages also plays a part in assisting the sealing of the apex member tips. Circulating lubricating oil also helps seal the rotor tips. The seal between the sides of the rotor and the casing side is of the piston ring-kind and it is installed between the eccentric and the rotor flank and between the eccentric and the bore in the side housings. Oil pressure of about 5 lbs. per sq. in. has been used to complete the sealing effect.

Some prolonged tests on a car fitted with an N.S.U. engine indicated a tendency for carbon to build up against the seals at low speeds and if the engine was operated above the recommended maximum speeds, for the springs to overheat and lose their strength.

The hollow rotor units are cooled by engine oil which is circulated through them and it is arranged, by means of a water-cooled heat interchanger, to maintain the maximum temperature at about 120°.

The Ignition System.—The earlier Wankel engines were prone to sparking plug trouble, which meant more frequent plug changing. The later engines with twin plugs, suitably located and fired, have overcome this trouble. Recent engines employ two 18 mm. plugs, each with its own coil and distributor. They are located as shown in Fig. 342. The plugs can be fired independently or together. When fired together the mixture heat

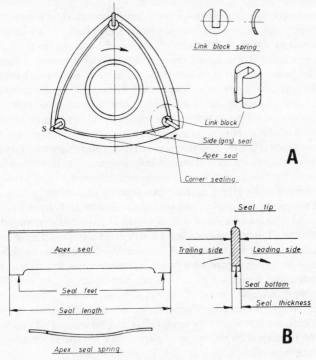

Fig. 343.—Methods for sealing the Rotor Tips and Corners.

release is shorter than with single plug ignition. It is mentioned later that from the viewpoint of minimum exhaust pollution, it is necessary to employ a different firing arrangement. In connection with the fitting of each plug it is arranged for the plug points to be inside their screwed holes, with a bronze plug having a central hole between the plug points and the chamber surface.

In the Mazda engine it has been found that the best performance was obtained by locating the plugs at 1·18 in. before T.D.C. and 0·7 in. after T.D.C. corresponding to about 5° at idling speed.

Carburation.—For most production, Wankel engines standard carburettors are used, but in the case of the twin-rotor engines two separate twin-choke or one four-barrel type are fitted. Thus, in the case of the Mazda engine, which develops a minimum output of 108 B.H.P. (net) at 7,000 r.p.m., a four-barrel Stromberg carburettor supplies the mixture to each chamber's double inlet ports—which then lead to a single port into the chamber. The primary stage of the carburettor supplies the mixture for starting and idling purposes, while both primary and secondary stages come into operation at medium to maximum engine speeds. In a more recent engine, direct fuel injection into the inlet ports is adopted; this ensures more uniform mixture strength and quantity into each chamber and also greater volumetric efficiency. Further the exhaust pollutants can then be better controlled.

The Twin-rotor Engine.—An interesting earlier application of the twin-rotor engine was that of the American Curtiss-Wright model which had a rated output of 310 B.H.P. and dry weight of 285 lb. The chambers were provided with cooling fins located in the cooling air stream. A smaller Wankel engine-the KKM 514—had twin-rotors of 330 c.c. chamber capacity; it was air-cooled and developed 38 B.H.P. at 5,000 r.p.m.

More recently, most of the smaller automobile engines, e.g. the N.S.U. Ro80 and Mazda types, are of the twin-rotor kind. In each case the rotors are mounted on a common double eccentric type of shaft with the rotors at 180° to one another, to give equal firing intervals.

The N.S.U. Ro80 Engine.—In connection with this well developed engine (Fig. 344) each chamber has a capacity of 497·5 c.c. and a displacement volume of 995 c.c. per shaft revolution. It can be shown that this displacement is equivalent to that of a 1990 c.c. four-stroke piston engine. The compression ratio is 9:1 and the maximum output, 113·5 B.H.P. (net) at 5,500 r.p.m.; the maximum torque is 117 lb. ft. at 4,500 r.p.m.

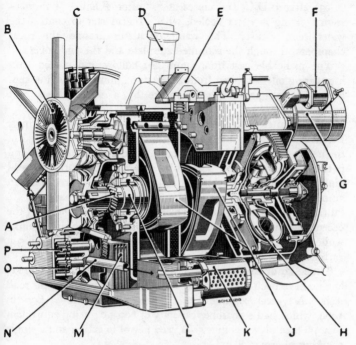

Fig. 344.—The Twin-Rotor Wankel Engine, showing some of its Features.
A—Oil pump and distributor drive. *B*—Viscous-type cooling fan and water impeller. *C*—Distributor. *D*—Oil filler pipe. *E*—Heater regulator housing. *F*—Starter motor solenoid. *G*—Starter motor. *H*—Transmission torque converter. *J*—Twin rotors. *K*—oil filter. *L*—Eccentric shaft. *M*—Oil cooler, with water jacket. *N*—Oil feed pipe (pump to filter). *O*—Oil gear-type pump for torque converter. *P*—Main gear-type oil pump for engine lubrication and cooling rotors.

It employs two sparking plugs, each with its own coil and distributor unit. Mixture is given by two 2-stage Solex cross-draught carburettors with starting and idling stages and an accelerator pump. The engine casing has two inlet ports leading, by a single entrance port, to the chamber.

The inlet port timing is such that the port opens at 108° before T.D.C. and closes at 40° after B.D.C. The exhaust port opens at 63° after B.D.C. and closes at 71° after T.D.C. The main engine casing is water cooled, with a thermostat to control the water temperature. The water circulates around the rotor chambers, through the intermediate plate and the end covers.

The malleable cast iron rotors are hollow and cooled with lubricating oil—supplied from a gear-type pump—which is kept down to its operating temperature by a water-cooled heat exchanger.

The eccentric shaft runs in two plain three-layer bearings. Balance weights for the rotors and eccentric parts are located in the belt pulley which drives the cooling fan and water pump, and in the rear shaft flange. The 6-bladed fan is of the variable speed kind, known as the variable torque type and it is controlled by an automatic clutch and the rubbing surfaces of the chambers are lubricated with an engine oil which is squirted from holes in the eccentric shaft. The rotor tips are sealed by cast iron sliding units while the working surfaces of the chambers are coated with silicon carbide particles in a nickel matrix.

Three-rotor Engines.—Several models of this type of Wankel engine have been built and subjected to both bench and car road tests. A typical example is the engine built by Daimler-Benz in 1969, which had a chamber capacity of 600 c.c., being equivalent to a 3·6 litre piston engine. It used petrol injection and a single sparking plug per chamber.

Four-rotor Engines.—Of the different engines of this type which have been built for experimental purposes the later Mercedes-Benz (1970) one known as the C.111 appears the more promising. It has a chamber capacity of 600 c.c.—equivalent to a 4·8 litre piston engine—and a maximum output of about 360 B.H.P. at 7,000 r.p.m. and torque of 290 lbs. ft. at 4,750 r.p.m. Special features of this engine are its fuel injection and electronic ignition systems, with a circulating oil system for lubrication and rotor cooling, etc. Its four rotors are mounted on a five main bearing-supported eccentric crankshaft.

In operation the engine is vibration free, particularly smooth and almost noiseless over the upper speed range and more flexible at the lower engine speeds. It weighs, dry, about 360 lb., i.e. about 1 lb. per horse power.

Exhaust Pollutants.—The Wankel engines mentioned earlier had a relatively high level of pollutants in the exhaust. Most of these consisted of hydrocarbons with low amounts of carbon monoxide and nitrogen oxides. This is due to the relatively large quench area of the chambers and to the action of the trailing edges of the seals scraping the quench layer of the hydrocarbons into the exhaust chamber. Again, there is sometimes a small leakage of air-petrol mixture across the leading apex seal into the exhaust port. Since U.S.A. legislation lays down definite limits to the amounts of pollutants in the exhausts of all cars manufactured or imported into that country and similar legislation is likely in the

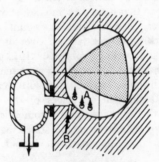

Fig. 345.—Methods for reducing Hydrocarbon and Carbonmonoxide Pollutants.

near future for many other countries it would be necessary for Wankel engines to be modified so as to conform to the legal requirements. The following are methods by which the pollutants can be reduced to the desired limits:—

(1) *By the Sparking Plugs.* It has been shown that with the single trailing plug pollutants occur higher loads and speeds, whereas the leading plug, alone reduces such pollutants. In recent engines an automatic control is fitted to switch the trailing plug in or out of action according to the emission requirements.

(2) *By Air Injection* into the exhaust ports, to oxidize the hydrocarbons and carbon monoxide. Fig. 345 shows three different air injection holes which have been tested. Of these, that shown

at (A) gave the lowest emission.   However, since such holes could be sooted up after a time it was decided to use the tangential port shown at B.

(3) *By means of an Exhaust Reactor.*—Tests have shown that the dual-cell recirculation kind of reactor was the most effective.

With all of the three systems in action together the exhaust pollutants can be reduced to the legal requirements. However, the addition of the extra devices adds to the complication of the Wankel engine.

Some General Considerations.—It will be evident that the original single rotor Wankel engine described previously has had to be much modified in design to reach the automobile engine production stage, so that it has become more complex, costly and heavier. Moreover, a considerable amount of research and development work was necessary to overcome the rotor and chamber sealing and cooling problems.   It is probable that still further experimental and bench testing work may have to be done in connection with the rotor seals, to ensure wear resistance with proper chamber sealing in all stages of the operating cycles, comparable with that of the modern automobile piston engine.

In regard to rotor chamber wear it is stated that this wear should not exceed about 1 mm. (0·0394 in.) or about 40/1000 inch per 60,000 road miles.

Concerning fuel consumption, the earlier single and twin-rotor engines had higher specific fuel values than their equivalent output piston engines.   The results of some tests on a twin rotor engine, given by Dr. Frode, a director of research for the N.S.U. Company, are reproduced in Fig. 346.   It will be seen from the lower graph that initial fuel consumption, at 1000 r.p.m. for the engine which developed 110 B.H.P. net at 7,000 r.p.m., was relatively high, but as the speed increased to 3,000 r.p.m. it fell continuously to about 0·53 lb. per B.H.P., after which it increased to 0·60 at 5,500 r.p.m.   The upper brake mean pressure graph shows an increasing value, as the fuel consumption decreased, until at about 2,500 r.p.m. it reached its maximum at about 137 lb. per sq. in.   Road test figures on certain production engines gave about 18 to 20 mp.g. compared with 24 to 25 for comparable output piston engine cars.

In regard to oil consumptions, the earlier twin-rotor engines were known to give appreciable greater values than for piston engines. Tests on some later car engines indicated appreciably greater consumptions than the 270 to 350 miles per pint of lubricant stated by the manufacturers.

Against these apparent shortcomings of the Wankel engines can be mentioned definite advantages, namely, (1) absence of

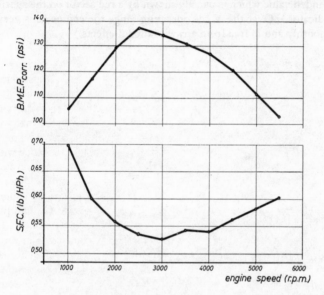

Fig. 346.—Fuel Consumption and B.M.E.P. Graphs for Twin-Rotor Engine.

reciprocating members, e.g. valve components, camshafts and their drives, (2) excellent engine balance at all speeds, (3) absence of vibrations over the speed range, (4) for a given power output the Wankel engine can be made much lighter than the equivalent piston engine. In the more recent higher output engines weights of only about 1·0 to 1·5 lb. per B.H.P. have been achieved, (5) in general the automobile engines are more compact than the equivalent piston engine and (6) due to the fact that the rotor and chamber units replace the relatively large number of components of the multi-cylinder piston engine it is anticipated that the cost of production of the engine can be much reduced, in the near future.

In operation, the car engine starts easily, with gradual acceleration up to about 2,500 to 3,000 r.p.m. and then accelerates more briskly up to its top speed—which should not exceed the recommended value which is usually shown by a red sector on the speed indicator.   Over the whole operating range the engine runs very smoothly and is free from any vibrational effects.

INDEX

* For convenience these are indexed as
Diesel Engines.